Lecture Notes
in Control and Information Sciences 240

Editor: M. Thoma

Springer
London
Berlin
Heidelberg
New York
Barcelona
Budapest
Hong Kong
Milan
Paris
Santa Clara
Singapore
Tokyo

Zongli Lin

Low Gain Feedback

To Professor Khalil.

With the compliments of
the author

Springer

Series Advisory Board

A. Bensoussan · M.J. Grimble · P. Kokotovic · H. Kwakernaak
J.L. Massey · Y.Z. Tsypkin

Author

Professor Zongli Lin
Department of Electrical Engineering, University of Virginia,
Charlottesville, Virginia 22903, USA

ISBN 1-85233-081-3 Springer-Verlag London Berlin Heidelberg

British Library Cataloguing in Publication Data
Lin, Zongli
 Low gain feedback. - (Lecture notes in control and
 information sciences ; 240)
 1. Feedback control systems
 I. Title
 629.8'312
ISBN 1852330813

Library of Congress Cataloging-in-Publication Data
Lin, Zongli, 1964-
 Low gain feedback / Zongli Lin
  p.   cm. -- (Lecture notes in control and information sciences
; 240)
 Includes bibliographical references and index.
 ISBN 1-85233-081-3 (pbk. : alk. paper)
 1. Feedback control systems. I. Title. II. Series.
 TJ216.L49 1998   98-39948
 629.8'3--dc21   CIP

Typesetting: Camera ready by author
Printed and bound at the Athenæum Press Ltd., Gateshead, Tyne & Wear
69/3830-543210 Printed on acid-free paper

To Jian, Tony and Vivian

Preface

Compared to that of high gain feedback, the powerfulness of low gain feedback is less recognized. Recently, low gain design techniques have been being developed for an increasing number of control problems, including control of linear systems with saturating actuators, semi-global stabilization of minimum phase input-output linearizable systems, H_2- and H_∞-suboptimal control, and nonlinear H_∞-control.

The purpose of the present monograph is to present some recent development in low gain feedback and its applications in a coherent manner. While most of the materials are drawn from recent papers published by the author and his co-authors in the past several years, many design and proof techniques presented in the monograph are greatly improved, simplified and unified. The monograph also contains some new results that have not been submitted for publication.

The intended audience of this monograph includes practicing control engineers and researchers in areas related to control engineering. An appropriate background for this monograph would be some first year graduate courses in linear systems and multivariable control. Some background in nonlinear control systems would greatly facilitate the reading of the monograph.

Our presentation is organized as follows. In Chapter 1, after a short introduction to low gain feedback, in comparison of high gain feedback, we include a list of notations and acronyms that are used throughout the monograph.

Chapter 2 presents two basic low gain feedback design methodologies, one based on a direct eigenstructure assignment and the other on the solution of a certain parameterized algebraic Riccati equation, and establishes some fundamental properties of low gain feedback. Both continuous-time and discrete-time systems are considered.

Chapter 3 contains the first application of low gain feedback: semi-global asymptotic stabilization of linear systems with saturating actuators via linear feedback. By semi-global asymptotic stabilizability via linear feedback, we mean that for any *a priori* given (arbitrarily large) bounded set of the state space, we find a linear feedback law such that the origin is a locally asymptotically stable equilibrium point with the given set contained in its basin of attraction. Both state feedback and output feedback case are considered. Our main results establish the fact that, if a linear system is asymptotically null controllable with bounded controls, then, when subject to actuator magnitude saturation, it is semi-globally asymptotically stabilizable by linear state feedback. If, in addition, the system is also detectable, then it is semi-globally asymptotically stabilizable via linear output feedback. Both continuous-time and discrete-time systems are considered.

Chapter 4 addresses several fundamental control problems beyond semi-global asymptotic stabilization for linear systems subject to actuator magnitude saturation. These problems include input-additive disturbance rejection and robust semi-global asymptotic stabilization in the presence of matched nonlinear uncertainties. To solve these problems, two new design techniques, referred to as low-and-high gain feedback and low gain based variable structure control, are developed. Both design techniques are based on the low gain feedback design technique. As in the low gain feedback design, there are two approaches to the low-and-high gain feedback design, basing respectively on the direct eigenstructure assignment and on the solution of a certain parameterized algebraic Riccati equation.

Chapter 5 utilizes the low gain feedback design techniques to construct feedback laws that solve semi-global output regulation problems for linear systems subject to actuator magnitude saturation. Solvability conditions as well as explicit construction of feedback laws that solve the problems are explicitly given. Our problem formulation follows the classical formulation of output regulation for linear systems and is a problem of controlling a linear system subject to actuator saturation in order to have its output track (or reject) a family of reference (or disturbance) signals produced by some external generator. The utilization of low-and-high gain feedback of Chapter 4 in the solution of semi-global output regulation problems is discussed.

Chapter 6 examines the problem of semi-global almost disturbance decoupling with internal stability for linear systems subject to actuator magnitude saturation and input additive disturbance. Here by semi-global we mean that the disturbances are bounded either in magnitude or in energy by any *a priori* given (arbitrarily large) bounded number. The low-and-high gain feedback laws as developed in Chapter 4 is utilized to establish that, semi-global almost disturbance decoupling with local asymptotic stability is always solvable via linear state feedback as long as the system in the absence of actuator saturation is stabilizable, no matter where the poles of the open loop system are, and the locations of these poles play a role only when semi-global asymptotic stabilization is required.

Chapter 7 develops some scheduling techniques that upgrade the low-and-high gain feedback laws to elevate some of the semi-global results into global ones. By scheduling the high gain component of the low-and-high gain feedback law, we show that, in a sharp contrast to global asymptotic stabilization problem as discussed in Chapter 3 where open loop poles are required to be in the closed left-half plane, global finite gain L_p-stabilization can always be achieved via state feedback, no matter where the open loop poles are. Moreover, the L_p-gain can be made arbitrarily small. By scheduling both the low gain and high gain component of the low-and-high gain feedback law as developed in Chapter 4, the semi-global results of Chapter 4 can be made global.

Chapter 8 studies the problem of semi-global asymptotic stabilization of linear systems subject to both actuator magnitude and rate saturation. By utilizing the low gain feedback design technique, it is shown that, if a linear system is asymptotically null controllable with bounded controls, then, when subject to both actuator magnitude and rate saturation, it is semi-globally asymptotically stabilizable by linear state feedback. If, in addition, the system is also detectable, then it is semi-globally asymptotically stabilizable via linear output feedback. Both continuous-time and discrete-time systems are considered.

Chapter 9 deals with minimum-phase input-output linearizable systems. Combining low gain feedback and the classical high gain feedback, we obtain a new family of low-and-high gain feedback laws and use it to establish semi-global asymptotic stabilizability and the semi-global practical stabilizability of such systems under some weak conditions, most of which are necessary. The main role low gain feedback plays here is the avoidance of the so-called peaking phenomenon.

Chapter 10 demonstrates how low gain feedback can be utilized to explicitly construct feedback laws that solve the problem of perfect regulation and H_2-suboptimal control problems for linear systems with invariant zeros on the $j\omega$ axis (unit circle for discrete-time systems). The complexity due to $j\omega$-axis (unit circle) invariant zeros have been well-understood in the literature and, as a result, they are always excluded from consideration in the explicit construction of feedback laws. Both continuous-time and discrete-time case are presented.

Chapter 11 illustrates how low gain feedback can be utilized to solve the general H_∞ almost disturbance decoupling problem. The role low gain feedback plays here is the treatment of $j\omega$-axis (unit circle) invariant zeros. As in the H_2-suboptimal control problems, the major challenge in explicit construction of suboptimal feedback laws comes from the presence of $j\omega$ (unit circle) zeros.

In the literature on the explicit construction of feedback laws, these invariant zeros have always been excluded from consideration. Both continuous-time and discrete-time case are considered. Generalization to nonlinear systems is also included.

The next two chapters include the applications of the low-and-high gain feedback design techniques to some physical systems. Chapter 12 considers the problem of balancing an inverted pendulum on a carriage, where the physical limitations impose a constraint on the maximum allowable motion of the carriage. Using low gain feedback design ideas, we provide robust *linear* controllers that balance the pendulum without violating the maximum allowable motion constraint.

In Chapter 13, the low-and-high gain design technique of Chapter 4 is combined with another design technique recently developed for linear systems with magnitude saturating actuators, the piecewise-linear LQ control, to yield a new design technique for linear systems with rate saturating actuators. The combined design takes advantages of both design techniques, while avoiding their disadvantages. An open loop exponentially unstable F-16 class fighter aircraft is used to demonstrate the effectiveness of the combined design method.

Finally, in the Appendix, we collect some technical tools that we have used in more than one place in the monograph.

I have been fortunate to have the benefit of the collaboration of several co-workers, from whom I have learned a great deal. Many of the results presented in this monograph are the results of our collaboration. Among these co-workers are Dr. Siva Banda of the Air Force Wright Laboratory, Dr. Ben M. Chen of National University of Singapore, Mr. Ravi Mantri of US Robotics, Inc., Dr. Meir Pachter of the Air Force Institute of Technology, Dr. Ali Saberi of Washington State University, Dr. Pedda Sannuti of Rutgers University, Dr. Yacov Shamash of State University of New York at Stony Brook, and Dr. Anton Stoorvogel of Eindhoven University of Technology, Dr. Andrew R. Teel of University of California, Santa Barbara.

I am especially grateful to Dr. Ben Chen, Dr. Ali Saberi, Dr. Yacov Shamash, and Dr. Gang Tao of University of Virginia for the fine examples they set and their continual support and friendship. I am also grateful to University of Virginia, in particular the Department of Electrical Engineering and other colleagues in the department, for making an excellent environment in which we live and work. I am indebted to Professor Petar Kokotovic for suggesting the inverted pendulum system of Chapter 12 as an example for low gain feedback. I would also like to acknowledge the Air Force Office of Scientific Research, whose support enabled me to participate in the 1996 AFOSR Summer Faculty Research Program and to be exposed to flight control problems.

Last, but not the least, I owe a great deal to my wife, Jian, and my children, Tony and Vivian, for all their sacrifice. It is their understanding, encouragement and love that encourage me to strive on.

This monograph was typeset by the author using LaTeX. All simulations and numerical computations were carried out in MATLAB. Diagrams were generated using xfig.

Zongli Lin
Charlottesville, Virginia
June, 1998

Contents

Chapter 1

Introduction

1.1. Introduction to Low Gain Feedback

In the classical control theory of single input single output systems, it is known that well-designed high gain feedback systems have the advantages of high steady-state accuracy consistent with stability, fast response, disturbance rejection, and insensitivity to parameter uncertainties and distortions (e.g., [22]). In the past two decades or so, many efforts have been devoted to the study of high gain feedback systems, leading to several high gain feedback design methodologies that extend the classical high gain feedback control theory to multivariable systems. These methods include multivariable asymptotic root loci (e.g., [23,34,35,81,96]), fast time scale assignment (e.g., [91]), perfect regulation or cheap optimal control regulators (e.g., [17,32,37,63]), geometric methods (e.g., [117,124,125]), semi-global stabilization of nonlinear systems (e.g., [6,42,107]), nonlinear H_∞-optimal control (e.g., [46,49,75]), and various other miscellaneous methods (e.g., [1,38,99]).

The concept underlying high gain feedback is that of asymptotics and, hence, by high gain feedback we mean a family of feedback laws in which a parameterized gain matrix, say $F(\varepsilon)$, approaches infinity as the parameter approaches its extreme value (typically zero or infinity). Consequently, the implementation of high gain feedback laws entails large control inputs (either in magnitude or in energy) and hence large actuation capacities. Another intricate feature of high gain feedback design methods is the restriction it sometimes places on the open loop system in order to achieve certain design objectives. One of most common such restrictions is the minimum-phase assumption, as demonstrated by the classical asymptotic root-loci design method.

1

As will be demonstrated throughout this monograph, low gain feedback has been conceived to either avoid or to complement high gain feedback whenever such "unpleasant" features of high gain feedback prevent certain control objectives from being achieved. In the past few years, we have developed several low gain design methods to achieve various control objectives that high gain feedback (or high gain feedback alone) could not achieve. These objectives include control of linear systems subject to actuator magnitude and/or rate saturation (e.g., [42–45,47,54,58,60,61,67–70,89]), semi-global stabilization of minimum-phase input-output linearizable nonlinear systems (e.g., [42,55,59]), nonlinear H_∞-control [46,49], H_2- and H_∞-suboptimal control [12,50,63–66], and stabilization of an inverted pendulum on a carriage with limited travel [62].

Similar to that of high gain feedback, the concept underlying low gain feedback is also that of asymptotics and, roughly speaking, by low gain feedback we mean a family of feedback laws in which a parameterized gain matrix, say $F(\varepsilon)$, approaches zero as the parameter ε approaches zero. In the development of low gain feedback design techniques, one observes that the closed-loop system properties induced by the low gain feedback are often mirror images of those induced by the high gain feedback, and are beautifully symmetric to each other. For example, high gain feedback induces fast time scales while low gain feedback induces slow ones.

There have also been cases where both low gain and high gain feedback are needed in order to achieve certain design objectives. In such cases, the design procedure is typically sequential. A family of low gain feedback laws, say, parameterized in ε, is first constructed. Based on the low gain design, a family of high gain feedback laws is then constructed. The high gain feedback laws could be parameterized in a new parameter, say ρ, or in the same parameter ε. The two families of feedback laws are then combined in a certain way to arrive at the desired final feedback laws, called low-and-high gain feedback laws.

As in any parameterization, the determination of the values for the low gain and high gain parameters is of great practical importance and often difficult. One way around this difficulty is to schedule these parameters as functions of the state of the system. The feedback laws are consequently referred to as scheduled low gain, scheduled high gain, or scheduled low-and-high gain feedback.

In comparison with that of high gain feedback, the development of low gain feedback design methodologies has attracted less attention. As mentioned earlier, we have recently started to systematically explore the effectiveness of low gain feedback (and of a combination of low and high gain feedback) and have developed several low gain and low-and-high gain design techniques for a number of control problems, including control of linear systems with actuator saturation, semi-global stabilization for minimum phase input output linearizable systems, H_2- and H_∞-suboptimal control problems, nonlinear H_∞-control problems and stabilization of some mechanical systems.

The purpose of this monograph is to systematically present various low gain based design techniques and to demonstrate their applications in solving various control problems.

1.2. Notations and Acronyms

Throughout this monograph we shall adopt the following notations and acronyms:

$\mathbb{R} :=$ the set of real numbers,

$\mathbb{R}_+ :=$ the set of nonnegative real numbers,

$\mathbb{N} :=$ the set of natural numbers,

$\mathbb{C} :=$ the entire complex plane,

$\mathbb{C}^- :=$ the open left-half complex plane,

$\mathbb{C}^+ :=$ the open right-half complex plane,

$\mathbb{C}^0 :=$ the imaginary axis in the complex plane,

$\mathbb{C}^\odot :=$ the set of complex numbers inside the unit circle,

$\mathbb{C}^\otimes :=$ the set of complex numbers outside the unit circle,

$\mathbb{C}^\mathrm{O} :=$ the unit circle in the complex plane,

$|x| :=$ the Euclidean norm, or 2-norm, of $x \in \mathbb{R}^n$,

$|x|_\infty := \max_i |x_i|$ for $x \in \mathbb{R}^n$,

$C^n :=$ the set of n times continuously differentiable functions,

$L_p^n :=$ the set of all measurable functions $x : [0, \infty) \to \mathbb{R}^n$ such that $\int_0^\infty |x(t)|^p dt < \infty$ for any $p \in [1, \infty)$,

$L_\infty^n :=$ the set of all measurable functions $x : [0, \infty) \to \mathbb{R}^n$ such that ess $\sup_{t \in [0,\infty)} |x(t)| < \infty$,

$\|x\|_{L_p} := \left(\int_0^\infty |x(t)|^2 dt \right)^{\frac{1}{2}}$, the L_p-norm of any $x \in L_p^n$,

$\|x\|_{L_\infty} :=$ ess $\sup_{t \in [0,\infty)} |x(t)|$, the L_∞-norm of any $x \in L_\infty^n$,

$\|x\|_{\infty,\mathrm{T}} :=$ ess $\sup_{t \in [T,\infty)} |x(t)|_\infty$, for any $T \geq 0$, and any $x \in L_\infty^n$,

$l_p^n :=$ the set of all sequences $\{x(k) \in \mathbb{R}^n\}_{k=0}^\infty$ such that $\sum_{k=0}^\infty |x(k)|^p dt < \infty$ for any $p \in [1, \infty)$,

$l_\infty^n :=$ the set of all sequences $\{x(k) \in \mathbb{R}^n\}_{k=0}^\infty$ such that

$$\sup_k |x(k)| < \infty,$$

$\|x\|_{l_p} := \left(\sum_{k=0}^{\infty} |x(k)|^2 dt \right)^{\frac{1}{2}}$, the l_p-norm of any $x \in l_p^n$,

$\|x\|_{l_\infty} := \sup_k |x(k)|$, the l_∞-norm of any $x \in l_\infty^n$,

$\|x\|_{\infty,K} := \sup_{k \geq K} |x(k)|$, for any $K \geq 0$, and any $x \in l_\infty^n$,

$A \otimes B :=$ Kronecker product of matrix A and B,

$I :=$ an identity matrix,

$I_k :=$ an identity matrix of dimension $k \times k$,

$|X| :=$ 2-norm of matrix X,

$X' :=$ the transpose of X,

$X^\dagger :=$ the Moore-Penrose (pseudo) inverse of X,

$\lambda(X) :=$ the set of eigenvalues of X,

$\lambda_{\max}(X) :=$ the maximum eigenvalues of X where $\lambda(X) \subset \mathbb{R}$,

$\lambda_{\min}(X) :=$ the minimum eigenvalues of X where $\lambda(X) \subset \mathbb{R}$,

$\mathrm{Im}\,(X) :=$ the range space of X,

$\boxed{A} :=$ the end of an algorithm,

$\boxed{C} :=$ the end of a corollary,

$\boxed{D} :=$ the end of a definition,

$\boxed{E} :=$ the end of an example,

$\boxed{H} :=$ the end of a hypothesis or an assumption,

$\boxed{L} :=$ the end of a lemma,

$\boxed{O} :=$ the end of an observation,

$\boxed{P} :=$ the end of a problem, a property or a proposition,

$\boxed{R} :=$ the end of a remark,

$\boxed{T} :=$ the end of a theorem,

$\boxed{X} :=$ the end of a proof,

ADDPMS := almost disturbance decoupling problem with measurement feedback and internal stability,

ADDPS := almost disturbance decoupling problem with state feedback and with internal stability,

ANCBC := asymptotically null controllable with bounded controls

ARE := algebraic Riccati equation,

LHG := low-and-high gain,

PLC := piecewise linear LQ control,

SCB := special coordinate basis.

Also,

- For the state variable x of a discrete-time system, we use the notation $x^+(k)$ to indicate $x(k+1)$,

- For $x \in \mathbb{R}^n$ and $y \in \mathbb{R}^m$, we often, by an abuse of notation, write (x, y) instead of $[x', y']'$,

- A function $f : \mathcal{W} \to \mathbb{R}_+$ is said to be positive definite on $\mathcal{W}_0 \subset \mathcal{W}$ if $f(x)$ is strictly positive for all $x \in \mathcal{W}_0$,

- For a continuous function $V : \mathbb{R}^n \to \mathbb{R}_+$, a level set $L_V(c)$ is defined as $L_V(c) := \{x \in \mathbb{R}^n : V(x) \le c\}$. Also, denote $L_V^o(c) := \{x \in \mathbb{R}^n : V(x) < c\}$,

- A function $\beta : \mathbb{R}_+ \to \mathbb{R}_+$ is said to be of class $\mathcal{K}$ if it is continuous, strictly increasing and satisfies $\beta(0) = 0$. It is of class $\mathcal{K}_\infty$ if in addition $\beta(s) \to \infty$ as $s \to \infty$.

Chapter 2

Basic Low Gain Feedback Design Techniques

2.1. Introduction

In this chapter, we present two basic low gain state feedback design techniques. The first low gain feedback design technique is developed using direct eigenstructure assignment. The other is based on the solutions of parameterized algebraic Riccati equations (ARE's). These two low gain design techniques yield two different families of state feedback laws, parameterized in a low-gain parameter ε. These families of state feedback laws play fundamental roles in establishing our results throughout this monograph.

We will consider both continuous-time and discrete-time systems. The continuous-time case is dealt with in Section 2.2 while its discrete-time counterpart is treated in Section 2.3. In both cases, we first explicitly construct these families of low gain state feedbacks and then establish their fundamental properties. Finally, we draw some concluding remarks on the comparison between the different low gain feedback design techniques in Section 2.4.

2.2. Continuous-Time Systems

Consider the linear system

$$\dot{x}(t) = Ax(t) + Bu(t), \ x(t) \in \mathbb{R}^n, u(t) \in \mathbb{R}^m, \tag{2.2.1}$$

where $x(t)$ is the state and $u(t)$ is the input.[1] We make the following assumption.

---
[1] For simplicity, we will suppress the independent variable t throughout the monograph.

7

Assumption 2.2.1. *The pair* (A, B) *is asymptotically null controllable with bounded controls (ANCBC), i.e.,*

1. (A, B) *is stabilizable;*

2. *All eigenvalues of* A *are in the closed left half* s-*plane.*   ⊞

The next two subsections each present a technique for designing low gain state feedback.

2.2.1. Eigenstructure Assignment Based Method

The eigenstructure assignment based low gain state feedback design is carried out in three steps.

Step 1. Find nonsingular transformation matrices T_s and T_I such that the pair (A, B) is transformed into the following block diagonal control canonical form,

$$T_s^{-1}AT_s = \begin{bmatrix} A_1 & 0 & \cdots & 0 & 0 \\ 0 & A_2 & \cdots & 0 & 0 \\ \vdots & \vdots & \ddots & \vdots & \vdots \\ 0 & 0 & \cdots & A_l & 0 \\ 0 & 0 & \cdots & 0 & A_0 \end{bmatrix}, \tag{2.2.2}$$

$$T_s^{-1}BT_I = \begin{bmatrix} B_1 & B_{12} & \cdots & B_{1l} & * \\ 0 & B_2 & \cdots & B_{2l} & * \\ \vdots & \vdots & \ddots & \vdots & \vdots \\ 0 & 0 & \cdots & B_l & * \\ B_{01} & B_{02} & \cdots & B_{0l} & * \end{bmatrix}, \tag{2.2.3}$$

where, A_0 contains all the open left-half plane eigenvalues of A, for each $i = 1$ to l, all eigenvalues of A_i are on the $j\omega$ axis and hence (A_i, B_i) is controllable as given by,

$$A_i = \begin{bmatrix} 0 & 1 & 0 & \cdots & 0 \\ 0 & 0 & 1 & \cdots & 0 \\ \vdots & \vdots & \vdots & \ddots & \vdots \\ 0 & 0 & 0 & \cdots & 1 \\ -a_{n_i}^i & -a_{n_i-1}^i & -a_{n_i-2}^i & \cdots & -a_1^i \end{bmatrix}, \quad B_i = \begin{bmatrix} 0 \\ 0 \\ \vdots \\ 0 \\ 1 \end{bmatrix},$$

and finally, $*$'s represent submatrices of less interest.

We note that the existence of the above canonical form was shown in [126]. The software realization can be found in [9].

Step 2. For each (A_i, B_i), let $F_i(\varepsilon) \in \mathbb{R}^{1 \times n_i}$ be the state feedback gain such that

$$\lambda(A_i + B_i F_i(\varepsilon)) = -\varepsilon + \lambda(A_i) \in \mathbb{C}^-, \quad \varepsilon \in (0, 1]. \qquad (2.2.4)$$

Note that $F_i(\varepsilon)$ is unique.

Step 3. Construct a family of low gain state feedback laws for system (2.2.1) as,

$$u = F(\varepsilon)x, \qquad (2.2.5)$$

where the low gain matrix $F(\varepsilon)$ is given by

$$F(\varepsilon) = T_{\mathrm{I}} \begin{bmatrix} F_1(\varepsilon) & 0 & \cdots & 0 & 0 & 0 \\ 0 & F_2(\varepsilon) & \cdots & 0 & 0 & 0 \\ \vdots & \vdots & \ddots & \vdots & \vdots & \vdots \\ 0 & 0 & \cdots & F_{l-1}(\varepsilon) & 0 & 0 \\ 0 & 0 & \cdots & 0 & F_l(\varepsilon) & 0 \\ 0 & 0 & \cdots & 0 & 0 & 0 \end{bmatrix} T_s^{-1}. \qquad (2.2.6)$$

Ⓐ

Remark 2.2.1. *It is clear from the construction of $F(\varepsilon)$ that $\lim_{\varepsilon \to 0} F(\varepsilon) = 0$. For this reason the family of state feedback laws (2.2.5) are referred to as low gain feedback, and ε the low gain parameter. We however also note that, as the following lemma shows, the low gain feedback laws (2.2.5) possess more intricate properties than simply having small feedback gains.* Ⓡ

Lemma 2.2.1. *Consider a single input pair (A, B) in the control canonical form,*

$$A = \begin{bmatrix} 0 & 1 & 0 & \cdots & 0 \\ 0 & 0 & 1 & \cdots & 0 \\ \vdots & \vdots & \vdots & \ddots & \vdots \\ 0 & 0 & 0 & \cdots & 1 \\ -a_n & -a_{n-1} & -a_{n-2} & \cdots & -a_1 \end{bmatrix}, \quad B = \begin{bmatrix} 0 \\ 0 \\ \vdots \\ 0 \\ 1 \end{bmatrix}. \qquad (2.2.7)$$

Assume that all the eigenvalues are in the closed left-half s-plane. Let $F(\varepsilon) \in \mathbb{R}^{1 \times n}$ be the state feedback gain such that $\lambda(A + BF(\varepsilon)) = -\varepsilon + \lambda(A)$. Then, there exists an $\varepsilon^* \in (0, 1]$ such that for all $\varepsilon \in (0, \varepsilon^*]$,

$$|F(\varepsilon)| \le \alpha\varepsilon, \qquad (2.2.8)$$

$$\left| e^{(A + BF(\varepsilon))t} \right| \le \frac{\beta}{\varepsilon^{r-1}} e^{-\varepsilon t/2}, \ \forall t \ge 0, \qquad (2.2.9)$$

$$\left| F(\varepsilon)(A + BF(\varepsilon))^\ell e^{(A + BF(\varepsilon))t} \right| \le \gamma_\ell \varepsilon e^{-\varepsilon t/2}, \ \forall \ell \in \mathbb{N}, \ \forall t \ge 0, \quad (2.2.10)$$

where r is the largest algebraic multiplicity of the eigenvalues of A, and α, β and γ_l's are all positive constants independent of ε. Ⓛ

Remark 2.2.2. *Property (2.2.8) indicates the asymptotic nature of the feed-back laws (2.2.5), i.e., $\lim_{\varepsilon \to 0} |F(\varepsilon)| = 0$.*

Recalling that $e^{(A+BF(\varepsilon))t}$ is the transition matrix of the closed-loop system under the low gain feedback, we see that Property (2.2.9) reveals that the closed-loop system under low gain feedback will peak slowly to a magnitude of order $O(1/\varepsilon^{r-1})$, with r being the largest algebraic multiplicity of the eigenvalues of A. Property (2.2.10) on the other hand implies that for any given bounded set of initial conditions, the control and all its derivatives can be made arbitrarily small by decreasing the value of the low gain parameter ε. This can also be seen in the simulation of the following example. ℝ

Example 2.2.1. Consider the single input system

$$\dot{x} = Ax + Bu, \qquad\qquad (2.2.11)$$

with

$$A = \begin{bmatrix} 0 & 1 & 0 & 0 \\ 0 & 0 & 1 & 0 \\ 0 & 0 & 0 & 1 \\ -1 & 0 & -2 & 0 \end{bmatrix}, \quad B = \begin{bmatrix} 0 \\ 0 \\ 0 \\ 1 \end{bmatrix}.$$

It is straightforward to verify that (A, B) is controllable with four controllable modes at $\{-j, -j, j, j\}$, where $j = \sqrt{-1}$. Following the eigenstructure assignment based low gain feedback design algorithm, we construct the following family of linear state feedback control laws,

$$u = [-\varepsilon^4 - 2\varepsilon^2 \quad -4\varepsilon^3 - 4\varepsilon \quad -6\varepsilon^2 \quad -4\varepsilon]x. \qquad (2.2.12)$$

Some simulation of the closed-loop system is given in Figs. 2.2.1-2.2.2. It can be easily seen from these figures that, for the same initial conditions, as the value of ε decreases, the state peaks slowly while the L_∞-norm of the control input decreases. 𝔼

Proof of Lemma 2.2.1. First, (2.2.8) follows trivially from the fact $F(\varepsilon)$ is a vector of polynomials in ε and $F(0) = 0$.

We next proceed to show (2.2.9) and (2.2.10). To this end, let

$$\det (sI - A) = \prod_{i=1}^{p} (s - \lambda_i)^{n_i},$$

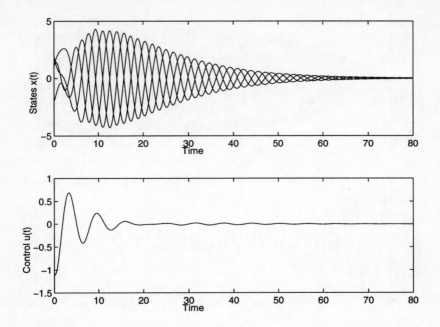

Figure 2.2.1: States and control input: $x(0) = (1, 2, -2, 1)$ and $\varepsilon = 0.1$.

where $\lambda_i \neq \lambda_j$, $i \neq j$. Then, for each $i = 1$ to p, the n_i generalized eigenvectors of A are given by ([27]),

$$
p_1^i = \begin{bmatrix} 1 \\ \lambda_i \\ \lambda_i^2 \\ \vdots \\ \lambda_i^{n-2} \\ \lambda_i^{n-1} \end{bmatrix}, \ p_2^i = \begin{bmatrix} 0 \\ 1 \\ 2\lambda_i \\ 3\lambda_i^2 \\ \vdots \\ (n-1)\lambda_i^{n-2} \end{bmatrix}, \cdots, p_{n_i}^i = \begin{bmatrix} 0 \\ 0 \\ 0 \\ \vdots \\ C_{n-2}^{n_i-1}\lambda_i^{n-n_i-1} \\ C_{n-1}^{n_i-1}\lambda_i^{n-n_i} \end{bmatrix}.
$$

Similarly, for each $i = 1$ to p, the n_i generalized eigenvectors of $A + BF(\varepsilon)$ are given by

$$
q_1^i = \begin{bmatrix} 1 \\ \lambda_{\varepsilon i} \\ \lambda_{\varepsilon i}^2 \\ \vdots \\ \lambda_{\varepsilon i}^{n-2} \\ \lambda_{\varepsilon i}^{n-1} \end{bmatrix}, \ q_2^i = \begin{bmatrix} 0 \\ 1 \\ 2\lambda_{\varepsilon i} \\ 3\lambda_{\varepsilon i}^2 \\ \vdots \\ (n-1)\lambda_{\varepsilon i}^{n-2} \end{bmatrix}, \cdots, q_{n_i}^i = \begin{bmatrix} 0 \\ 0 \\ 0 \\ \vdots \\ C_{n-2}^{n_i-1}\lambda_{\varepsilon i}^{n-n_i-1} \\ C_{n-1}^{n_i-1}\lambda_{\varepsilon i}^{n-n_i} \end{bmatrix},
$$

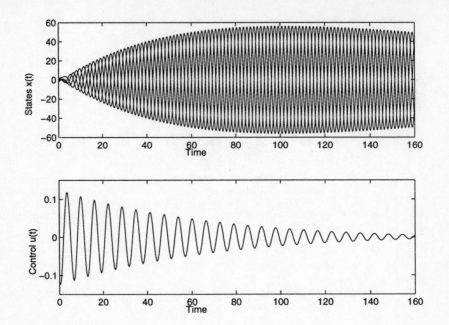

Figure 2.2.2: States and control input: $x(0) = (1, 2, -2, 1)$ and $\varepsilon = 0.01$.

where $\lambda_{\varepsilon i} = -\varepsilon + \lambda_i$ and C_n^i is defined as

$$C_n^i = \begin{cases} \dfrac{n!}{(n-i)!i!} & 0 \le i \le n, \\ 0 & \text{otherwise.} \end{cases}$$

We next form the following two nonsingular transformation matrices,

$$P = [p_1^1, p_2^1, \cdots, p_{n_1}^1, \cdots, p_1^p, p_2^p, \cdots, p_{n_p}^p],$$

which is independent of ε, and

$$Q(\varepsilon) = [q_1^1, q_2^1, \cdots, q_{n_1}^1, \cdots, q_1^p, q_2^p, \cdots, q_{n_p}^p].$$

It then follows that

$$P^{-1}AP = \begin{bmatrix} J_1 & 0 & \cdots & 0 \\ 0 & J_2 & \cdots & 0 \\ \vdots & \vdots & \ddots & \vdots \\ 0 & 0 & \cdots & J_p \end{bmatrix}, \quad J_i = \begin{bmatrix} \lambda_i & 1 & 0 & \cdots & 0 \\ 0 & \lambda_i & 1 & \cdots & 0 \\ \vdots & \vdots & \vdots & \cdots & \vdots \\ 0 & 0 & 0 & \cdots & \lambda_i \end{bmatrix}, \quad (2.2.13)$$

and

$$Q^{-1}(\varepsilon)(A + BF(\varepsilon))Q(\varepsilon) = \begin{bmatrix} J_{\varepsilon 1} & 0 & \cdots & 0 \\ 0 & J_{\varepsilon 2} & \cdots & 0 \\ \vdots & \vdots & \ddots & \vdots \\ 0 & 0 & \cdots & J_{\varepsilon p} \end{bmatrix},$$

$$J_{\varepsilon i} = \begin{bmatrix} \lambda_{\varepsilon i} & 1 & 0 & \cdots & 0 \\ 0 & \lambda_{\varepsilon i} & 1 & \cdots & 0 \\ \vdots & \vdots & \vdots & \cdots & \vdots \\ 0 & 0 & 0 & \cdots & \lambda_{\varepsilon i} \end{bmatrix}. \qquad (2.2.14)$$

By the definition of generalized eigenvectors, we have the following relationships,

$$Ap_1^i = \lambda_i p_1^i,$$
$$Ap_j^i = p_{j-1}^i + \lambda_i p_j^i, \quad j = 2, 3, \cdots, n_i, \ i = 1, 2, \cdots, p, \qquad (2.2.15)$$

and

$$(A + BF(\varepsilon))q_1^i = \lambda_{\varepsilon i} q_1^i,$$
$$(A + BF(\varepsilon))q_j^i = q_{j-1}^i + \lambda_{\varepsilon i} q_j^i, \ j = 2, 3, \cdots, n_i, \ i = 1, 2, \cdots, p. \ (2.2.16)$$

Denoting $A_n = [-a_n \ -a_{n-1} \ \cdots \ -a_1]$, it then follows form (2.2.15) and (2.2.16) that, for each $i = 1$ to p, we have, for $j = 1, 2, \cdots, n_i$,

$$A_n p_j^i = C_{n-1}^{j-1}\lambda_i^{n-j+1} + C_{n-1}^{j-2}\lambda_i^{n-j+1} = C_n^{j-1}\lambda_i^{n-j+1},$$
$$(A_n + F(\varepsilon))q_j^i = C_{n-1}^{j-1}\lambda_{\varepsilon i}^{n-j+1} + C_{n-1}^{j-2}\lambda_{\varepsilon i}^{n-j+1} = C_n^{j-1}\lambda_{\varepsilon i}^{n-j+1},$$

from which we have

$$F(\varepsilon)q_j^i = -A_n q_j^i + C_n^{j-1}\lambda_{\varepsilon i}^{n-j+1}$$

$$= -A_n \begin{bmatrix} 0 \\ \vdots \\ 0 \\ C_{j-1}^{j-1} \\ C_j^{j-1}(-\varepsilon + \lambda_i) \\ \vdots \\ C_{n-1}^{j-1}(-\varepsilon + \lambda_i)^{n-j} \end{bmatrix} + C_n^{j-1}(-\varepsilon + \lambda_i)^{n-j+1}$$

$$= -A_n \begin{bmatrix} 0 \\ \vdots \\ 0 \\ C_{j-1}^{j-1} \\ C_j^{j-1} \sum_{k=0}^{1} (-\varepsilon)^k \lambda_i^{1-k} \\ \vdots \\ C_{n-1}^{j-1} \sum_{k=0}^{n-j} C_{n-j}^k (-\varepsilon)^k \lambda_i^{n-j-k} \end{bmatrix} + C_n^{j-1} (-\varepsilon + \lambda_i)^{n-j+1}$$

$$= -A_n \sum_{k=0}^{n_i-j} (-\varepsilon)^k C_{k+j-1}^k p_{j+k}^i + C_n^{j-1} \sum_{k=0}^{n_i-j} (-\varepsilon)^k C_{n-j+1}^k \lambda_i^{n-j-k+1}$$

$$+ \Delta\left(\varepsilon^{n_i-j+1}\right)$$

$$= -\sum_{k=0}^{n_i-j} (-\varepsilon)^k C_{k+j-1}^k C_n^{j+k-1} \lambda_i^{n-j-k+1}$$

$$+ C_n^{j-1} \sum_{k=0}^{n_i-j} (-\varepsilon)^k C_{n-j+1}^k \lambda_i^{n-j-k+1} + \Delta\left(\varepsilon^{n_i-j+1}\right)$$

$$= \Delta\left(\varepsilon^{n_i-j+1}\right),$$

where $\Delta(\varepsilon^{n_i-j+1})$ is a polynomial in ε whose coefficients of terms of order lower than $n_i - j + 1$ are all zero. Hence, we have, for $\varepsilon \in (0, 1]$,

$$\left|F(\varepsilon)q_j^i\right| \leq \delta_j^i \varepsilon^{n_i-j+1}, \tag{2.2.17}$$

where δ_j^i is some positive constant independent of ε.

We next note that both $Q(\varepsilon)$ and $Q^{-1}(\varepsilon)$ are continuous functions of ε satisfying $Q(0) = P$ and $Q^{-1}(0) = P^{-1}$. Now from the continuity of norm functions it follows that there exists an $\varepsilon^* \in (0, 1]$, such that for all $\varepsilon \in (0, \varepsilon^*]$,

$$|Q(\varepsilon)| \leq |P| + 1, \quad \left|Q^{-1}(\varepsilon)\right| \leq |P^{-1}| + 1. \tag{2.2.18}$$

We are now ready to show (2.2.9) and (2.2.10). It follows from (2.2.17) and (2.2.18) that, for all $\varepsilon \in (0, \varepsilon^*]$,

$$\left|F(\varepsilon)e^{(A+BF(\varepsilon))t}\right| = \left|F(\varepsilon)Q(\varepsilon)e^{Q^{-1}(\varepsilon)(A+BF(\varepsilon))Q(\varepsilon)t}Q^{-1}(\varepsilon)\right|$$

$$\leq (|P^{-1}| + 1) \sum_{i=1}^{p} \left|F(\varepsilon)[q_1^i, q_2^i, \cdots, q_{n_i}^i]e^{J_{\varepsilon i}t}\right|$$

$$\leq (|P^{-1}| + 1) \sum_{i=1}^{p} \left|F(\varepsilon)[q_1^i, q_2^i, \cdots, q_{n_i}^i]\right|$$

$$\times \begin{bmatrix} e^{\lambda_\varepsilon i t} & t e^{\lambda_\varepsilon i t} & t^2 e^{\lambda_\varepsilon i t}/2! & \cdots & t^{n_i-1} e^{\lambda_\varepsilon i t}/(n_i-1)! \\ 0 & e^{\lambda_\varepsilon i t} & t e^{\lambda_\varepsilon i t} & \cdots & t^{n_i-2} e^{\lambda_\varepsilon i t}/(n_i-2)! \\ \vdots & \vdots & \vdots & \ddots & \vdots \\ 0 & 0 & 0 & \cdots & e^{\lambda_\varepsilon i t} \end{bmatrix} \Bigg]$$

$$\leq (|P^{-1}|+1) \sum_{i=1}^{p} \sum_{j=1}^{n_i} \sum_{l=1}^{j} \delta_l^i \varepsilon^{n_i-l+1} t^{j-l} e^{-\varepsilon t}/(j-l)!$$

$$\leq (|P^{-1}|+1) \sum_{i=1}^{p} \sum_{j=1}^{n_i} \sum_{l=1}^{j} \frac{2^{j-l} \delta_l^i (j-l)^{j-l} \varepsilon^{n_i-j+1}}{(j-l)!} e^{-\varepsilon t/2}$$

$$\leq (|P^{-1}|+1) \sum_{i=1}^{p} \sum_{j=1}^{n_i} \sum_{l=1}^{j} \frac{2^{j-l} \delta_l^i (j-l)^{j-l}}{(j-l)!} \varepsilon e^{-\varepsilon t/2}$$

$$= \gamma_0 \varepsilon e^{-\varepsilon t/2},$$

where

$$\gamma_0 = (|P^{-1}|+1) \sum_{i=1}^{p} \sum_{j=1}^{n_i} \sum_{l=1}^{j} \frac{2^{j-l} \delta_l^i (j-l)^{j-l}}{(j-l)!}$$

is independent of ε. This shows (2.2.10) for $\ell = 0$, which together with (2.2.8) and the commutability between $(A + BF(\varepsilon))^\ell$ and $e^{(A+BF(\varepsilon))t}$ imply (2.2.10) for $\ell \geq 1$.

Finally, using (2.2.18), (2.2.9) can be shown as follows. For all $\varepsilon \in (0, \varepsilon^*]$,

$$\left| e^{(A+BF(\varepsilon))t} \right| = \left| Q(\varepsilon) e^{Q^{-1}(\varepsilon)(A+BF(\varepsilon))Q(\varepsilon)t} Q^{-1}(\varepsilon) \right|$$

$$\leq (|P|+1)(|P^{-1}|+1) \sum_{i=1}^{p} \left| e^{J_\varepsilon i t} \right|$$

$$\leq (|P|+1)(|P^{-1}|+1)$$

$$\times \sum_{i=1}^{p} \Bigg[\begin{bmatrix} e^{\lambda_\varepsilon i t} & t e^{\lambda_\varepsilon i t} & t^2 e^{\lambda_\varepsilon i t}/2! & \cdots & t^{n_i-1} e^{\lambda_\varepsilon i t}/(n_i-1)! \\ 0 & e^{\lambda_\varepsilon i t} & t e^{\lambda_\varepsilon i t} & \cdots & t^{n_i-2} e^{\lambda_\varepsilon i t}/(n_i-2)! \\ \vdots & \vdots & \vdots & \ddots & \vdots \\ 0 & 0 & 0 & \cdots & e^{\lambda_\varepsilon i t} \end{bmatrix} \Bigg]$$

$$\leq (|P|+1)(|P^{-1}\|+1) \sum_{i=1}^{p} \sum_{j=1}^{n_i} \sum_{l=1}^{j} t^{j-l} e^{-\varepsilon t}/(j-l)!$$

$$\leq (|P|+1)(|P^{-1}\|+1) \sum_{i=1}^{p} \sum_{j=1}^{n_i} \sum_{l=1}^{j} \frac{2^{j-l}(j-l)^{j-l}}{(j-l)!\varepsilon^{j-l}} e^{-\varepsilon t/2}$$

$$\leq (|P|+1)(|P^{-1}|+1) \sum_{i=1}^{p} \sum_{j=1}^{n_i} \sum_{l=1}^{j} \frac{2^{j-l}(j-l)^{j-l}}{(j-l)!\varepsilon^{r-1}} e^{-\varepsilon t/2}$$

$$= \beta \frac{e^{-\varepsilon t/2}}{\varepsilon^{r-1}},$$

where $r = \max\{n_1, n_2, \cdots, n_p\}$ and

$$\beta = (|P| + 1)\,(|P^{-1}| + 1) \sum_{i=1}^{p} \sum_{j=1}^{n_i} \sum_{l=1}^{j} \frac{2^{j-l}(j-l)^{j-l}}{(j-l)!}$$

is independent of ε.   ⊠

The next three lemmas, Lemmas 2.2.2-2.2.4, establish a framework for Lyapunov analysis of the closed-loop system under low gain feedback. As can be seen throughout this monograph, this framework simplifies and unifies the proofs of various results due to low gain feedback.

Lemma 2.2.2. *Consider a single input pair* (A, B) *in the form of (2.2.7) with all eigenvalues of A on the $j\omega$ axis. Let $F(\varepsilon) \in \mathbb{R}^{1 \times n}$ be the unique matrix such that $\lambda(A + BF(\varepsilon)) = -\varepsilon + \lambda(A)$, $\varepsilon \in (0, 1]$. Then, there exists a nonsingular transformation matrix $Q(\varepsilon) \in \mathbb{R}^{n \times n}$ such that*

$$Q^{-1}(\varepsilon)(A + BF(\varepsilon))Q(\varepsilon) = J(\varepsilon) := \mathrm{blkdiag}\,\{J_0(\varepsilon), J_1(\varepsilon), \cdots, J_l(\varepsilon)\}, \quad (2.2.19)$$

where

$$J_0(\varepsilon) = \begin{bmatrix} -\varepsilon & 1 & & \\ & \ddots & \ddots & \\ & & -\varepsilon & 1 \\ & & & -\varepsilon \end{bmatrix}_{n_0 \times n_0}, \quad\quad (2.2.20)$$

and for each $i = 1$ to l,

$$J_i(\varepsilon) = \begin{bmatrix} J_i^\star(\varepsilon) & I_2 & & \\ & \ddots & \ddots & \\ & & J_i^\star(\varepsilon) & I_2 \\ & & & J_i^\star(\varepsilon) \end{bmatrix}_{2n_i \times 2n_i}, \quad J_i^\star(\varepsilon) = \begin{bmatrix} -\varepsilon & \beta_i \\ -\beta_i & -\varepsilon \end{bmatrix},$$

$$(2.2.21)$$

with $\beta_i > 0$ for all $i = 1$ to l and $\beta_i \neq \beta_j$ for $i \neq j$.   Ⓛ

Remark 2.2.3. *For each $i = 1$ to l, $J_i(0)$ can be viewed as a chain of "generalized integrators" of length n_i, with each "generalized integrator corresponding to a pair of $j\omega$ axis poles. As the next lemma shows, such a notion of chains of generalized integrators allows us to develop a slow time scale assignment through low gain feedback.*   Ⓡ

Proof of Lemma 2.2.2. Let

$$\det\left(sI - A - BF(\varepsilon)\right) = (s+\varepsilon)^{n_0} \prod_{i=1}^{l} (s+\varepsilon - j\beta_i)^{n_i}(s+\varepsilon + j\beta_i)^{n_i}. \quad (2.2.22)$$

Then, the n_0 generalized eigenvectors $A + BF(\varepsilon)$ corresponding to the eigenvalue $\lambda_{\varepsilon 0} = -\varepsilon$ are ([27]),

$$q_1^0(\varepsilon) = \begin{bmatrix} 1 \\ \lambda_{\varepsilon 0} \\ \lambda_{\varepsilon 0}^2 \\ \vdots \\ \lambda_{\varepsilon 0}^{n-2} \\ \lambda_{\varepsilon 0}^{n-1} \end{bmatrix}, \quad q_2^0(\varepsilon) = \begin{bmatrix} 0 \\ 1 \\ 2\lambda_{\varepsilon 0} \\ 3\lambda_{\varepsilon 0}^2 \\ \vdots \\ (n-1)\lambda_{\varepsilon 0}^{n-2} \end{bmatrix}, \cdots,$$

$$q_{n_0}^0(\varepsilon) = \begin{bmatrix} 0 \\ 0 \\ 0 \\ \vdots \\ C_{n-2}^{n_0-1}\lambda_{\varepsilon 0}^{n-n_0-1} \\ C_{n-1}^{n_0-1}\lambda_{\varepsilon 0}^{n-n_0} \end{bmatrix}. \quad (2.2.23)$$

Similarly, for $i = 1$ to l, the n_i generalized eigenvectors of $A + BF(\varepsilon)$ corresponding to eigenvalues of $\lambda_{\varepsilon i} = -\varepsilon + j\beta_i$ and $\bar{\lambda}_{\varepsilon i} = -\varepsilon - j\beta_i$ are given, respectively, by

$$q_1^i(\varepsilon) = \begin{bmatrix} 1 \\ \lambda_{\varepsilon i} \\ \lambda_{\varepsilon i}^2 \\ \vdots \\ \lambda_{\varepsilon i}^{q-2} \\ \lambda_{\varepsilon i}^{n-1} \end{bmatrix}, \quad q_2^i(\varepsilon) = \begin{bmatrix} 0 \\ 1 \\ 2\lambda_{\varepsilon i} \\ 3\lambda_{\varepsilon i}^2 \\ \vdots \\ (n-1)\lambda_{\varepsilon i}^{n-2} \end{bmatrix}, \cdots,$$

$$q_{n_i}^i(\varepsilon) = \begin{bmatrix} 0 \\ 0 \\ 0 \\ \vdots \\ C_{n-2}^{n_i-1}\lambda_{\varepsilon i}^{n-n_i-1} \\ C_{n-1}^{n_i-1}\lambda_{\varepsilon i}^{n-n_i} \end{bmatrix}, \quad (2.2.24)$$

and their complex conjugates $\bar{q}_j^i(\varepsilon)$, $j = 1$ to n_i.

We then form the following real nonsingular transformation matrix,

$$Q(\varepsilon) = [\,Q_0(\varepsilon) \quad Q_1(\varepsilon) \quad Q_2(\varepsilon) \quad \cdots \quad Q_l(\varepsilon)\,], \quad (2.2.25)$$

where

$$Q_0(\varepsilon) = [\,q_1^0(\varepsilon) \quad q_2^0(\varepsilon) \quad \cdots \quad q_{n_0}^0(\varepsilon)\,],$$

and for $i = 1$ to l,

$$Q_i(\varepsilon) = \left[\frac{q_1^i + \bar{q}_1^i}{2} \quad \frac{q_1^i - \bar{q}_1^i}{2j} \quad \frac{q_2^i + \bar{q}_2^i}{2} \quad \frac{q_2^i - \bar{q}_2^i}{2j} \right.$$
$$\left. \cdots \quad \frac{q_{n_i}^i + \bar{q}_{n_i}^i}{2} \quad \frac{q_{n_i}^i - \bar{q}_{n_i}^i}{2j} \right].$$

It can now be readily verified that

$$Q^{-1}(\varepsilon)(A + BF(\varepsilon))Q(\varepsilon) = J(\varepsilon), \qquad (2.2.26)$$

where $J(\varepsilon)$ is as defined in (2.2.19). This completes the proof.   ⊠

Lemma 2.2.3. *Consider a single input pair (A, B) in the form of (2.2.7) with all eigenvalues of A on the $j\omega$ axis. Let $J(\varepsilon)$ be as given in Lemma 2.2.2. Let*

$$S(\varepsilon) = \text{blkdiag}\{S_0(\varepsilon), S_1(\varepsilon), S_2(\varepsilon), \cdots, S_l(\varepsilon)\}, \qquad (2.2.27)$$

where $S_0(\varepsilon) = \text{diag}\{\varepsilon^{n_0-1}, \varepsilon^{n_0-2}, \cdots, \varepsilon, 1\}$ and for each $i = 1$ to l, $S_i(\varepsilon) = $ blkdiag $\{\varepsilon^{n_i-1}I_2, \varepsilon^{n_i-2}I_2, \cdots, \varepsilon I_2, I_2\}$.

   Then,

   1.

$$S(\varepsilon)J(\varepsilon)S^{-1}(\varepsilon) = \varepsilon\tilde{J}(\varepsilon) := \varepsilon\text{blkdiag}\left\{\tilde{J}_0, \tilde{J}_1(\varepsilon), \cdots, \tilde{J}_l(\varepsilon)\right\}, \qquad (2.2.28)$$

where

$$\tilde{J}_0 = \begin{bmatrix} -1 & 1 & & \\ & \ddots & \ddots & \\ & & -1 & 1 \\ & & & -1 \end{bmatrix}_{n_0 \times n_0}, \qquad (2.2.29)$$

and for each $i = 1$ to l,

$$\tilde{J}_i(\varepsilon) = \begin{bmatrix} \tilde{J}_i^\star(\varepsilon) & I_2 & & \\ & \ddots & \ddots & \\ & & \tilde{J}_i^\star(\varepsilon) & I_2 \\ & & & \tilde{J}_i^\star(\varepsilon) \end{bmatrix}_{2n_i \times 2n_i}, \quad \tilde{J}_i^\star(\varepsilon) = \begin{bmatrix} -1 & \beta_i/\varepsilon \\ -\beta_i/\varepsilon & -1 \end{bmatrix},$$
$$(2.2.30)$$

with $\beta_i > 0$ for all $i = 1$ to l and $\beta_i \neq \beta_j$ for $i \neq j$;

   2. The unique positive definite solution $\tilde{P}$ to the Lyapunov equation

$$\tilde{J}(\varepsilon)'\tilde{P} + \tilde{P}\tilde{J}(\varepsilon) = -I \qquad (2.2.31)$$

is independent of ε.   Ⓛ

Proof of Lemma 2.2.3. Item 1 is the result of some straightforward matrix manipulation. To show Item 2, we observe that the solution $\tilde{P}$ to the Lyapunov equation (2.2.31) is of block diagonal form,

$$\tilde{P} = \text{blkdiag}\{\tilde{P}_0, \tilde{P}_1(\varepsilon), \tilde{P}_2(\varepsilon), \cdots, \tilde{P}_l(\varepsilon)\}, \qquad (2.2.32)$$

where $\tilde{P}_0$ is the unique positive definite solution to the Lyapunov equation

$$\tilde{J}_0' \tilde{P}_0 + \tilde{P}_0 \tilde{J}_0 = -I, \qquad (2.2.33)$$

and, for $i = 1$ to l, $\tilde{P}_i(\varepsilon)$ is the positive definite solution to the Lyapunov equation

$$\tilde{J}_i'(\varepsilon)\tilde{P}_i + \tilde{P}_i \tilde{J}_i(\varepsilon) = -I. \qquad (2.2.34)$$

Clearly, $\tilde{P}_0$ is independent of ε. It remains to show that for each $i = 1$ to l, $\tilde{P}_i(\varepsilon)$ is also independent of ε. To this end, we notice that

$$T_i^{-1} \tilde{J}_i(\varepsilon) T_i = \text{blkdiag}\{J_i^+(\varepsilon), J_i^-(\varepsilon)\}, \qquad (2.2.35)$$

where

$$J_i^+(\varepsilon) = \begin{bmatrix} -1 + j\frac{\beta_i}{\varepsilon} & 1 & & \\ & \ddots & \ddots & \\ & & \ddots & 1 \\ & & & -1 + j\frac{\beta_i}{\varepsilon} \end{bmatrix}, \qquad (2.2.36)$$

$$J_i^-(\varepsilon) = \bar{J}_i^+(\varepsilon) = \begin{bmatrix} -1 - j\frac{\beta_i}{\varepsilon} & 1 & & \\ & \ddots & \ddots & \\ & & \ddots & 1 \\ & & & -1 - j\frac{\beta_i}{\varepsilon} \end{bmatrix}, \qquad (2.2.37)$$

and the nonsingular transformation matrix T_i is given by

$$T_i = \begin{bmatrix} 1 & & & & 1 & & \\ j & & & & -j & & \\ & 1 & & & & 1 & \\ & j & & & & -j & \\ & & \ddots & & & & \ddots \\ & & & 1 & & & & 1 \\ & & & j & & & & -j \end{bmatrix}_{2n_i \times 2n_i}. \qquad (2.2.38)$$

Noting that

$$e^{J_i^+(\varepsilon)t} = e^{-t + j\beta_i t/\varepsilon} \begin{bmatrix} 1 & t & \frac{t^2}{2!} & \cdots & \\ & 1 & t & \cdots & \\ & & 1 & \cdots & \\ & & & \ddots & \\ & & & & 1 \end{bmatrix},$$

$$e^{(J_i^+(\varepsilon))^* t} = e^{-t - j\beta_i t/\varepsilon} \begin{bmatrix} 1 & & & & \\ t & 1 & & & \\ \frac{t^2}{2} & t & 1 & & \\ \vdots & \vdots & \vdots & \ddots & \\ & & & & 1 \end{bmatrix},$$

we see that

$$e^{(J_i^+(\varepsilon))^* t} e^{J_i^+(\varepsilon) t} = e^{-2t} \begin{bmatrix} 1 & t & \frac{t^2}{2!} & \cdots \\ t & 1+t^2 & \cdots & \cdots \\ \frac{t^2}{2!} & \vdots & \ddots & \cdots \\ \vdots & \vdots & \vdots & \ddots \end{bmatrix}$$

is independent of ε. Similarly,

$$e^{(J_i^-(\varepsilon))^* t} e^{J_i^-(\varepsilon) t} = e^{-2t} \begin{bmatrix} 1 & t & \frac{t^2}{2!} & \cdots \\ t & 1+t^2 & \cdots & \cdots \\ \frac{t^2}{2!} & \vdots & \ddots & \cdots \\ \vdots & \vdots & \vdots & \ddots \end{bmatrix}$$

is also independent of ε.

Hence, using the fact that $T_i^* T_i = 2I_{2n_i}$, we have

$$\tilde{P}_i(\varepsilon) = \int_0^\infty e^{\tilde{J}_i \mathsf{T}(\varepsilon) t} e^{\tilde{J}_i(\varepsilon) t} dt$$

$$= 2 \int_0^\infty (T_i^{-1})^* \text{blkdiag}\{e^{(J_i^+(\varepsilon))^* t} e^{J_i^+(\varepsilon) t}, e^{(J_i^-(\varepsilon))^* t} e^{J_i^-(\varepsilon) t}\} T_i^{-1} dt$$

and is independent of ε. ⊠

Lemma 2.2.4. Let A, B, $F(\varepsilon)$, $Q(\varepsilon)$, l, and n_i for $i = 0$ to l, be as defined in Lemma 2.2.2 and its proof. Let the scaling matrix $S(\varepsilon)$ be as defined in Lemma 2.2.3. Then, there exists $\alpha, \beta, \theta \geq 0$ independent of ε such that, for all $\varepsilon \in (0, 1]$,

$$|F(\varepsilon) Q(\varepsilon) S^{-1}(\varepsilon)| \leq \alpha \varepsilon, \tag{2.2.39}$$

$$|F(\varepsilon) A Q(\varepsilon) S^{-1}(\varepsilon)| \leq \beta \varepsilon, \tag{2.2.40}$$

$$|Q(\varepsilon)| \leq \theta, \ |Q^{-1}(\varepsilon)| \leq \theta. \tag{2.2.41}$$

▣

Proof of Lemma 2.2.4. Observe that

$$F(\varepsilon) Q(\varepsilon) S^{-1}(\varepsilon) = [\, F(\varepsilon) Q_0(\varepsilon) S_0^{-1}(\varepsilon) \ \ F(\varepsilon) Q_1(\varepsilon) S_1^{-1}(\varepsilon) \ \cdots \ F(\varepsilon) Q_l(\varepsilon) S_l^{-1}(\varepsilon) \,],$$
$$\tag{2.2.42}$$

where $Q_i(\varepsilon)$, $i = 0, 1, 2, \cdots, l$, are defined in (2.2.25). We next recall that (see (2.2.17)) for each $i = 0$ to l and for each $j = 1$ to n_i, there exists a $\delta_j^i \geq 0$, independent of ε, such that,

$$\left| F(\varepsilon) q_j^i(\varepsilon) \right| \leq \delta_j^i \varepsilon^{n_i - j + 1}, \ \ \forall \varepsilon \in (0, 1]. \tag{2.2.43}$$

It is now clear that there exists a $\delta_0 \geq 0$ such that

$$\left| F(\varepsilon) Q_0(\varepsilon) S_0^{-1}(\varepsilon) \right| \leq \delta_0 \varepsilon, \ \ \forall \varepsilon \in (0, 1]. \tag{2.2.44}$$

For each $i = 1$ to l, noting the definition of $Q_i(\varepsilon)$, it is also straightforward to verify that there exists a $\delta_i \geq 0$, independent of ε, such that,

$$\left| F(\varepsilon) Q_i(\varepsilon) S_i^{-1}(\varepsilon) \right| \leq \delta_i \varepsilon, \ \ \forall \varepsilon \in (0, 1]. \tag{2.2.45}$$

The existence of α now follows readily. The existence of β follows with similar arguments from the combination of (2.2.17) and (2.2.16). The existence of θ follows from the facts that $Q(\varepsilon)$ is a polynomial in ε and that $Q(0)$, being the transformation matrix that takes A into its real Jordan form, is nonsingular (and hence $Q^{-1}(\varepsilon)$ is continuously differentiable in ε). $\boxtimes$

Finally, we establish a lemma that is useful in examining the properties of the closed-loop system under low gain feedback and in the presence of external inputs. For example, we will need this lemma in Chapter 10 when we utilize the low gain feedback design technique to construct feedback laws that solve the H_∞-ADDPMS.

Lemma 2.2.5. *Let A and $Q(\varepsilon)$ be as given in the proof of Lemma 2.2.2. Let $E \in \mathbb{R}^{n \times q}$ is such that*

$$\text{Im}(E) \subset \cap_{w \in \lambda(A)} \text{Im}(wI - A), \tag{2.2.46}$$

where q is any integer. Then, there exists a $\delta \geq 0$, independent of ε, such that

$$|Q^{-1}(\varepsilon)E| \leq \delta, \ \ \varepsilon \in (0, 1], \tag{2.2.47}$$

and, if we partition $Q^{-1}(\varepsilon)E$ according to that of $J(\varepsilon)$ of Lemma 2.2.2 as,

$$Q^{-1}(\varepsilon)E = \begin{bmatrix} E_0(\varepsilon) \\ E_1(\varepsilon) \\ \vdots \\ E_l(\varepsilon) \end{bmatrix}, \ \ E_0(\varepsilon) = \begin{bmatrix} E_{01}(\varepsilon) \\ E_{02}(\varepsilon) \\ \vdots \\ E_{0n_0}(\varepsilon) \end{bmatrix}_{n_0 \times q}, \ \ E_i(\varepsilon) = \begin{bmatrix} E_{i1}(\varepsilon) \\ E_{i2}(\varepsilon) \\ \vdots \\ E_{in_i}(\varepsilon) \end{bmatrix}_{2n_i \times q},$$

$$\tag{2.2.48}$$

then, there exists a $\beta \geq 0$, independent of ε, such that, for each $i = 0$, to l,

$$|E_{in_i}(\varepsilon)| \leq \beta\varepsilon. \tag{2.2.49}$$

□

Proof of Lemma 2.2.5. The existence of a $\delta \geq 0$ that satisfies (2.2.47) follows readily from Lemma 2.2.4.

To show the existence of $\beta \geq 0$ that satisfies (2.2.49), we note that Assumption (2.2.46) implies that, for each $i = 0$ to l,

$$|E_{in_i}(0)| = 0. \tag{2.2.50}$$

The existence of such a β now follows trivially from the continuous differentiability of $Q^{-1}(\varepsilon)$.　　　　　　　　　　　　　　　　　　　　　　　　　　　　　⊠

2.2.2. ARE Based Method

The ARE based low gain state feedback design for the system (2.2.1) is carried out in two steps.

Step 1. Solve the following algebraic Riccati equation

$$A'P + PA - PBB'P + \varepsilon I = 0, \quad \varepsilon \in (0, 1], \tag{2.2.51}$$

for the unique positive definite solution $P(\varepsilon)$. The existence of such a solution is established in Lemma 2.2.6 below. We also note here that the choice of εI in the above ARE, and other ARE's in this monograph, is purely for simplicity in the presentation. In fact, it can be replaced by any $Q(\varepsilon) : (0, 1] \to \mathbb{R}^{n \times n}$ as long as it is positive definite for all $\varepsilon \in (0, 1]$ and satisfies

$$\lim_{\varepsilon \to 0} Q(\varepsilon) = 0. \tag{2.2.52}$$

Step 2. Construct a family of low gain state feedback laws as

$$u = F(\varepsilon)x, \tag{2.2.53}$$

where

$$F(\varepsilon) = -B'P(\varepsilon). \tag{2.2.54}$$

Ａ

The following lemma establishes the basic property of this family of ARE based low gain feedback laws.

Lemma 2.2.6. *Let Assumption 2.2.1 hold. Then, for each $\varepsilon \in (0, 1]$, there exists a unique matrix $P(\varepsilon) > 0$ that solves the ARE (2.2.51). Moreover, such a $P(\varepsilon)$ satisfies,*

1. $\lim_{\varepsilon \to 0} P = 0$;

2. *There exists a constant $\alpha > 0$, independent of ε, such that,*

$$\left| P^{\frac{1}{2}} A^\ell P^{-\frac{1}{2}} \right| \le \alpha^\ell, \, \forall \ell \in \mathbb{N}, \, \forall \varepsilon \in (0, 1]. \tag{2.2.55}$$

Note that for notational convenience, we have denoted and will often denote $P = P(\varepsilon)$. ⬜

Remark 2.2.4. *Lemma 2.2.6 reveals that for any given bounded set of initial conditions, the control and all its derivatives can be made arbitrarily small by decreasing the value of the low gain parameter ε.* ⓡ

Proof of Lemma 2.2.6. The existence of a unique positive definite solution $P(\varepsilon)$ for all $\varepsilon > 0$ has been established in [123]. The same paper established that for $\varepsilon = 0$ there is a unique solution $P(0) = 0$ for which $A + BB'P(0)$ has all eigenvalues in the closed left-half plane. Item 1 of the lemma thus follows readily from the continuity of the solution of the algebraic Riccati equation for $\varepsilon = 0$, which was established in [116].

To show Item 2 of the lemma, post-multiply both sides of (2.2.51) by P^{-1} to obtain,

$$\varepsilon P^{-1} = PBB' - A' - PAP^{-1}, \tag{2.2.56}$$

which, by $\text{tr}(PAP^{-1}) = \text{tr}(A)$, implies that

$$\left| \varepsilon P^{-1} \right| = \lambda_{\max}(\varepsilon P^{-1}) \le \text{tr}(\varepsilon P^{-1}) = \text{tr}(PBB') - 2\text{tr}(A). \tag{2.2.57}$$

Also, pre- and post-multiply (2.2.51) by $P^{-\frac{1}{2}}$ and $P^{-1}A'P^{\frac{1}{2}}$ to obtain,

$$P^{-\frac{1}{2}} A' A' P^{\frac{1}{2}} + P^{\frac{1}{2}} AP^{-1} A' P^{\frac{1}{2}} - P^{\frac{1}{2}} BB' A' P^{\frac{1}{2}} + \varepsilon P^{-\frac{3}{2}} A' P^{\frac{1}{2}} = 0, \tag{2.2.58}$$

which, by the fact that $\text{tr}(XY) = \text{tr}(YX)$ for any two matrices X and Y, implies that

$$\left| P^{\frac{1}{2}} AP^{-\frac{1}{2}} \right|^2 \le \text{tr}(P^{\frac{1}{2}} AP^{-1} A' P^{\frac{1}{2}}) = \text{tr}(PBB'A') - \text{tr}(A'A') - \text{tr}(\varepsilon P^{-1}A'). \tag{2.2.59}$$

By (2.2.57) and Item 1, we have

$$\left| P^{\frac{1}{2}} AP^{-\frac{1}{2}} \right| \le \alpha, \quad \forall \varepsilon \in (0, 1], \tag{2.2.60}$$

for some $\alpha > 0$ independent of ε. Finally, we have

$$\left| P^{\frac{1}{2}} A^\ell P^{-\frac{1}{2}} \right| \le \left| P^{\frac{1}{2}} A P^{-\frac{1}{2}} \right|^\ell \le \alpha^\ell. \tag{2.2.61}$$

⌧

2.3. Discrete-Time Systems

Consider the discrete-time linear system

$$x^+(k) = Ax(k) + Bu(k), \quad x(k) \in \mathbb{R}^n, \ u(k) \in \mathbb{R}^m, \tag{2.3.1}$$

where $x(k)$ is the state and $u(k)$ is the input.[2] We make the following assumption.

Assumption 2.3.1. *The pair (A, B) is asymptotically null controllable with bounded controls (ANCBC), i.e.,*

1. *(A, B) is stabilizable;*

2. *All eigenvalues of A are inside or on the unit circle.*   ⊞

The next two subsections each present a technique for designing low gain state feedback.

2.3.1. Eigenstructure Assignment Based Method

The eigenstructure assignment based low gain state feedback design is carried out in three steps.

Step 1. Find nonsingular transformation matrices T_s and T_I such that the pair (A, B) is transformed into the following block diagonal control canonical form,

$$T_s^{-1} A T_s = \begin{bmatrix} A_1 & 0 & \cdots & 0 & 0 \\ 0 & A_2 & \cdots & 0 & 0 \\ \vdots & \vdots & \ddots & \vdots & \vdots \\ 0 & 0 & \cdots & A_l & 0 \\ 0 & 0 & 0 & 0 & A_0 \end{bmatrix}, \tag{2.3.2}$$

$$T_s^{-1} B T_I = \begin{bmatrix} B_1 & B_{12} & \cdots & B_{1l} & * \\ 0 & B_2 & \cdots & B_{2l} & * \\ \vdots & \vdots & \ddots & \vdots & \vdots \\ 0 & 0 & \cdots & B_l & * \\ B_{01} & B_{02} & \cdots & B_{0l} & * \end{bmatrix}, \tag{2.3.3}$$

---

[2]For simplicity, we will suppress the independent variable k throughout the monograph.

where, A_0 contains all the eigenvalues of A that are inside the unit circle, for each $i = 1$ to l, all eigenvalues of A_i are on the unit circle and hence (A_i, B_i) is controllable as given by,

$$A_i = \begin{bmatrix} 0 & 1 & 0 & \cdots & 0 \\ 0 & 0 & 1 & \cdots & 0 \\ \vdots & \vdots & \vdots & \ddots & \vdots \\ 0 & 0 & 0 & \cdots & 1 \\ -a_{n_i}^i & -a_{n_i-1}^i & -a_{n_i-2}^i & \cdots & -a_1^i \end{bmatrix}, \quad B_i = \begin{bmatrix} 0 \\ 0 \\ \vdots \\ 0 \\ 1 \end{bmatrix},$$

and finally, *'s represent submatrices of less interest.

We note that the existence of the above canonical form was shown in [126]. The software realization can be found in [9].

Step 2. For each (A_i, B_i), let $F_i(\varepsilon) \in \mathbb{R}^{1 \times n_i}$ be the state feedback gain such that

$$\lambda(A_i + B_i F_i(\varepsilon)) = (1 - \varepsilon)\lambda(A_i) \in \mathbb{C}^\circ, \quad \varepsilon \in (0, 1]. \tag{2.3.4}$$

Note that $F_i(\varepsilon)$ is unique.

Step 3. Construct a family of low gain state feedback laws for system (2.3.1) as,

$$u = F(\varepsilon)x, \tag{2.3.5}$$

where the low gain matrix $F(\varepsilon)$ is given by

$$F(\varepsilon) = T_I \begin{bmatrix} F_1(\varepsilon) & 0 & \cdots & 0 & 0 & 0 \\ 0 & F_2(\varepsilon) & \cdots & 0 & 0 & 0 \\ \vdots & \vdots & \ddots & \vdots & \vdots & \vdots \\ 0 & 0 & \cdots & F_{l-1}(\varepsilon) & 0 & 0 \\ 0 & 0 & \cdots & 0 & F_l(\varepsilon) & 0 \\ 0 & 0 & \cdots & 0 & 0 & 0 \end{bmatrix} T_s^{-1}. \tag{2.3.6}$$

Ⓐ

Remark 2.3.1. *As in the continuous-time case, it is clear from the construction of $F(\varepsilon)$ that $\lim_{\varepsilon \to 0} F(\varepsilon) = 0$. For this reason the family of state feedback laws (2.3.5) are referred to as low gain feedback, and ε the low gain parameter. We however also note that, as the following lemma shows, the low gain feedback laws (2.3.5) possess more intricate properties than simply having small feedback gains.* Ⓡ

Lemma 2.3.1. *Consider a single input pair* (A, B) *in the control canonical form,*

$$A = \begin{bmatrix} 0 & 1 & 0 & \cdots & 0 \\ 0 & 0 & 1 & \cdots & 0 \\ \vdots & \vdots & \vdots & \ddots & \vdots \\ 0 & 0 & 0 & \cdots & 1 \\ -a_n & -a_{n-1} & -a_{n-2} & \cdots & -a_1 \end{bmatrix}, \quad B = \begin{bmatrix} 0 \\ 0 \\ \vdots \\ 0 \\ 1 \end{bmatrix}. \tag{2.3.7}$$

Assume that all the eigenvalues are in the closed left half s-plane. Let $F(\varepsilon) \in \mathbb{R}^{1 \times n}$ be the state feedback gain such that $\lambda(A + BF(\varepsilon)) = (1 - \varepsilon)\lambda(A)$. Then, there exists an $\varepsilon^* \in (0, 1]$ such that for all $\varepsilon \in (0, \varepsilon^*]$,

$$|F(\varepsilon)| \leq \alpha \varepsilon, \tag{2.3.8}$$

$$\left|(A + BF(\varepsilon))^k\right| \leq \frac{\beta}{\varepsilon^{r-1}}(1 - \varepsilon)^{k/2}, \, \forall k \geq 0, \tag{2.3.9}$$

$$\left|F(\varepsilon)(A + BF(\varepsilon))^{k+\ell}\right| \leq \gamma_\ell \varepsilon (1 - \varepsilon)^{k/2}, \, \forall \ell \in \mathbb{N}, \, \forall k \geq 0, \tag{2.3.10}$$

where r is the largest algebraic multiplicity of the eigenvalues of A, and α, β and γ_ℓ's are all positive constants independent of ε. $\boxed{\text{L}}$

Remark 2.3.2. *Property (2.3.8) indicates the asymptotic nature of the feedback laws (2.3.5), i.e., $\lim_{\varepsilon \to 0} |F(\varepsilon)| = 0$.*

Recalling that $(A + BF(\varepsilon))^k$ is the transition matrix of the closed-loop system under the low gain feedback, we see that Property (2.3.9) reveals that the closed-loop system under low gain feedback will peak slowly to a magnitude of order $O(1/\varepsilon^{r-1})$, with r being the largest algebraic multiplicity of the eigenvalues of A. Property (2.3.10) on the other hand implies that for any given bounded set of initial conditions, the control and all its derivatives can be made arbitrarily small by decreasing the value of the low gain parameter ε. This can also be seen in the simulation of the following example. $\boxed{\text{R}}$

Example 2.3.1. Consider the single input system

$$x^+ = Ax + Bu, \tag{2.3.11}$$

with

$$A = \begin{bmatrix} 0 & 1 & 0 & 0 \\ 0 & 0 & 1 & 0 \\ 0 & 0 & 0 & 1 \\ -1 & 2\sqrt{2} & -4 & 2\sqrt{2} \end{bmatrix}, \quad B = \begin{bmatrix} 0 \\ 0 \\ 0 \\ 1 \end{bmatrix}.$$

It is straightforward to verify that the open loop system has repeated poles at $\left\{\frac{\sqrt{2}}{2} + j\frac{\sqrt{2}}{2}, \frac{\sqrt{2}}{2} + j\frac{\sqrt{2}}{2}, \frac{\sqrt{2}}{2} - j\frac{\sqrt{2}}{2}, \frac{\sqrt{2}}{2} - j\frac{\sqrt{2}}{2}\right\}$, where $j = \sqrt{-1}$. Following the eigenstructure assignment based low gain feedback design algorithm, we construct the following family of linear state feedback control laws,

$$u = [-\varepsilon^4 + 4\varepsilon^3 - 6\varepsilon^2 + 4\varepsilon \quad -2\sqrt{2}\varepsilon^3 + 6\sqrt{2}\varepsilon^2 - 6\sqrt{2}\varepsilon \quad -4\varepsilon^2 + 8\varepsilon \quad -2\sqrt{2}\varepsilon]x. \quad (2.3.12)$$

Some simulation of the closed-loop system is given in Figs. 2.3.1-2.3.2. It can be easily seen from these figures that, for the same initial conditions, as the value of ε decreases, the state peaks slowly while the l_∞-norm of control input decreases. $\boxed{\text{E}}$

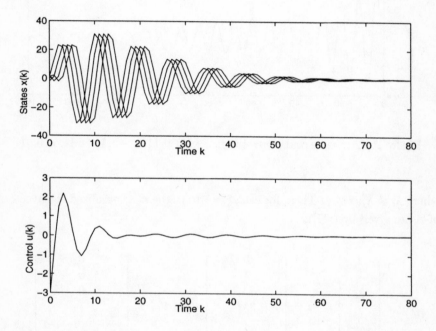

Figure 2.3.1: States and control input: $x(0) = (1, 2, -2, 1)$ and $\varepsilon = 0.1$.

Proof of Lemma 2.3.1. First, (2.3.8) follows trivially from the fact $F(\varepsilon)$ is a vector of polynomials in ε and $F(0) = 0$.

We next proceed to show (2.3.9) and (2.3.10). To this end, let

$$\det(sI - A) = \prod_{i=1}^{p} (s - \lambda_i)^{n_i},$$

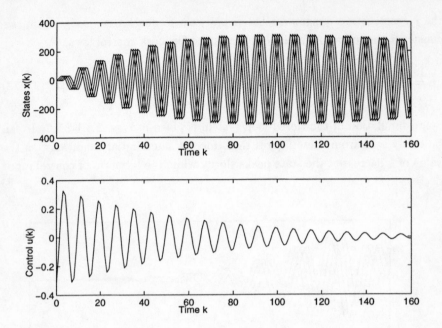

Figure 2.3.2: States and control input: $x(0) = (1, 2, -2, 1)$ and $\varepsilon = 0.01$.

where $\lambda_i \neq \lambda_j$, $i \neq j$. Then, for each $i = 1$ to p, the n_i generalized eigenvectors of A are given by ([27]),

$$
p_1^i = \begin{bmatrix} 1 \\ \lambda_i \\ \lambda_i^2 \\ \vdots \\ \lambda_i^{n-2} \\ \lambda_i^{n-1} \end{bmatrix}, \quad
p_2^i = \begin{bmatrix} 0 \\ 1 \\ 2\lambda_i \\ 3\lambda_i^2 \\ \vdots \\ (n-1)\lambda_i^{n-2} \end{bmatrix}, \cdots,
p_{n_i}^i = \begin{bmatrix} 0 \\ 0 \\ 0 \\ \vdots \\ C_{n-2}^{n_i-1} \lambda_i^{n-n_i-1} \\ C_{n-1}^{n_i-1} \lambda_i^{n-n_i} \end{bmatrix}.
$$

Similarly, for each $i = 1$ to p, the n_i generalized eigenvectors of $A + BF(\varepsilon)$ are given by

$$
q_1^i = \begin{bmatrix} 1 \\ \lambda_{\varepsilon i} \\ \lambda_{\varepsilon i}^2 \\ \vdots \\ \lambda_{\varepsilon i}^{n-2} \\ \lambda_{\varepsilon i}^{n-1} \end{bmatrix}, \quad
q_2^i = \begin{bmatrix} 0 \\ 1 \\ 2\lambda_{\varepsilon i} \\ 3\lambda_{\varepsilon i}^2 \\ \vdots \\ (n-1)\lambda_{\varepsilon i}^{n-2} \end{bmatrix}, \cdots,
q_{n_i}^i = \begin{bmatrix} 0 \\ 0 \\ 0 \\ \vdots \\ C_{n-2}^{n_i-1} \lambda_{\varepsilon i}^{n-n_i-1} \\ C_{n-1}^{n_i-1} \lambda_{\varepsilon i}^{n-n_i} \end{bmatrix},
$$

where $\lambda_{\varepsilon i} = (1 - \varepsilon)\lambda_i$ and C_n^i is defined as

$$C_n^i = \begin{cases} \dfrac{n!}{(n-i)!i!} & 0 \le i \le n, \\[2mm] 0 & \text{otherwise.} \end{cases}$$

We next form the following two nonsingular transformation matrices

$$P = [p_1^1, p_2^1, \cdots, p_{n_1}^1, \cdots, p_1^p, p_2^p, \cdots, p_{n_p}^p],$$

which is independent of ε, and

$$Q(\varepsilon) = [q_1^1, q_2^1, \cdots, q_{n_1}^1, \cdots, q_1^p, q_2^p, \cdots, q_{n_p}^p].$$

It then follows that

$$P^{-1}AP = \begin{bmatrix} J_1 & 0 & \cdots & 0 \\ 0 & J_2 & \cdots & 0 \\ \vdots & \vdots & \ddots & \vdots \\ 0 & 0 & \cdots & J_p \end{bmatrix}, \quad J_i = \begin{bmatrix} \lambda_i & 1 & 0 & \cdots & 0 \\ 0 & \lambda_i & 1 & \cdots & 0 \\ \vdots & \vdots & \vdots & \cdots & \vdots \\ 0 & 0 & 0 & \cdots & \lambda_i \end{bmatrix}, \quad (2.3.13)$$

and

$$Q^{-1}(\varepsilon)(A + BF(\varepsilon))Q(\varepsilon) = \begin{bmatrix} J_{\varepsilon 1} & 0 & \cdots & 0 \\ 0 & J_{\varepsilon 2} & \cdots & 0 \\ \vdots & \vdots & \ddots & \vdots \\ 0 & 0 & \cdots & J_{\varepsilon p} \end{bmatrix},$$

$$J_{\varepsilon i} = \begin{bmatrix} \lambda_{\varepsilon i} & 1 & 0 & \cdots & 0 \\ 0 & \lambda_{\varepsilon i} & 1 & \cdots & 0 \\ \vdots & \vdots & \vdots & \cdots & \vdots \\ 0 & 0 & 0 & \cdots & \lambda_{\varepsilon i} \end{bmatrix}. \quad (2.3.14)$$

By the definition of generalized eigenvectors, we have the following relationships,

$$Ap_1^i = \lambda_i p_1^i,$$
$$Ap_j^i = p_{j-1}^i + \lambda_i p_j^i, \quad j = 2, 3, \cdots, n_i, \ i = 1, 2, \cdots, p, \quad (2.3.15)$$

and

$$(A + BF(\varepsilon))q_1^i = \lambda_{\varepsilon i} q_1^i,$$
$$(A + BF(\varepsilon))q_j^i = q_{j-1}^i + \lambda_{\varepsilon i} q_j^i, \quad j = 2, 3, \cdots, n_i, \ i = 1, 2, \cdots, p. \ (2.3.16)$$

Denoting $A_n = [-a_n \quad -a_{n-1} \quad \cdots \quad -a_1]$, it then follows form (2.3.15) and (2.3.16) that, for each $i = 1$ to p, we have that, for $j = 1, 2, \cdots, n_i$,

$$A_n p_j^i = C_{n-1}^{j-1} \lambda_i^{n-j+1} + C_{n-1}^{j-2} \lambda_i^{n-j+1} = C_n^{j-1} \lambda_i^{n-j+1},$$
$$(A_n + F(\varepsilon))q_j^i = C_{n-1}^{j-1} \lambda_{\varepsilon i}^{n-j+1} + C_{n-1}^{j-2} \lambda_{\varepsilon i}^{n-j+1} = C_n^{j-1} \lambda_{\varepsilon i}^{n-j+1},$$

from which we have

$$F(\varepsilon)q_j^i = -A_n q_j^i + C_n^{j-1}\lambda_{\varepsilon i}^{n-j+1}$$

$$= -A_n \begin{bmatrix} 0 \\ \vdots \\ 0 \\ C_{j-1}^{j-1} \\ C_j^{j-1}(1-\varepsilon)\lambda_i \\ \vdots \\ C_{n-1}^{j-1}(1-\varepsilon)^{n-j}\lambda_i^{n-j} \end{bmatrix} + C_n^{j-1}(1-\varepsilon)^{n-j+1}\lambda_i^{n-j+1}$$

$$= -A_n \begin{bmatrix} 0 \\ \vdots \\ 0 \\ C_{j-1}^{j-1} \\ C_j^{j-1}\sum_{k=0}^{1}(-\varepsilon)^k\lambda_i \\ \vdots \\ C_{n-1}^{j-1}\sum_{k=0}^{n-j}C_{n-j}^k(-\varepsilon)^k\lambda_i^{n-j} \end{bmatrix} + C_n^{j-1}(1-\varepsilon)^{n-j+1}\lambda_i^{n-j+1}$$

$$= -A_n \sum_{k=0}^{n_i-j}(-\varepsilon)^k\lambda_i^k C_{k+j-1}^k p_{j+k}^i + C_n^{j-1}\sum_{k=0}^{n_i-j}(-\varepsilon)^k C_{n-j+1}^k \lambda_i^{n-j+1}$$

$$+ \Delta\left(\varepsilon^{n_i-j+1}\right)$$

$$= -\sum_{k=0}^{n_i-j}(-\varepsilon)^k C_{k+j-1}^k C_n^{j+k-1}\lambda_i^{n-j+1} + C_n^{j-1}\sum_{k=0}^{n_i-j}(-\varepsilon)^k C_{n-j+1}^k \lambda_i^{n-j+1}$$

$$+ \Delta\left(\varepsilon^{n_i-j+1}\right)$$

$$= \Delta\left(\varepsilon^{n_i-j+1}\right),$$

where $\Delta\left(\varepsilon^{n_i-j+1}\right)$ is a polynomial in ε whose coefficients of terms of order lower than $n_i - j + 1$ are all zero. Hence, we have, for $\varepsilon \in (0,1]$,

$$\left|F(\varepsilon)q_j^i\right| \le \delta_j^i \varepsilon^{n_i-j+1}, \qquad (2.3.17)$$

where δ_j^i is some positive constant independent of ε.

We next note that both $Q(\varepsilon)$ and $Q^{-1}(\varepsilon)$ are continuous functions of ε satisfying $Q(0) = P$ and $Q^{-1}(0) = P^{-1}$. Now from the continuity of norm functions it follows that there exists an $\varepsilon^* \in (0, 1/2]$, such that for all $\varepsilon \in (0, \varepsilon^*]$,

$$|Q(\varepsilon)| \le |P| + 1, \quad \left|Q^{-1}(\varepsilon)\right| \le \left|P^{-1}\right| + 1. \qquad (2.3.18)$$

We are now ready to show (2.3.9) and (2.3.10). It follows from (2.3.17) and (2.3.18) that, for all $\varepsilon \in (0, \varepsilon^*]$,

$$\left|F(\varepsilon)(A + BF(\varepsilon))^k\right| = \left|F(\varepsilon)Q(\varepsilon)\left(Q^{-1}(\varepsilon)(A + BF(\varepsilon))Q(\varepsilon)\right)^k Q^{-1}(\varepsilon)\right|$$

$$\leq \left(|P^{-1}|+1\right) \sum_{i=1}^{p} \left|F(\varepsilon)[q_1^i, q_2^i, \cdots, q_{n_i}^i] J_{\varepsilon i}^k\right|$$

$$\leq \left(|P^{-1}|+1\right) \sum_{i=1}^{p} \left| F(\varepsilon)[q_1^i, q_2^i, \cdots, q_{n_i}^i] \right.$$

$$\times \left.\begin{bmatrix} \lambda_{\varepsilon i}^k & k\lambda_{\varepsilon i}^{k-1} & \frac{k(k-1)}{2!}\lambda_{\varepsilon i}^{k-2} & \cdots & \frac{k(k-1)\cdots(k-n_i+2)}{(n_i-1)!}\lambda_{\varepsilon i}^{k-n_i+1} \\ 0 & \lambda_{\varepsilon i}^k & k\lambda_{\varepsilon i}^{k-1} & \cdots & \frac{k(k-1)\cdots(k-n_i+1)}{(n_i-2)!}\lambda_{\varepsilon i}^{k-n_i+2} \\ \vdots & \vdots & \vdots & \ddots & \vdots \\ 0 & 0 & 0 & \cdots & \lambda_{\varepsilon i}^k \end{bmatrix}\right|$$

$$\leq \left(|P^{-1}|+1\right) \sum_{i=1}^{p}\sum_{j=1}^{n_i}\sum_{l=1}^{j} \frac{\delta_l^i k! \varepsilon^{n_i-l+1}(1-\varepsilon)^{k-j+l}}{(k-j+l)!(j-l)!}$$

$$\leq \left(|P^{-1}|+1\right) \sum_{i=1}^{p}\sum_{j=1}^{n_i}\sum_{l=1}^{j} \frac{\delta_l^i k! \varepsilon^{n_i-l+1}(1-\varepsilon)^{k-j+l+n}}{(1-\varepsilon)^n(k-j+l)!(j-l)!}, \tag{2.3.19}$$

where, for the consistency of the notations, $k(k-1)\cdots(k-j+l+1) = \frac{k!}{(k-j+l)!}$ is taken to be zero for $k < j-l$. Noting that

$$\frac{k!}{(k-j+l)!}(1-\varepsilon)^{k-j+l+n} \leq 2^{j-l}\left(\frac{j-l}{\varepsilon}\right)^{j-l}(1-\varepsilon)^{k/2}, \tag{2.3.20}$$

(2.3.19) can be continued as

$$|F(\varepsilon)(A+BF(\varepsilon))^k| \leq \left(|P^{-1}|+1\right)\sum_{i=1}^{p}\sum_{j=1}^{n_i}\sum_{l=1}^{j}\frac{2^{j-l}\delta_l^i(j-l)^{j-l}}{(1-\varepsilon)^n(j-l)!}\varepsilon^{n_i-j+1}(1-\varepsilon)^{k/2}$$

$$\leq \left(|P^{-1}|+1\right)\sum_{i=1}^{p}\sum_{j=1}^{n_i}\sum_{l=1}^{j}\frac{2^{j-l}\delta_l^i(j-l)^{j-l}}{\frac{1}{2^n}(j-l)!}\varepsilon(1-\varepsilon)^{k/2}$$

$$= \gamma_0\varepsilon(1-\varepsilon)^{k/2}, \tag{2.3.21}$$

where

$$\gamma_0 = 2^n \left(|P^{-1}|+1\right)\sum_{i=1}^{p}\sum_{j=1}^{n_i}\sum_{l=1}^{j}\frac{2^{j-l}\delta_l^i(j-l)^{j-l}}{(j-l)!}$$

is independent of ε. This shows (2.3.10) for $\ell = 0$, which together with (2.3.8) imply (2.3.10) for $\ell > 0$.

Finally, using (2.3.18) and (2.3.9) can be shown as follows. For all $\varepsilon \in (0, \varepsilon^*]$,

$$|(A+BF(\varepsilon))^k| = \left|Q(\varepsilon)\left(Q^{-1}(\varepsilon)(A+BF(\varepsilon))Q(\varepsilon)\right)^k Q^{-1}(\varepsilon)\right|$$

$$\leq (|P| + 1)(|P^{-1}| + 1) \sum_{i=1}^{p} |J_{\varepsilon i}^{k}|$$

$$\leq (|P| + 1)(|P^{-1}| + 1)$$

$$\times \sum_{i=1}^{p} \left| \left| \begin{bmatrix} \lambda_{\varepsilon i}^{k} & k\lambda_{\varepsilon i}^{k-1} & \frac{k(k-1)}{2!}\lambda_{\varepsilon i}^{k-2} & \cdots & \frac{k(k-1)\cdots(k-n_i+2)}{(n_i-1)!}\lambda_{\varepsilon i}^{k-n_i+1} \\ 0 & \lambda_{\varepsilon i}^{k} & k\lambda_{\varepsilon i}^{k-1} & \cdots & \frac{k(k-1)\cdots(k-n_i+1)}{(n_i-2)!}\lambda_{\varepsilon i}^{k-n_i+2} \\ \vdots & \vdots & \vdots & \ddots & \vdots \\ 0 & 0 & 0 & \cdots & \lambda_{\varepsilon i}^{k} \end{bmatrix} \right| \right|$$

$$\leq (|P| + 1)(|P^{-1}| + 1) \sum_{i=1}^{p} \sum_{j=1}^{n_i} \sum_{l=1}^{j} \frac{k!}{(k-j+l)!(j-l)!}(1-\varepsilon)^{k-j+l}$$

$$\leq (|P|+1)(|P^{-1}|+1) \sum_{i=1}^{p} \sum_{j=1}^{n_i} \sum_{l=1}^{j} \frac{k!}{(1-\varepsilon)^n(k-j+l)!(j-l)!}(1-\varepsilon)^{k-j+l+n}$$

$$\leq (|P| + 1)(|P^{-1}| + 1) \sum_{i=1}^{p} \sum_{j=1}^{n_i} \sum_{l=1}^{j} \frac{2^{j-l}(j-l)^{j-l}}{(1-\varepsilon)^n(j-l)!\varepsilon^{j-l}}(1-\varepsilon)^{k/2}$$

$$\leq (|P| + 1)(|P^{-1}| + 1) \sum_{i=1}^{p} \sum_{j=1}^{n_i} \sum_{l=1}^{j} \frac{2^{j-l}(j-l)^{j-l}}{\frac{1}{2^n}(j-l)!\varepsilon^{r-1}}(1-\varepsilon)^{k/2}$$

$$= \frac{\beta}{\varepsilon^{r-1}}(1-\varepsilon)^{k/2}, \tag{2.3.22}$$

where $r = \max\{n_1, n_2, \cdots, n_p\}$ and

$$\beta = 2^n(|P| + 1)(|P^{-1}| + 1) \sum_{i=1}^{p} \sum_{j=1}^{n_i} \sum_{l=1}^{j} \frac{2^{j-l}(j-l)^{j-l}}{(j-l)!}$$

is independent of ε.   ⊠

The next three lemmas, Lemmas 2.3.2-2.3.4, establish a framework for Lyapunov analysis of the closed-loop system under low gain feedback. As can be seen throughout this monograph, this framework simplifies and unifies the proofs of various results due to low gain feedback.

Lemma 2.3.2. *Consider a single input pair (A, B) in the form of (2.3.7) with all eigenvalues of A on the unit circle. Let $F(\varepsilon) \in \mathbb{R}^{1 \times n}$ be the unique matrix such that $\lambda(A + BF(\varepsilon)) = (1 - \varepsilon)\lambda(A)$, $\varepsilon \in (0, 1]$. Then, there exists a nonsingular transformation matrix $Q(\varepsilon) \in \mathbb{R}^{n \times n}$ such that*

$$Q^{-1}(\varepsilon)(A + BF(\varepsilon))Q(\varepsilon) = J(\varepsilon) := \text{blkdiag}\left\{J_{-1}(\varepsilon), J_{+1}(\varepsilon), J_1(\varepsilon), \cdots, J_l(\varepsilon)\right\},$$

$$\tag{2.3.23}$$

where

$$
J_{-1}(\varepsilon) = \begin{bmatrix} -(1-\varepsilon) & 1 & & \\ & \ddots & \ddots & \\ & & -(1-\varepsilon) & 1 \\ & & & -(1-\varepsilon) \end{bmatrix}_{n_{-1} \times n_{-1}} , \qquad (2.3.24)
$$

$$
J_{+1}(\varepsilon) = \begin{bmatrix} 1-\varepsilon & 1 & & \\ & \ddots & \ddots & \\ & & 1-\varepsilon & 1 \\ & & & 1-\varepsilon \end{bmatrix}_{n_{+1} \times n_{+1}} , \qquad (2.3.25)
$$

and for each $i = 1$ to l,

$$
J_i(\varepsilon) = \begin{bmatrix} J_i^\star(\varepsilon) & I_2 & & \\ & \ddots & \ddots & \\ & & J_i^\star(\varepsilon) & I_2 \\ & & & J_i^\star(\varepsilon) \end{bmatrix}_{2n_i \times 2n_i} , \quad J_i^\star(\varepsilon) = (1-\varepsilon) \begin{bmatrix} \alpha_i & \beta_i \\ -\beta_i & \alpha_i \end{bmatrix} ,
$$

$$(2.3.26)$$

with $\alpha_i^2 + \beta_i^2 = 1$ for all $i = 1$ to l and $\alpha_i \neq \alpha_j$ for $i \neq j$. $\quad\boxed{}$

Proof of Lemma 2.3.2. Let

$$
\det(sI - A - BF(\varepsilon)) = (s+1-\varepsilon)^{n-1}(s-1+\varepsilon)^{n+1}
$$
$$
\times \prod_{i=1}^{l} [s - (1-\varepsilon)(\alpha + j\beta_i)]^{n_i} [s - (1-\varepsilon)(\alpha - j\beta_i)]^{n_i}. \quad (2.3.27)
$$

Then, the n_{-1} generalized eigenvectors $A + BF(\varepsilon)$ corresponding to the eigenvalue $\lambda_{\varepsilon,-1} = -(1-\varepsilon)$ are ([27]),

$$
q_1^{-1}(\varepsilon) = \begin{bmatrix} 1 \\ \lambda_{\varepsilon,-1} \\ \lambda_{\varepsilon,-1}^2 \\ \vdots \\ \lambda_{\varepsilon,-1}^{n-2} \\ \lambda_{\varepsilon,-1}^{n-1} \end{bmatrix}, \; q_2^{-1}(\varepsilon) = \begin{bmatrix} 0 \\ 1 \\ 2\lambda_{\varepsilon,-1} \\ 3\lambda_{\varepsilon,-1}^2 \\ \vdots \\ (n-1)\lambda_{\varepsilon,-1}^{n-2} \end{bmatrix}, \cdots,
$$

$$
q_{n-1}^{-1}(\varepsilon) = \begin{bmatrix} 0 \\ 0 \\ 0 \\ \vdots \\ C_{n-2}^{n_{-1}-1} \lambda_{\varepsilon,-1}^{n-n_{-1}-1} \\ C_{n-1}^{n_{-1}-1} \lambda_{\varepsilon,-1}^{n-n_{-1}} \end{bmatrix}. \qquad (2.3.28)
$$

The n_{+1} generalized eigenvectors $A + BF(\varepsilon)$ corresponding to the eigenvalue $\lambda_{\varepsilon,+1} = (1 - \varepsilon)$ are,

$$q_1^{+1}(\varepsilon) = \begin{bmatrix} 1 \\ \lambda_{\varepsilon,+1} \\ \lambda_{\varepsilon,+1}^2 \\ \vdots \\ \lambda_{\varepsilon,+1}^{n-2} \\ \lambda_{\varepsilon,+1}^{n-1} \end{bmatrix}, \; q_2^{+1}(\varepsilon) = \begin{bmatrix} 0 \\ 1 \\ 2\lambda_{\varepsilon,+1} \\ 3\lambda_{\varepsilon,+1}^2 \\ \vdots \\ (n-1)\lambda_{\varepsilon,+1}^{n-2} \end{bmatrix}, \cdots,$$

$$q_{n+1}^{+1}(\varepsilon) = \begin{bmatrix} 0 \\ 0 \\ 0 \\ \vdots \\ C_{n-2}^{n+1-1}\lambda_{\varepsilon,+1}^{n-n+1-1} \\ C_{n-1}^{n+1-1}\lambda_{\varepsilon,+1}^{n-n+1} \end{bmatrix}. \qquad (2.3.29)$$

Similarly, for $i = 1$ to l, the n_i generalized eigenvectors of $A + BF(\varepsilon)$ corresponding to eigenvalues of $\lambda_{\varepsilon i} = (1 - \varepsilon)(\alpha_i + j\beta_i)$ and $\bar{\lambda}_{\varepsilon i} = (1 - \varepsilon)(\alpha_i - j\beta_i)$ are given, respectively, by

$$q_1^i(\varepsilon) = \begin{bmatrix} 1 \\ \lambda_{\varepsilon i} \\ \lambda_{\varepsilon i}^2 \\ \vdots \\ \lambda_{\varepsilon i}^{q-2} \\ \lambda_{\varepsilon i}^{n-1} \end{bmatrix}, \; q_2^i(\varepsilon) = \begin{bmatrix} 0 \\ 1 \\ 2\lambda_{\varepsilon i} \\ 3\lambda_{\varepsilon i}^2 \\ \vdots \\ (n-1)\lambda_{\varepsilon i}^{n-2} \end{bmatrix}, \cdots,$$

$$q_{n_i}^i(\varepsilon) = \begin{bmatrix} 0 \\ 0 \\ 0 \\ \vdots \\ C_{n-2}^{n_i-1}\lambda_{\varepsilon i}^{n-n_i-1} \\ C_{n-1}^{n_i-1}\lambda_{\varepsilon i}^{n-n_i} \end{bmatrix} \qquad (2.3.30)$$

and their complex conjugates $\bar{q}_j^i(\varepsilon)$, $j = 1$ to n_i.

We then form the following real nonsingular transformation matrix,

$$Q(\varepsilon) = [Q_{-1}(\varepsilon) \; Q_{+1}(\varepsilon) \; Q_1(\varepsilon) \; Q_2(\varepsilon) \; \cdots \; Q_l(\varepsilon)], \qquad (2.3.31)$$

where

$$Q_{-1}(\varepsilon) = [q_1^{-1}(\varepsilon) \; q_2^{-1}(\varepsilon) \; \cdots \; q_{n_{-1}}^{-1}(\varepsilon)],$$

$$Q_{+1}(\varepsilon) = [q_1^{+1}(\varepsilon) \; q_2^{+1}(\varepsilon) \; \cdots \; q_{n_{+1}}^{+1}(\varepsilon)],$$

and for $i = 1$ to l,

$$Q_i(\varepsilon) = \left[\frac{q_1^i + \bar{q}_1^i}{2} \quad \frac{q_1^i - \bar{q}_1^i}{2j} \quad \frac{q_2^i + \bar{q}_2^i}{2} \quad \frac{q_2^i - \bar{q}_2^i}{2j} \right.$$

$$\left. \cdots \quad \frac{q_{n_i}^i + \bar{q}_{n_i}^i}{2} \quad \frac{q_{n_i}^i - \bar{q}_{n_i}^i}{2j} \right].$$

It can now be readily verified that

$$Q^{-1}(\varepsilon)(A + BF(\varepsilon))Q(\varepsilon) = J(\varepsilon), \tag{2.3.32}$$

where $J(\varepsilon)$ is as defined in (2.2.19). This completes the proof. ⌗

Lemma 2.3.3. *Consider a single input pair (A, B) in the form of (2.3.7) with all eigenvalues of A on the unit circle. Let $J(\varepsilon)$ be as given in Lemma 2.3.2. Let*

$$S(\varepsilon) = \mathrm{blkdiag}\{S_{-1}(\varepsilon), S_{+1}(\varepsilon), S_1(\varepsilon), S_2(\varepsilon), \cdots, S_l(\varepsilon)\}, \tag{2.3.33}$$

where $S_{-1}(\varepsilon) = \mathrm{diag}\{\varepsilon^{n_{-1}-1}, \varepsilon^{n_{-1}-2}, \cdots, \varepsilon, 1\}$, $S_{+1}(\varepsilon) = \mathrm{diag}\{\varepsilon^{n_{+1}-1}, \varepsilon^{n_{+1}-2}, \cdots, \varepsilon, 1\}$, and for each $i = 1$ to l, $S_i(\varepsilon) = \mathrm{blkdiag}\{\varepsilon^{n_i-1}I_2, \varepsilon^{n_i-2}I_2, \cdots, \varepsilon I_2, I_2\}$.

Then,

1.

$$S(\varepsilon)J(\varepsilon)S^{-1}(\varepsilon) = \tilde{J}(\varepsilon) := \mathrm{blkdiag}\left\{\tilde{J}_{-1}(\varepsilon), \tilde{J}_{+1}(\varepsilon), \tilde{J}_1(\varepsilon), \cdots, \tilde{J}_l(\varepsilon)\right\}, \tag{2.3.34}$$

where

$$\tilde{J}_{-1}(\varepsilon) = \begin{bmatrix} -(1-\varepsilon) & \varepsilon & & \\ & \ddots & \ddots & \\ & & -(1-\varepsilon) & \varepsilon \\ & & & -(1-\varepsilon) \end{bmatrix}_{n_{-1} \times n_{-1}}, \tag{2.3.35}$$

$$\tilde{J}_{+1}(\varepsilon) = \begin{bmatrix} (1-\varepsilon) & \varepsilon & & \\ & \ddots & \ddots & \\ & & (1-\varepsilon) & \varepsilon \\ & & & (1-\varepsilon) \end{bmatrix}_{n_{+1} \times n_{+1}}, \tag{2.3.36}$$

and for each $i = 1$ to l,

$$\tilde{J}_i(\varepsilon) = \begin{bmatrix} J_i^\star(\varepsilon) & \varepsilon I_2 & & \\ & \ddots & \ddots & \\ & & J_i^\star(\varepsilon) & \varepsilon I_2 \\ & & & J_i^\star(\varepsilon) \end{bmatrix}_{2n_i \times 2n_i}, \quad J_i^\star(\varepsilon) = (1-\varepsilon)\begin{bmatrix} \alpha_i & \beta_i \\ -\beta_i & \alpha_i \end{bmatrix}, \tag{2.3.37}$$

with $\beta_i > 0$ for all $i = 1$ to l and $\beta_i \neq \beta_j$ for $i \neq j$;

2. *There exists an $\varepsilon^* \in (0, 1]$ such that the unique positive definite solution $\tilde{P}(\varepsilon)$ to the Lyapunov equation*

$$\tilde{J}'(\varepsilon)\tilde{P}\tilde{J}(\varepsilon) - \tilde{P} = -\varepsilon I \qquad (2.3.38)$$

is bounded over $\varepsilon \in (0, \varepsilon^*]$, i.e., there exist positive definite matrices $\tilde{P}_1$ and $\tilde{P}_2$ such that

$$\tilde{P}_1 \leq \tilde{P}(\varepsilon) \leq \tilde{P}_2, \quad \forall \varepsilon \in (0, \varepsilon^*]. \qquad (2.3.39)$$

□

Proof of Lemma 2.3.3. Item 1 is the result of some straightforward matrix manipulation. To show Item 2, we observe that the solution $\tilde{P}(\varepsilon)$ to the Lyapunov equation (2.3.38) is of block diagonal form

$$\tilde{P}(\varepsilon) = \text{blkdiag}\{\tilde{P}_{-1}(\varepsilon), \tilde{P}_{+1}(\varepsilon), \tilde{P}_1(\varepsilon), \tilde{P}_2(\varepsilon), \cdots, \tilde{P}_l(\varepsilon)\}, \qquad (2.3.40)$$

where $\tilde{P}_{-1}(\varepsilon)$ and $\tilde{P}_{+1}(\varepsilon)$ are respectively the unique positive definite solutions to the Lyapunov equations

$$\tilde{J}'_{-1}(\varepsilon)\tilde{P}_{-1}J_{-1}(\varepsilon) - \tilde{P}_{-1} = -\varepsilon I, \qquad (2.3.41)$$

and

$$\tilde{J}'_{+1}(\varepsilon)\tilde{P}_{+1}J_{+1}(\varepsilon) - \tilde{P}_{+1} = -\varepsilon I, \qquad (2.3.42)$$

and, for $i = 1$ to l, $\tilde{P}_i(\varepsilon)$ is the positive definite solution to the Lyapunov equation

$$\tilde{J}'_i(\varepsilon)\tilde{P}_i\tilde{J}_i(\varepsilon) - \tilde{P}_i = -\varepsilon I. \qquad (2.3.43)$$

Hence we will only need to show that each of these blocks in $\tilde{P}(\varepsilon)$ is bounded for sufficiently small ε. Let us first examine $\tilde{P}_{-1}(\varepsilon)$, which is given by,

$$\tilde{P}_{-1}(\varepsilon) = \varepsilon \sum_{i=0}^{\infty} \left(\tilde{J}'_{-1}(\varepsilon)\tilde{J}_{-1}(\varepsilon)\right)^i, \qquad (2.3.44)$$

where

$$\tilde{J}'_{-1}(\varepsilon)\tilde{J}_{-1}(\varepsilon) = (1-\varepsilon)^2 I_{n-1} + \varepsilon(1-\varepsilon)\begin{bmatrix} 0 & 1 & 0 & \cdots & 0 \\ 1 & 0 & 1 & \cdots & 0 \\ 0 & 1 & 0 & \ddots & 0 \\ \vdots & \vdots & \ddots & \ddots & \vdots \\ 0 & 0 & 0 & \cdots & 0 \end{bmatrix}$$

$$\geq (1-\varepsilon)^2 I_{n-1} - \theta_{-1}\varepsilon(1-\varepsilon)I_{n-1}$$

$$\geq [1 - (2 + \theta_{-1})\varepsilon]I_{n-1}, \qquad (2.3.45)$$

and

$$\tilde{J}'_{-1}(\varepsilon)\tilde{J}_{-1}(\varepsilon) = (1-\varepsilon)^2 I_{n_{-1}} + \varepsilon(1-\varepsilon)\begin{bmatrix} 0 & 1 & 0 & \cdots & 0 \\ 1 & 0 & 1 & \cdots & 0 \\ 0 & 1 & 0 & \ddots & 0 \\ \vdots & \vdots & \ddots & \ddots & \vdots \\ 0 & 0 & 0 & \cdots & 0 \end{bmatrix}$$

$$\leq (1-\varepsilon)^2 I_{n_{-1}} + \theta_{-1}\varepsilon(1-\varepsilon)I_{n_{-1}}$$
$$= [1 - (2-\theta_{-1})\varepsilon - (\theta_{-1}-1)\varepsilon^2]I_{n_{-1}}, \qquad (2.3.46)$$

for some $\theta_{-1} \in [1,2)$ independent of ε. Hence, for sufficiently small ε, we have

$$\tilde{P}_{-1}(\varepsilon) \geq \varepsilon \sum_{i=0}^{\infty}[1 - (2+\theta_{-1})\varepsilon]^i I_{n_{-1}} = \frac{1}{2+\theta_{-1}}I_{n_{-1}}, \qquad (2.3.47)$$

and

$$\tilde{P}_{-1}(\varepsilon) \leq \varepsilon \sum_{i=0}^{\infty}[1 - (2-\theta_{-1})\varepsilon]^i I_{n_{-1}} = \frac{1}{2-\theta_{-1}}I_{n_{-1}}. \qquad (2.3.48)$$

Similarly, we can show that

$$\tilde{P}_{+1}(\varepsilon) \geq \frac{1}{2+\theta_{+1}}I_{n_{+1}}, \qquad (2.3.49)$$

and

$$\tilde{P}_{+1}(\varepsilon) \leq \frac{1}{2-\theta_{+1}}I_{n_{+1}}, \quad \theta_{+1} = \theta_{-1}. \qquad (2.3.50)$$

It remains to show that, for each $i = 1$ to l, $\tilde{P}_i(\varepsilon)$ is also bounded over $\varepsilon \in (0,1]$. Recall that

$$\tilde{P}_i(\varepsilon) = \varepsilon \sum_{i=0}^{\infty}\left(\tilde{J}'_i(\varepsilon)\tilde{J}_i(\varepsilon)\right)^i. \qquad (2.3.51)$$

Note that

$$\tilde{J}'_i(\varepsilon)\tilde{J}_i(\varepsilon) = (1-\varepsilon)^2 I_{2n_i} + \varepsilon(1-\varepsilon)\begin{bmatrix} 0 & 1 & 0 & \cdots & 0 \\ 1 & 0 & 1 & \cdots & 0 \\ 0 & 1 & 0 & \ddots & 0 \\ \vdots & \vdots & \ddots & \ddots & \vdots \\ 0 & 0 & 0 & \cdots & 0 \end{bmatrix} \otimes \begin{bmatrix} \alpha_i & \beta_i \\ -\beta_i & \alpha_i \end{bmatrix}$$

$$\geq (1-\varepsilon)^2 I_{2n_i} - \theta_i\varepsilon(1-\varepsilon)I_{2n_i}$$
$$\geq [1 - (2+\theta_i)\varepsilon]I_{2n_i}, \qquad (2.3.52)$$

and

$$\tilde{J}'_i(\varepsilon)\tilde{J}_i(\varepsilon) = (1-\varepsilon)^2 I_{2n_i} + \varepsilon(1-\varepsilon) \begin{bmatrix} 0 & 1 & 0 & \cdots & 0 \\ 1 & 0 & 1 & \cdots & 0 \\ 0 & 1 & 0 & \ddots & 0 \\ \vdots & \vdots & \ddots & \ddots & \vdots \\ 0 & 0 & 0 & \cdots & 0 \end{bmatrix} \otimes \begin{bmatrix} \alpha_i & \beta_i \\ -\beta_i & \alpha_i \end{bmatrix}$$

$$\leq (1-\varepsilon)^2 I_{2n_i} + \theta_i \varepsilon(1-\varepsilon) I_{2n_i}$$

$$= [1 - (2 - \theta_i)\varepsilon - (\theta_i - 1)\varepsilon^2] I_{2n_i}, \tag{2.3.53}$$

for some $\theta_i \in [1,2)$ independent of ε. Hence, for sufficiently small ε, we have

$$\tilde{P}_i(\varepsilon) \geq \varepsilon \sum_{i=0}^{\infty} [1 - (2 + \theta_i)\varepsilon] I_{2n_i} = \frac{1}{2+\theta_i} I_{2n_i}, \tag{2.3.54}$$

and

$$\tilde{P}_i(\varepsilon) \leq \varepsilon \sum_{i=0}^{\infty} [1 - (2 - \theta_i)]^i I_{2n_i} = \frac{1}{2-\theta_i} I_{2n_i}. \tag{2.3.55}$$

This concludes our proof of Lemma 2.3.3.   ⊠

Lemma 2.3.4. *Let A, B, $F(\varepsilon)$, $Q(\varepsilon)$, l, and n_i for $i = 0$ to l, be as defined in Lemma 2.2.2 and its proof. Let the scaling matrix $S(\varepsilon)$ be as defined in Lemma 2.3.3. Then, there exists an $\alpha, \beta, \theta \geq 0$ independent of ε such that, for all $\varepsilon \in (0,1]$,*

$$|F(\varepsilon)Q(\varepsilon)S^{-1}(\varepsilon)| \leq \alpha\varepsilon, \tag{2.3.56}$$

$$|F(\varepsilon)AQ(\varepsilon)S^{-1}(\varepsilon)| \leq \beta\varepsilon, \tag{2.3.57}$$

$$|Q(\varepsilon)| \leq \theta, \ |Q^{-1}(\varepsilon)| \leq \theta. \tag{2.3.58}$$

                                                                                ⊡

Proof of Lemma 2.3.4. Observe that

$$F(\varepsilon)Q(\varepsilon)S^{-1}(\varepsilon) = [\, F(\varepsilon)Q_{-1}(\varepsilon)S_{-1}^{-1}(\varepsilon) \quad F(\varepsilon)Q_{+1}(\varepsilon)S_{+1}^{-1}(\varepsilon)$$
$$F(\varepsilon)Q_1 S_1^{-1}(\varepsilon) \quad \cdots \quad F(\varepsilon)Q_l(\varepsilon)S_l^{-1}(\varepsilon)\,], \tag{2.3.59}$$

where $Q_i(\varepsilon)$, $i = 0, 1, 2, \cdots, l$, are defined in (2.3.31). We next recall that (see (2.3.17)) for each $i \in \{-1, +1, 1, 2, \cdots, l\}$ and for each $j = 1$ to n_i, there exists a $\delta_j^i \geq 0$, independent of ε, such that,

$$\left|F(\varepsilon)q_j^i(\varepsilon)\right| \leq \delta_j^i \varepsilon^{n_i - j + 1}, \ \forall \varepsilon \in (0,1]. \tag{2.3.60}$$

It is now clear that there exist δ_{-1}, $\delta_{+1} \geq 0$ such that

$$\left|F(\varepsilon)Q_{-1}(\varepsilon)S_{-1}^{-1}(\varepsilon)\right| \leq \delta_{-1}\varepsilon, \ \ \forall \varepsilon \in (0,1], \tag{2.3.61}$$

and

$$\left|F(\varepsilon)Q_{-1}(\varepsilon)S_{+1}^{-1}(\varepsilon)\right| \leq \delta_{+1}\varepsilon, \ \ \forall \varepsilon \in (0,1]. \tag{2.3.62}$$

For each $i = 1$ to l, noting the definition of $Q_i(\varepsilon)$, it is also straightforward to verify that there exists a $\delta_i \geq 0$, independent of ε, such that,

$$\left|F(\varepsilon)Q_i(\varepsilon)S_i^{-1}(\varepsilon)\right| \leq \delta_i\varepsilon, \ \ \forall \varepsilon \in (0,1]. \tag{2.3.63}$$

The existence of α now follows readily. The existence of β follows with similar arguments from the combination of (2.3.17) and (2.3.16). The existence of θ follows from the facts that $Q(\varepsilon)$ is a polynomial in ε and that $Q(0)$, being the transformation matrix that takes A into its real Jordan form, is nonsingular (and hence $Q^{-1}(\varepsilon)$ is continuously differentiable in ε).   ⊠

Finally, we establish a lemma that is useful in examining the properties of the closed-loop system under low gain feedback and in the presence of external inputs. For example, we will need this lemma in Chapter 10 when we utilize the low gain feedback design technique to construct feedback laws that solve the H_∞-ADDPMS for discrete-time linear systems.

Lemma 2.3.5. *Let A and $Q(\varepsilon)$ be as given in the proof of Lemma 2.3.2. Let $E \in \mathbb{R}^{n \times q}$ is such that*

$$\mathrm{Im}(E) \subset \cap_{w \in \lambda(A)}\mathrm{Im}(wI - A), \tag{2.3.64}$$

where q is any integer. Then, there exists a $\delta \geq 0$, independent of ε, such that

$$\left|Q^{-1}(\varepsilon)E\right| \leq \delta, \ \ \varepsilon \in (0,1], \tag{2.3.65}$$

and, if we partition $Q^{-1}(\varepsilon)E$ according to that of $J(\varepsilon)$ of Lemma 2.3.2 as,

$$Q^{-1}(\varepsilon)E = \begin{bmatrix} E_0(\varepsilon) \\ E_1(\varepsilon) \\ \vdots \\ E_l(\varepsilon) \end{bmatrix}, \ \ E_0(\varepsilon) = \begin{bmatrix} E_{01}(\varepsilon) \\ E_{02}(\varepsilon) \\ \vdots \\ E_{0n_0}(\varepsilon) \end{bmatrix}_{n_0 \times q}, \ \ E_i(\varepsilon) = \begin{bmatrix} E_{i1}(\varepsilon) \\ E_{i2}(\varepsilon) \\ \vdots \\ E_{in_i}(\varepsilon) \end{bmatrix}_{2n_i \times q}, \tag{2.3.66}$$

then, there exists a $\beta \geq 0$, independent of ε, such that, for each $i = 0$, to l,

$$\left|E_{in_i}(\varepsilon)\right| \leq \beta\varepsilon. \tag{2.3.67}$$

Ⓛ

Proof of Lemma 2.3.5. The existence of a $\delta \geq 0$ that satisfies (2.3.65) follows readily from Lemma 2.3.4.

To show the existence of $\beta \geq 0$ that satisfies (2.3.67), we note that Assumption (2.3.64) implies that, for each $i = 0$ to l,

$$|E_{in_i}(0)| = 0. \qquad (2.3.68)$$

The existence of such a β now follows trivially from the continuous differentiability of $Q^{-1}(\varepsilon)$.   ☒

2.3.2. ARE Based Method

The ARE based low gain state feedback design for the discrete-time system (2.3.1) is carried out in two steps.

Step 1. Solve the following algebraic Riccati equation

$$P = A'PA - A'PB(B'PB + I)^{-1}B'PA + \varepsilon I, \qquad (2.3.69)$$

for the unique positive definite solution $P(\varepsilon)$. The existence of such a solution is established in Lemma 2.3.6 below.

Step 2. Construct a family of low gain state feedback laws as

$$u = F(\varepsilon)x, \qquad (2.3.70)$$

where

$$F(\varepsilon) = -(B'P(\varepsilon)B + I)^{-1}B'P(\varepsilon)A. \qquad (2.3.71)$$

                                                                                        🅐

The following lemma establishes the basic property of this family of ARE based low gain feedback laws.

Lemma 2.3.6. *Let Assumption 2.3.1 hold. Then, for each $\varepsilon \in (0,1]$, there exists a unique matrix $P(\varepsilon) > 0$ that solves the ARE (2.3.69). Moreover, such a $P(\varepsilon)$ satisfies,*

1. $\lim_{\varepsilon \to 0} P = 0$;

2. *There exists an $\varepsilon^* \in (0,1]$ such that,*

$$\left| P^{\frac{1}{2}} A^\ell P^{-\frac{1}{2}} \right| \leq \sqrt{2}^\ell, \; \forall \ell \in \mathbf{N}, \; \forall \varepsilon \in (0, \varepsilon^*]. \qquad (2.3.72)$$

Note that for notational convenience, we have denoted and will often denote $P = P(\varepsilon)$. $\boxed{\text{L}}$

Remark 2.3.3. *Lemma 2.3.6 reveals that for any given bounded set of initial conditions, the control and all its derivatives can be made arbitrarily small by decreasing the value of the low gain parameter* ε. $\boxed{\text{R}}$

Proof of Lemma 2.3.6. Existence and uniqueness of such a solution for $\varepsilon > 0$ follows from [94]. For $\varepsilon = 0$, it is trivial to see that (2.3.69) has a solution $P = 0$ which is semi-stabilizing. Again from [94], this semi-stabilizing solution is unique. Finally, the fact that $\lim_{\varepsilon \to 0} P(\varepsilon) = 0$ follows readily from the standard continuity arguments.

To show Item 2 of the lemma, pre- and post-multiplying both sides of the ARE (2.3.69) by $P^{-\frac{1}{2}}(\varepsilon)$, we obtain

$$P^{-\frac{1}{2}} A' P^{\frac{1}{2}} \left[I - P^{\frac{1}{2}} B (B'PB + I)^{-1} B' P^{\frac{1}{2}} \right] P^{\frac{1}{2}} A P^{-\frac{1}{2}} = I - \varepsilon P^{-1}. \quad (2.3.73)$$

By Item 1, $\lim_{\varepsilon \to 0} P = 0$, hence it follows from (2.3.73) that there exists an $\varepsilon^* > 0$ such that for all $\varepsilon \in (0, \varepsilon^*]$,

$$P^{-\frac{1}{2}} A' P A P^{-\frac{1}{2}} \le 2I - 2\varepsilon P^{-1} \le 2I, \quad (2.3.74)$$

which implies that

$$\lambda_{\max} \left(P^{-\frac{1}{2}} A' P A P^{-\frac{1}{2}} \right) \le 2. \quad (2.3.75)$$

This completes the proof of Item 2 for $\ell = 1$. For the case of $\ell > 1$, we have

$$\left| P^{\frac{1}{2}} A^{\ell} P^{-\frac{1}{2}} \right| \le \left| P^{\frac{1}{2}} A P^{-\frac{1}{2}} \right|^{\ell} \le \sqrt{2}^{\ell}. \quad (2.3.76)$$

$\boxed{\text{※}}$

2.4. Comparison Between Different Design Techniques

The biggest advantage of the eigenstructure assignment method for low gain feedback design is that it results in feedback gains that are polynomial matrix in the low gain parameter ε. Thus the design is "one-shot" in the sense that if the value of the low gain parameter ε is required to change, the design process need not be repeated.

The ARE based method for low gain feedback design is conceptually appealing. As will be seen throughout the monograph, they also result in feedback laws that are more robust with respect to actuator nonlinearities than the eigenstructure assignment based low gain feedback laws. The disadvantage of the

ARE based method is that the resulting feedback gain is indirectly dependent on the low gain parameter ε. For different values of ε, the solution of a parameterized ARE is required. The solution of these parameterized ARE's may become numerically ill-conditioned as ε goes to zero.

Chapter 3

Semi-Global Stabilization of Linear Systems with Saturating Actuators

3.1. Introduction

It is known that a linear time-invariant system subject to actuator saturation can be globally asymptotically stabilized if and only if it is asymptotically null controllable with bounded controls (ANCBC) (see [103]). It is also shown by Fuller [18] and Sussmann and Yang [109] that in general one must use non-linear feedback laws to achieve global asymptotic stabilization and only some special cases can be handled by linear feedback laws. A nested feedback design technique for designing nonlinear globally asymptotically stabilizing feedback laws was proposed in [110] for a chain of integrators of any length and was fully generalized in [108] for general linear ANCBC systems. Alternative solutions to the global stabilization by ARE based feedback laws were later proposed in [77,104,112].

In this chapter, we present a result that complements the "negative results" of [18] and [109]. We consider semi-global exponential stabilization of general linear ANCBC systems subject to actuator magnitude saturation and show that, in contrast to the situation in global asymptotic stabilization, one can utilize low gain feedback to semi-globally exponentially stabilize such systems via linear feedback laws. Here by semi-global exponential stabilization we mean local exponential stabilization of the system with a basin of attraction that contains an *a priori* given bounded set as a subset. (For a precise definition of semi-global exponential stabilization, see problem statements in Sections

43

3.3.1 and 3.4.1.) Relaxing the requirement of global stabilization to that of semi-global stabilization, from engineering point of view, makes sense, since in general a plant's model is usually valid in some region of the state space. Such relaxation gives us simple linear feedback laws and stronger stability property for the closed-loop system, that is, the exponential stability of the closed-loop system, rather than asymptotic stability.

In designing low gain feedback laws, we can either use eigenstructure assignment algorithm or the ARE based design algorithm. As will become clear shortly, the ARE based design algorithm results in feedback laws that are robust with respect to the actuator nonlinearities.

We will consider both continuous-time and discrete-time systems. Section 3.2 defines families of saturation functions to be encountered in the chapter and the remainder of this monograph. Section 3.3 deals with continuous-time systems. Section 3.4 deals with discrete-time systems. Section 3.5 contains some concluding remarks.

3.2. Definition of Saturations Functions

We define three classes of saturation functions, class $\mathcal{S}_1$, class $\mathcal{S}_2$ and class $\mathcal{S}_3$. These are the types of saturation functions we will encounter in this chapter and the remainder of this monograph.

Definition 3.2.1. *A function $\sigma : \mathbb{R}^m \to \mathbb{R}^m$ is called a class $\mathcal{S}_1$ saturation function if*

   1. $\sigma(u)$ is decentralized, i.e., $\sigma(u) = [\,\sigma_1(u_1) \quad \sigma_2(u_2) \quad \cdots \quad \sigma_m(u_m)\,]'$;

   2. There exists a $\Delta > 0$ such that, for each $i = 1$ to m,

$$\sigma_i(u_i) \begin{cases} = u_i & \text{if } |u_i| \leq \Delta, \\ < -\Delta & \text{if } u_i < -\Delta, \\ > \Delta & \text{if } u_i > \Delta. \end{cases} \tag{3.2.1}$$

Moreover, the set of all class $\mathcal{S}_1$ saturation functions that satisfy (3.2.1) with a fixed $\Delta > 0$ is denoted by $\mathcal{S}_1(\Delta)$. ▫

The class $\mathcal{S}_3$ saturation functions are characterized by the following properties.

Definition 3.2.2. *A function $\sigma : \mathbb{R}^m \to \mathbb{R}^m$ is called a class $\mathcal{S}_3$ saturation function if*

1. $\sigma(u)$ is decentralized, i.e., $\sigma(u) = [\,\sigma_1(u_1) \quad \sigma_2(u_2) \quad \cdots \quad \sigma_m(u_m)\,]'$; and for each $i = 1$ to m,

2. $s\sigma_i(s) > 0$ whenever $s \neq 0$;

3. There exist a $\Delta > 0$ and $b > b_1 > 0$ such that

$$b_1 \leq \frac{\sigma_i(s)}{s} \leq b, \ 0 < |s| \leq \Delta.$$

Ⓓ

Remark 3.2.1.   1. Graphically, the saturation function resides in the first and third quadrants and there exist $\Delta > 0$ and $b > b_1 > 0$ such that for $|s| \leq \Delta$, the saturation function lies in the linear sector between the graphs graphs $(s, b_1 s)$ and (s, bs), which implies that,

$$s[\sigma_i(s) - b_1 s] \geq 0, \tag{3.2.2}$$

$$|\sigma_i(s) - b_1 s| \leq (b - b_1)|s|. \tag{3.2.3}$$

For notational simplicity, but without loss of generality, we will assume that $b_1 = 1$.

Ⓡ

Definition 3.2.3. The set of all class $\mathcal{S}_3$ saturation functions that satisfy Definition 3.2.2 with a fixed triple of constants Δ, $b_1 = 1$ and b is denoted by $\mathcal{S}_3(\Delta, b)$. The subset of locally Lipschitz members of $\mathcal{S}_3(\Delta, b)$ is denoted as $\mathcal{S}_2(\Delta, b)$. The subset of all globally Lipschitz members of $\mathcal{S}_3(\Delta, b)$ with a Lipschitz constant δ is denoted as $\mathcal{S}_2^{\text{GL}}(\Delta, b, \delta)$.

Ⓓ

Remark 3.2.2. It follows directly from Definition 3.2.3 that the functions $\sigma(t) = t$, $\arctan(t)$, $\tanh(t)$ and the standard saturation function $\sigma(t) = \text{sign}(t)$ $\min\{|t|, 1\}$ are all class $\mathcal{S}_2$ saturation functions. Moreover, functions like $\sigma(s) = 2s + s\sin(1/s)$ with $\sigma(0) = 0$, which are not even one-sided differentiable at the origin, are also class $\mathcal{S}_3$ saturation functions.

Ⓡ

3.3. Continuous-Time Systems

3.3.1. Problem Statement

Consider a linear system subject to actuator magnitude saturation described by

$$\begin{cases} \dot{x} = Ax + B\sigma(u), \\ y = Cx, \end{cases} \tag{3.3.1}$$

where $x \in \mathbb{R}^n$ is the state, $u \in \mathbb{R}^m$ is the control input to the actuators, $y \in \mathbb{R}^p$ is the measurement output, and $\sigma : \mathbb{R}^m \to \mathbb{R}^m$ is a saturation function of either class $\mathcal{S}_1$ or class $\mathcal{S}_2$, as defined in Section 3.2.

We make the following assumptions on system (3.3.1).

Assumption 3.3.1. *The pair (A, B) is asymptotically null controllable with bounded controls (ANCBC), i.e.,*

1. *(A, B) is stabilizable;*

2. *All eigenvalues of A are in the closed left-half s-plane.* ⊞

Assumption 3.3.2. *The pair (A, C) is detectable.* ⊞

The problems that we are to solve using low gain feedback are the following.

Problem 3.3.1. (Semi-global exponential stabilization via linear static state feedback) *Consider system (3.3.1) with $\sigma \in \mathcal{S}_1(\Delta)$ [or $\sigma \in \mathcal{S}_2(\Delta, b)$], where $\Delta > 0$ and $b \geq 1$. For any a priori given bounded set of initial conditions $\mathcal{X} \subset \mathbb{R}^n$, find a state feedback law $u = F_{\mathcal{X}}x$ such that, for any $\sigma \in \mathcal{S}_1(\Delta)$ [or $\sigma \in \mathcal{S}_2(\Delta, b)$], the equilibrium $x = 0$ of the closed-loop system is locally exponentially stable with $\mathcal{X}$ contained in its basin of attraction.* ▣

Problem 3.3.2. (Semi-global exponential stabilization via linear dynamic output feedback) *Consider system (3.3.1) with $\sigma \in \mathcal{S}_1(\Delta)$ [or $\sigma \in \mathcal{S}_2(\Delta, b).$], where $\Delta > 0$ and $b \geq 1$. For any a priori given bounded set $\mathcal{W} \subset \mathbb{R}^{2n}$, find a linear dynamic output feedback control law of dynamical order n,*

$$\begin{cases} \dot{z} = Gz + Hy, \quad z \in \mathbb{R}^n, \\ u = Mz + Ny, \end{cases} \tag{3.3.2}$$

such that, for any $\sigma \in \mathcal{S}_1(\Delta)$ [or $\sigma \in \mathcal{S}_2(\Delta, b)$], the equilibrium $(x, z) = (0, 0)$ of the closed-loop system is locally exponentially stable with $\mathcal{W}$ contained in its basin of attraction. ▣

The solution of the above two problems requires a family of feedback laws. The objective here is to show that low gain feedback design methods can be utilized to construct families of linear feedback laws that solve the above two problems for system (3.3.1) with any $\sigma \in \mathcal{S}_1(\Delta)$. And, moreover, the ARE based low gain feedback laws can solve these problems for system (3.3.1) with any $\sigma \in \mathcal{S}_2(\Delta, b)$. Our design of low gain feedback laws does not require any specific knowledge of σ as long as it belongs to $\mathcal{S}_1(\Delta)$ or $\mathcal{S}_2(\Delta, b)$. In this sense, the low gain feedback laws we are to construct are robust with respect to the actuator nonlinearities.

3.3.2. Eigenstructure Assignment Design

Design Algorithm:

Step 1 - State Feedback Design. For the matrix pair (A, B), construct a family of low gain state feedback laws as,

$$u = F(\varepsilon)x, \tag{3.3.3}$$

where $F(\varepsilon)$ is as given by (2.2.6). The existence of such an $F(\varepsilon)$ is guaranteed by Assumption 3.3.1.

Step 2 - Output Feedback Design. Construct a family of low gain output feedback laws as,

$$\begin{cases} \dot{\hat{x}} = A\hat{x} + BF(\varepsilon)\hat{x} - L(y - C\hat{x}), \\ u = F(\varepsilon)\hat{x}, \end{cases} \tag{3.3.4}$$

where L is chosen such that $A+LC$ is asymptotically stable. The existence of such an L is due to Assumption 3.3.2. Ⓐ

With these two families of feedback laws, we have the following results.

Theorem 3.3.1. *Consider system (3.3.1) with $\sigma \in \mathcal{S}_1(\Delta)$, where $\Delta > 0$ is any (arbitrarily small) constant. If Assumption 3.3.1 is satisfied, then the family of state feedback laws (3.3.3) solves Problem 3.3.1. If Assumptions 3.3.1 and 3.3.2 are satisfied, then the family of output feedback laws (3.3.4) solves Problem 3.3.2.* Ⓣ

Proof of Theorem 3.3.1. The idea of the proof is as follows. We first consider the given system (3.3.1) in the absence of saturation, i.e.,

$$\dot{x} = Ax + Bu, \tag{3.3.5}$$

and show that the closed-loop system is exponentially stable for all sufficiently small ε, and, for any *a priori* given bounded set $\mathcal{W}$, there exists an $\varepsilon^* > 0$ such that, for any $\varepsilon \in (0, \varepsilon^*]$, each channel of the control input u is uniformly bounded by Δ.

Without loss of generality, assume that the pair (A, B) is already in the block diagonal control canonical form of (2.2.2)-(2.2.3).

State Feedback Case:

Under the family of state feedback laws (3.3.3), the closed-loop system is given by,

$$\begin{cases} \dot{x}_1 = (A_1 + B_1 F_i(\varepsilon))x_1 + \sum_{j=2}^{l} B_{1j} F_j(\varepsilon)x_j, \\ \dot{x}_2 = (A_2 + B_2 F_2(\varepsilon))x_2 + \sum_{j=3}^{l} B_{2j} F_j(\varepsilon)x_j, \\ \quad\vdots \\ \dot{x}_l = (A_l + B_l F_l(\varepsilon))x_l, \\ \dot{x}_0 = A_0 x_0 + \sum_{j=1}^{l} B_{0j} F_j(\varepsilon)x_j, \end{cases} \tag{3.3.6}$$

where $x = [x_1', x_2', \cdots, x_l', x_0']'$.

Now for each $i = 1$ to l, let $Q_i(\varepsilon)$, $S_i(\varepsilon)$, $J_i(\varepsilon)$, $\tilde{J}_i(\varepsilon)$ and $\tilde{P}_i$ be as defined in Lemmas 2.2.2-2.2.4 for the triple $(A_i, B_i, F_i(\varepsilon))$. Define a state transformation as,

$$\tilde{x} = [\tilde{x}_1', \tilde{x}_2', \cdots, \tilde{x}_l', \tilde{x}_0']', \tag{3.3.7}$$

where $\tilde{x}_0 = x_0$ and for each $i = 1$ to l, $\tilde{x}_i = S_i(\varepsilon)Q_i^{-1}(\varepsilon)x_i$.

The closed-loop system in the new state variable is then given by,

$$\begin{cases} \dot{\tilde{x}}_1 = \varepsilon \tilde{J}_1(\varepsilon)\tilde{x}_1 + \sum_{j=2}^{l} S_1(\varepsilon)Q_1^{-1}(\varepsilon)B_{1j}F_j(\varepsilon)Q_j(\varepsilon)S_j^{-1}(\varepsilon)\tilde{x}_j, \\ \dot{\tilde{x}}_2 = \varepsilon \tilde{J}_2(\varepsilon)\tilde{x}_2 + \sum_{j=3}^{l} S_2(\varepsilon)Q_2^{-1}(\varepsilon)B_{2j}F_j(\varepsilon)Q_j(\varepsilon)S_j^{-1}(\varepsilon)\tilde{x}_j, \\ \quad\vdots \\ \dot{\tilde{x}}_l = \varepsilon \tilde{J}_l(\varepsilon)\tilde{x}_l, \\ \dot{x}_0 = A_0 x_0 + \sum_{j=1}^{l} B_{0j}F_j(\varepsilon)Q_j(\varepsilon)S_j^{-1}(\varepsilon)\tilde{x}_j, \end{cases} \tag{3.3.8}$$

for which we consider the Lyapunov function

$$V(\tilde{x}) = \sum_{i=1}^{l} \kappa^i \tilde{x}_i' \tilde{P}_i \tilde{x}_i + \tilde{x}_0' \tilde{P}_0 \tilde{x}_0, \tag{3.3.9}$$

where $\tilde{P}_0 > 0$ is such that $A_0'\tilde{P}_0 + \tilde{P}_0 A_0 = -I$ and $\kappa > 0$ is constant whose value is to be determined later. The existence of such a $\tilde{P}_0$ is due to the fact that A_0 is asymptotically stable.

The derivative of V along the trajectories of the closed-loop system can be evaluated as

$$\dot{V} = \sum_{i=1}^{l} \left[-\varepsilon\kappa^i \tilde{x}_i' \tilde{x}_i + 2 \sum_{j=i+1}^{l} \kappa^i \tilde{x}_i' \tilde{P}_i S_i(\varepsilon)Q_i^{-1}(\varepsilon)B_{ij}F_j(\varepsilon)Q_j(\varepsilon)S_j^{-1}(\varepsilon)\tilde{x}_j \right]$$
$$-\tilde{x}_0'\tilde{x}_0 + 2\sum_{j=1}^{n} \tilde{x}_0' \tilde{P}_0 B_{0j}F_j(\varepsilon)Q_j(\varepsilon)S_j^{-1}(\varepsilon)\tilde{x}_j. \tag{3.3.10}$$

In view of Lemmas 2.2.3 and 2.2.4, it is straightforward to verify that there exist a constant $\kappa > 0$ and an $\varepsilon_1^* \in (0,1]$ such that,

$$\dot{V} \le -\frac{\varepsilon}{2}\tilde{x}'\tilde{x}, \ \ \forall \varepsilon \in (0, \varepsilon_1^*], \tag{3.3.11}$$

which shows that the closed-loop system is exponentially stable for all $\varepsilon \in (0, \varepsilon_1^*]$. It remains to show that, for the given $\mathcal{W}$, there exists an $\varepsilon^* \in (0, \varepsilon_1^*]$ such that, for any $\varepsilon \in (0, \varepsilon^*]$,

$$|F(\varepsilon)x(t)|_\infty = \max_j |F_j(\varepsilon)Q_j(\varepsilon)S_j^{-1}(\varepsilon)\tilde{x}_j(t)| \leq \Delta, \ \ \forall t \geq 0 \text{ and } \forall x(0) \in \mathcal{W}.$$

$$(3.3.12)$$

To this end, let $c > 0$ be such that

$$c \geq \sup_{x \in \mathcal{W}, \varepsilon \in (0, \varepsilon_1^*]} V(\tilde{x}). \qquad (3.3.13)$$

The right-hand side is well-defined since $Q_i^{-1}(\varepsilon)$ is bounded by Lemma 2.2.4. Let $\varepsilon^* \in (0, \varepsilon_1^*]$ be such that $\tilde{x} \in L_V(c)$ implies $|F(\varepsilon)x|_\infty \leq \Delta$. The existence of such an ε^* is due to Lemma 2.2.4 and the fact that $V(\tilde{x})$ is independent of ε. It then follows from (3.3.11) that any trajectory that starts from $\mathcal{W} \subset L_V(c)$ will remain inside $L_V(c)$ and hence (3.3.12) always holds for any $\varepsilon \in (0, \varepsilon^*]$. This completes the proof of the state feedback result.

Output Feedback Case:

From the proof of the state feedback case, it is straightforward to see that under the family of output feedback laws (3.3.4), the closed-loop system can be written as,

$$\begin{cases} \dot{\tilde{x}}_1 = \varepsilon \tilde{J}_1(\varepsilon)\tilde{x}_1 + \sum_{j=2}^l S_1(\varepsilon)Q_1^{-1}(\varepsilon)B_{1j}F_j(\varepsilon)[Q_j(\varepsilon)S_j^{-1}(\varepsilon)\tilde{x}_j - e_j] \\ \qquad\quad -S_1(\varepsilon)Q_1^{-1}(\varepsilon)B_1F_1(\varepsilon)e_1, \\ \dot{\tilde{x}}_2 = \varepsilon \tilde{J}_2(\varepsilon)\tilde{x}_2 + \sum_{j=3}^l S_2(\varepsilon)Q_2^{-1}(\varepsilon)B_{2j}F_j(\varepsilon)[Q_j(\varepsilon)S_j^{-1}(\varepsilon)\tilde{x}_j - e_j] \\ \qquad\quad -S_2(\varepsilon)Q_2^{-1}(\varepsilon)B_2F_2(\varepsilon)e_2, \\ \qquad\qquad\qquad\qquad\qquad \vdots \\ \dot{\tilde{x}}_l = \varepsilon \tilde{J}_l(\varepsilon)\tilde{x}_l - S_l(\varepsilon)Q_l^{-1}(\varepsilon)B_lF_l(\varepsilon)e_l, \\ \dot{x}_0 = A_0x_0 + \sum_{j=1}^l B_{0j}F_j(\varepsilon)[Q_j(\varepsilon)S_j^{-1}(\varepsilon)\tilde{x}_j - e_j], \\ \dot{e} = (A + LC)e, \end{cases} \qquad (3.3.14)$$

where

$$e = \begin{bmatrix} e_1 \\ e_2 \\ \vdots \\ e_l \end{bmatrix} = x - \hat{x}.$$

For this closed-loop system, let us consider the Lyapunov function

$$V(\tilde{x}, e) = \sum_{i=1}^l \kappa^i \tilde{x}_i' \tilde{P}_i \tilde{x}_i + \tilde{x}_0' \tilde{P}_0 \tilde{x}_0 + \kappa^{l+1} e' P_e e, \qquad (3.3.15)$$

where $\tilde{P}_0 > 0$ is the same as in (3.3.9) and P_e is such that $(A+LC)'P_e + P_e(A+LC) = -I$ and $\kappa > 0$ is constant whose value is to be determined later. Here the existence of such a P_e is due to the fact that $A + LC$ is asymptotically stable.

The derivative of V along the trajectories of the closed-loop system can be evaluated as

$$
\dot{V} = \sum_{i=1}^{l} \left[-\varepsilon \kappa^i \tilde{x}_i' \tilde{x}_i - 2\kappa^i \tilde{x}_i' \tilde{P}_i S_i(\varepsilon) Q_i^{-1}(\varepsilon) B_i F_i(\varepsilon) e_i \right.
$$

$$
\left. +2 \sum_{j=i+1}^{l} \kappa^i \tilde{x}_i' \tilde{P}_i S_i(\varepsilon) Q_i^{-1}(\varepsilon) B_{ij} F_j(\varepsilon) [Q_j(\varepsilon) S_j^{-1}(\varepsilon) \tilde{x}_j - e_j] \right]
$$

$$
-\tilde{x}_0' \tilde{x}_0 + 2 \sum_{j=1}^{n} \tilde{x}_0' \tilde{P}_0 B_{0j} F_j(\varepsilon) [Q_j(\varepsilon) S_j^{-1}(\varepsilon) \tilde{x}_j - e_j] - \kappa^{l+1} e' e. \quad (3.3.16)
$$

In view of Lemmas 2.2.3 and 2.2.4, it is straightforward to verify that there exist a constant $\kappa > 0$ and an $\varepsilon_1^* \in (0, 1]$ such that,

$$
\dot{V} \le -\frac{\varepsilon}{2} \tilde{x}' \tilde{x} - \frac{1}{2} e' e, \quad \forall \varepsilon \in (0, \varepsilon_1^*], \quad (3.3.17)
$$

which shows that the closed-loop system is exponentially stable for all $\varepsilon \in (0, \varepsilon_1^*]$. It remains to show that, for the given $\mathcal{W}$, there exists an $\varepsilon^* \in (0, \varepsilon_1^*]$ such that, for any $\varepsilon \in (0, \varepsilon^*]$,

$$
|F(\varepsilon) \hat{x}(t)|_\infty = \max_j \left| F_j(\varepsilon) [Q_j(\varepsilon) S_j^{-1}(\varepsilon) \tilde{x}_j(t) - e_j(t)] \right| \le \Delta,
$$

$$
\forall t \ge 0 \text{ and } \forall (x(0), \hat{x}(0)) \in \mathcal{W}. \quad (3.3.18)
$$

To this end, let $c > 0$ be such that

$$
c \ge \sup_{(x, \hat{x}) \in \mathcal{W}, \varepsilon \in (0, \varepsilon_1^*]} V(\tilde{x}, e). \quad (3.3.19)
$$

The right-hand side is well-defined since $Q_i^{-1}(\varepsilon)$ is bounded by Lemma 2.2.4. Let $\varepsilon^* \in (0, \varepsilon_1^*]$ be such that $(\tilde{x}, e) \in L_V(c)$ implies $|F(\varepsilon) \hat{x}| \le \Delta$. The existence of such an ε^* is due to Lemmas 2.2.1 and 2.2.4 and the fact that $V(\tilde{x}, e)$ is independent of ε. It then follows from (3.3.17) that any trajectory that starts from $\mathcal{W} \subset L_V(c)$ will remain inside $L_V(c)$ and hence (3.3.18) always holds for any $\varepsilon \in (0, \varepsilon^*]$. This completes the proof of the output feedback result and of Theorem 3.3.1. $\boxtimes$

3.3.3. ARE Based Design

Design Algorithm:

Step 1 - State Feedback Design. For the matrix pair (A, B), construct a family of low gain state feedback laws as,

$$u = -B'P(\varepsilon)x, \qquad (3.3.20)$$

where $P(\varepsilon) > 0$ is the solution to the H_2-ARE

$$A'P + PA - PBB'P + \varepsilon I = 0, \ \varepsilon \in (0, 1]. \qquad (3.3.21)$$

The existence of such a $P(\varepsilon)$ is guaranteed by Assumption 3.3.1.

We recall that the above is the ARE based low gain feedback design of Section 2.2.2. Hence, Lemma 2.2.6 applies.

Step 2 - Output Feedback Design. Construct a family of low gain output feedback laws as,

$$\begin{cases} \dot{\hat{x}} = A\hat{x} - BB'P(\varepsilon)\hat{x} - L(y - C\hat{x}), \\ u = -B'P(\varepsilon)\hat{x}, \end{cases} \qquad (3.3.22)$$

where L is chosen such that $A+LC$ is asymptotically stable. The existence of such an L is due to Assumption 3.3.2. 　　　　　 Ⓐ

With these two families of feedback laws, we have the following results.

Theorem 3.3.2. *Consider system (3.3.1) with $\sigma \in \mathcal{S}_2(\Delta, b)$, where $\Delta > 0$ is any (arbitrarily small) constant and b is any (arbitrarily large) constant. If Assumption 3.3.1 is satisfied, then the family of state feedback laws (3.3.20) solves Problem 3.3.1. If Assumptions 3.3.1 and 3.3.2 are satisfied, then the family of output feedback laws (3.3.22) solves Problem 3.3.2.* 　　　 Ⓣ

Proof of Theorem 3.3.2. We will separate the proof into state feedback and output feedback case. For each case, we will show that there exists an $\varepsilon^* \in (0, 1]$ such that for each $\varepsilon \in (0, \varepsilon^*]$, the closed-loop system is locally exponentially stable with $\mathcal{W}$ contained in its basin of attraction.

State Feedback Case:

Under the family of state feedback laws (3.3.20), the closed-loop system is then given by,

$$\dot{x} = Ax + B\sigma(-B'P(\varepsilon)x), \qquad (3.3.23)$$

for which let us consider the Lyapunov function

$$V(x) = x'P(\varepsilon)x, \qquad (3.3.24)$$

and let $c > 0$ be a constant such that

$$c \geq \sup_{x \in W, \varepsilon \in (0,1]} x'P(\varepsilon)x. \tag{3.3.25}$$

Such a c exists since $\lim_{\varepsilon \to 0} P(\varepsilon) = 0$ by Lemma 2.2.6 and W is bounded. Let $\varepsilon^* \in (0,1]$ be such that, for each $\varepsilon \in (0, \varepsilon^*]$, $x \in L_V(c)$ implies that $|B'P(\varepsilon)x|_\infty \leq \Delta$. The existence of such an ε^* is again due to the fact that $\lim_{\varepsilon \to 0} P(\varepsilon) = 0$.

The evaluation of the derivatives of V along the trajectories of the closed-loop system, using Remark 3.2.1, gives that for all $x \in L_V(c)$,

$$\dot{V} = -\varepsilon x'x + 2x'P(\varepsilon)B[\sigma(-B'P(\varepsilon)x) + B'P(\varepsilon)x] - x'P(\varepsilon)BB'P(\varepsilon)x$$
$$\leq -\varepsilon x'x - 2\sum_{i=1}^{m} u_i[\sigma_i(u_i) - \mathrm{sat}_\Delta(u_i)]$$
$$\leq -\varepsilon x'x, \tag{3.3.26}$$

where u_i is the ith element of u.

The above shows that, for any $\varepsilon \in (0, \varepsilon^*]$, the equilibrium $x = 0$ of the closed-loop system (3.3.23) locally exponentially stable with $L_V(c) \supset W$ contained in its basin of attraction. This completes the proof of the state feedback case.

Output Feedback Case:

Under the family of state feedback laws (3.3.22), the closed-loop system is then given by,

$$\begin{cases} \dot{x} = Ax + B\sigma(-B'P(\varepsilon)\hat{x}), \\ \dot{\hat{x}} = A\hat{x} - BB'P(\varepsilon)\hat{x} - L(y - C\hat{x}). \end{cases} \tag{3.3.27}$$

Letting $e = x - \hat{x}$, we can rewritten (3.3.27) as

$$\begin{cases} \dot{x} = Ax + B\sigma(-B'P(\varepsilon)x + B'P(\varepsilon)e), \\ \dot{e} = (A + LC)e + B[\sigma(-B'P(\varepsilon)x + B'P(\varepsilon)e) + B'P(\varepsilon)x - B'P(\varepsilon)e]. \end{cases} \tag{3.3.28}$$

We now consider the Lyapunov function

$$V(x) = x'P(\varepsilon)x + \lambda_{\max}^{\frac{1}{2}}(P(\varepsilon))e'P_e e, \tag{3.3.29}$$

where $P_e > 0$ is such that

$$(A + LC)'P_e + P_e(A + LC) = -I. \tag{3.3.30}$$

The existence of such a P_e is due to the fact that $A + LC$ is asymptotically stable.

Let $c > 0$ be a constant such that

$$c \geq \sup_{(x,\hat{x}) \in \mathcal{W}, \varepsilon \in (0,1]} \left(x'P(\varepsilon))x + \lambda_{\max}^{\frac{1}{2}}(P(\varepsilon)e'P_e e) \right). \tag{3.3.31}$$

Such a c exists since $\lim_{\varepsilon \to 0} P(\varepsilon) = 0$ by Lemma 2.2.6 and $\mathcal{W}$ is bounded. Let $\varepsilon^* \in (0,1]$ be such that, for each $\varepsilon \in (0,\varepsilon^*]$, $(x,e) \in L_V(c)$ implies that $|u|_\infty = |B'P(\varepsilon)x - B'P(\varepsilon)e|_\infty \leq \Delta$. The existence of such an ε^* is again due to the fact that $\lim_{\varepsilon \to 0} P(\varepsilon) = 0$.

The evaluation of the derivatives of V along the trajectories of the closed-loop system (3.3.28), using Remark 3.2.1, gives that for all $(x,e) \in L_V(c)$,

$$
\begin{aligned}
\dot{V} &= -\varepsilon x'x + 2x'P(\varepsilon)B[\sigma(-B'P(\varepsilon)x + B'P(\varepsilon)e) + B'P(\varepsilon)x] \\
&\quad -x'P(\varepsilon)BB'P(\varepsilon)x - \lambda_{\max}^{\frac{1}{2}}(P(\varepsilon))e'e \\
&\quad +2\lambda_{\max}^{\frac{1}{2}}(P(\varepsilon))e'P_e B[\sigma(-B'P(\varepsilon)x + B'P(\varepsilon)e) + B'P(\varepsilon)x - B'P(\varepsilon)e] \\
&\leq -\varepsilon x'x - u_x'u_x - \lambda_{\max}^{\frac{1}{2}}(P(\varepsilon))e'e - 2u'[\sigma(u) - u] - 2u_e'[\sigma(u) - u] - 2u_x'u_e \\
&\quad -2\lambda_{\max}^{\frac{1}{2}}(P(\varepsilon))e'P_e B[\sigma(u) - u] \\
&\leq -\varepsilon x'x - u_x'u_x - \lambda_{\max}^{\frac{1}{2}}(P(\varepsilon))e'e + 2(b-1)|u_e|(|u_x| + |u_e|) + 2|u_x||u_e| \\
&\quad +2(b-1)\lambda_{\max}^{\frac{1}{2}}(P(\varepsilon))|P_e B||e|(|u_x| + |u_e|) \\
&\leq -\varepsilon x'x - \lambda_{\max}^{\frac{1}{2}}(P(\varepsilon)) \Big[1 - 4(b-1)^2|B|^2\lambda_{\max}^{\frac{3}{2}}(P(\varepsilon)) - 2b|B|\lambda_{\max}^{\frac{3}{2}}P(\varepsilon)) \\
&\quad -4(b-1)^2|P_e B|^2|B|^2\lambda_{\max}^{\frac{5}{2}}(P(\varepsilon)) - 2(b-1)|P_e B||B|\lambda_{\max}(P(\varepsilon)) \Big] e'e,
\end{aligned}
\tag{3.3.32}
$$

where we have defined $u_x = -B'P(\varepsilon)x$ and $u_e = -B'P(\varepsilon)e$, with $u = u_x - u_e$.

Recalling that $\lim_{\varepsilon \to 0} P(\varepsilon) = 0$, we see that there exists an $\varepsilon^* \in (0,\varepsilon_1^*]$ such that for all $\varepsilon \in (0,\varepsilon^*]$,

$$\dot{V} \leq -\varepsilon x'x - \frac{1}{2}\lambda_{\max}^{\frac{1}{2}}(P(\varepsilon))e'e. \tag{3.3.33}$$

The above shows that, for any $\varepsilon \in (0,\varepsilon^*]$, the equilibrium $(x,e) = (0,0)$ of the closed-loop system (3.3.28) is locally exponentially stable with $L_V(c)$ contained in its basin of attraction. This completes the proof of the output feedback case and of Theorem 3.3.2, since $(x,\hat{x}) \in \mathcal{W}$ implies $(x,e) \in L_V(c)$. $\boxtimes$

3.4. Discrete-Time Systems

3.4.1. Problem Statement

Consider a discrete-time linear system subject to actuator magnitude saturation described by

$$\begin{cases} x^+ = Ax + B\sigma(u), \\ y = Cx, \end{cases} \tag{3.4.1}$$

where $x \in \mathbb{R}^n$ is the state, $u \in \mathbb{R}^m$ is the control input to the actuators, $y \in \mathbb{R}^p$ is the measurement output, and $\sigma : \mathbb{R}^m \to \mathbb{R}^m$ is a saturation function of either class S_1 or class S_3, as defined in Section 3.2.

We make the following assumptions on system (3.4.1).

Assumption 3.4.1. *The pair (A, B) is asymptotically null controllable with bounded controls (ANCBC), i.e.,*

1. *(A, B) is stabilizable;*

2. *All eigenvalues of A are inside or on the unit circle.* ⊞

Assumption 3.4.2. *The pair (A, C) is detectable.* ⊞

The problems that we are to solve using low gain feedback are the following.

Problem 3.4.1. (Semi-global exponential stabilization via linear static state feedback) *Consider system (3.4.1) with $\sigma \in S_1(\Delta)$ [or $\sigma \in S_3(\Delta, b)$], where $\Delta > 0$ and $b \geq 1$. For any a priori given bounded set of initial conditions $\mathcal{X} \subset \mathbb{R}^n$, find a state feedback law $u = F_{\mathcal{X}}x$ such that, for any $\sigma \in S_1(\Delta)$ [or $\sigma \in S_3(\Delta, b)$], the equilibrium $x = 0$ of the closed-loop system is locally exponentially stable with $\mathcal{X}$ contained in its basin of attraction.* ℗

Problem 3.4.2. (Semi-global exponential stabilization via linear dynamic output feedback) *Consider system (3.3.1) with $\sigma \in S_1(\Delta)$ [or $\sigma \in S_3(\Delta, b)$], where $\Delta > 0$ and $b \geq 1$. For any a priori given bounded set $\mathcal{W} \subset \mathbb{R}^{2n}$, find a linear dynamic output feedback control law of dynamical order n,*

$$\begin{cases} z^+ = Gz + Hy, \quad z \in \mathbb{R}^n, \\ u = Mz + Ny, \end{cases} \tag{3.4.2}$$

such that, for any $\sigma \in S_1(\Delta)$ [or $\sigma \in S_3(\Delta, b)$], the equilibrium $(x, z) = (0, 0)$ of the closed-loop system is locally exponentially stable with $\mathcal{W}$ contained in its basin of attraction. ℗

The solution of the above two problems requires a family of feedback laws. The objective here is to show that low gain feedback design methods can be utilized to construct families of linear feedback laws that solve the above two problems for system (3.4.1) with any $\sigma \in \mathcal{S}_1(\Delta)$. And, moreover, the ARE based low gain feedback laws can solve these problems for system (3.4.1) with any $\sigma \in \mathcal{S}_3(\Delta, b)$. Our design of low gain feedback laws does not require any specific knowledge of σ as long as it belongs to $\mathcal{S}_1(\Delta)$ or $\mathcal{S}_3(\Delta, b)$. In this sense, the low gain feedback laws we are to construct are robust with respect to the actuator nonlinearities.

3.4.2. Eigenstructure Assignment Design

Design Algorithm:

Step 1 - State Feedback Design. For the matrix pair (A, B), construct a family of low gain state feedback laws as,

$$u = F(\varepsilon)x, \tag{3.4.3}$$

where $F(\varepsilon)$ is as given by (2.3.6). The existence of such an $F(\varepsilon)$ is guaranteed by Assumption 3.4.1.

Step 2 - Output Feedback Design. Construct a family of low gain output feedback laws as,

$$\begin{cases} \hat{x}^+ = A\hat{x} + BF(\varepsilon)\hat{x} - L(y - C\hat{x}), \\ u = F(\varepsilon)\hat{x}, \end{cases} \tag{3.4.4}$$

where L is chosen such that $A+LC$ is asymptotically stable. The existence of such an L is due to Assumption 3.4.2. $\quad\boxed{\text{A}}$

With these two families of feedback laws, we have the following results.

Theorem 3.4.1. *Consider system (3.4.1) with $\sigma \in \mathcal{S}_1(\Delta)$, where $\Delta > 0$ is any (arbitrarily small) constant. If Assumption 3.4.1 is satisfied, then the family of state feedback laws (3.4.3) solves Problem 3.4.1. If Assumptions 3.4.1 and 3.4.2 are satisfied, then the family of output feedback laws (3.4.4) solves Problem 3.4.2.* $\quad\boxed{\text{T}}$

Proof of Theorem 3.4.1. The idea of the proof is as follows. We first consider the given system (3.4.1) in the absence of saturation, i.e.,

$$x^+ = Ax + Bu, \tag{3.4.5}$$

and show that the closed-loop system is exponentially stable for all sufficiently small ε, and, for any *a priori* given bounded set $\mathcal{W}$, there exists an $\varepsilon^* > 0$ such that, for any $\varepsilon \in (0, \varepsilon^*]$, each channel of the control input u is uniformly bounded by Δ.

Without loss of generality, assume that the pair (A, B) is already in the block diagonal control canonical form of (2.3.2)-(2.3.3).

State Feedback Case:

Under the family of state feedback laws (3.4.3), the closed-loop system is then given by,

$$\begin{cases} x_1^+ = (A_1 + B_1 F_i(\varepsilon))x_1 + \sum_{j=2}^{l} B_{1j} F_j(\varepsilon) x_j, \\ x_2^+ = (A_2 + B_2 F_2(\varepsilon))x_2 + \sum_{j=3}^{l} B_{2j} F_j(\varepsilon) x_j, \\ \qquad \vdots \\ x_l^+ = (A_l + B_l F_l(\varepsilon))x_l, \\ x_0^+ = A_0 x_0 + \sum_{j=1}^{l} B_{0j} F_j(\varepsilon) x_j, \end{cases} \tag{3.4.6}$$

where $x = [x_1', x_2', \cdots, x_l', x_0']'$.

Now for each $i = 1$ to l, let $Q_i(\varepsilon)$, $S_i(\varepsilon)$, $J_i(\varepsilon)$, $\tilde{J}_i(\varepsilon)$ and $\tilde{P}_i(\varepsilon)$ be as defined in Lemmas 2.3.2-2.3.4 for the triple $(A_i, B_i, F_i(\varepsilon))$. Define a state transformation as,

$$\tilde{x} = [\tilde{x}_1', \tilde{x}_2', \cdots, \tilde{x}_l', \tilde{x}_0']', \tag{3.4.7}$$

where $\tilde{x}_0 = x_0$ and for each $i = 1$ to l, $\tilde{x}_i = S_i(\varepsilon) Q_i^{-1}(\varepsilon) x_i$.

The closed-loop system in the new state variable is then given by,

$$\begin{cases} \tilde{x}_1^+ = \tilde{J}_1(\varepsilon)\tilde{x}_1 + \sum_{j=2}^{l} S_1(\varepsilon) Q_1^{-1}(\varepsilon) B_{1j} F_j(\varepsilon) Q_j(\varepsilon) S_j^{-1}(\varepsilon) \tilde{x}_j, \\ \tilde{x}_2^+ = \tilde{J}_2(\varepsilon)\tilde{x}_2 + \sum_{j=3}^{l} S_2(\varepsilon) Q_2^{-1}(\varepsilon) B_{2j} F_j(\varepsilon) Q_j(\varepsilon) S_j^{-1}(\varepsilon) \tilde{x}_j, \\ \qquad \vdots \\ \tilde{x}_l^+ = \tilde{J}_l(\varepsilon)\tilde{x}_l, \\ x_0^+ = A_0 x_0 + \sum_{j=1}^{l} B_{0j} F_j(\varepsilon) Q_j(\varepsilon) S_j^{-1}(\varepsilon) \tilde{x}_j, \end{cases} \tag{3.4.8}$$

for which we consider the Lyapunov function

$$V(\tilde{x}) = \sum_{i=1}^{l} \kappa^i \tilde{x}_i' \tilde{P}_i \tilde{x}_i + \tilde{x}_0' \tilde{P}_0 \tilde{x}_0, \tag{3.4.9}$$

where $\tilde{P}_0 > 0$ is such that $A_0' \tilde{P}_0 \tilde{P}_0 - \tilde{P}_0 = -I$ and $\kappa > 0$ is constant whose value is to be determined later. The existence of such a $\tilde{P}_0$ is due to the fact that A_0 is asymptotically stable.

The difference of V along the trajectories of the closed-loop system can be evaluated as

$$
\Delta V = \sum_{i=1}^{l} \left[-\varepsilon \kappa^i \tilde{x}_i' \tilde{x}_i + 2 \sum_{j=i+1}^{l} \kappa^i \tilde{x}_i' \tilde{J}_i'(\varepsilon) \tilde{P}_i(\varepsilon) S_i(\varepsilon) Q_i^{-1}(\varepsilon) B_{ij} F_j(\varepsilon) Q_j(\varepsilon) S_j^{-1}(\varepsilon) \tilde{x}_j \right.
$$
$$
+ \kappa^i \left(\sum_{j=i+1}^{l} S_i(\varepsilon) Q_i^{-1}(\varepsilon) B_{ij} F_j(\varepsilon) Q_j(\varepsilon) S_j^{-1}(\varepsilon) \tilde{x}_j \right)' \tilde{P}_i(\varepsilon)
$$
$$
\left. \times \left(\sum_{j=i+1}^{l} S_i(\varepsilon) Q_i^{-1}(\varepsilon) B_{ij} F_j(\varepsilon) Q_j(\varepsilon) S_j^{-1}(\varepsilon) \tilde{x}_j \right) \right] - \tilde{x}_0' \tilde{x}_0
$$
$$
+ 2 \sum_{j=1}^{n} \tilde{x}_0' A_0' \tilde{P}_0 B_{0j} F_j(\varepsilon) Q_j(\varepsilon) S_j^{-1}(\varepsilon) \tilde{x}_j
$$
$$
+ \left(\sum_{i=1}^{m} B_{0j} F_j(\varepsilon) Q_j(\varepsilon) S_j^{-1}(\varepsilon) \tilde{x}_j \right)' \tilde{P}_0 \left(\sum_{i=1}^{m} B_{0j} F_j(\varepsilon) Q_j(\varepsilon) S_j^{-1}(\varepsilon) \tilde{x}_j \right).
$$
$$(3.4.10)$$

In view of Lemmas 2.3.3 and 2.3.4, it is straightforward to verify that there exist a constant $\kappa > 0$ and an $\varepsilon_1^* \in (0, 1]$ such that,

$$
\Delta V \leq -\frac{\varepsilon}{2} \tilde{x}' \tilde{x}, \quad \forall \varepsilon \in (0, \varepsilon_1^*], \tag{3.4.11}
$$

which shows that the closed-loop system is exponentially stable for all $\varepsilon \in (0, \varepsilon_1^*]$. It remains to show that, for the given $\mathcal{W}$, there exists an $\varepsilon^* \in (0, \varepsilon_1^*]$ such that, for any $\varepsilon \in (0, \varepsilon^*]$,

$$
|F(\varepsilon) x(k)|_\infty = \max_j \left| F_j(\varepsilon) Q_j(\varepsilon) S_j^{-1}(\varepsilon) \tilde{x}_j(k) \right| \leq \Delta, \quad \forall k \geq 0 \text{ and } \forall x(0) \in \mathcal{W}. \tag{3.4.12}
$$

To this end, let $c > 0$ be such that

$$
c \geq \sup_{x \in \mathcal{W}, \varepsilon \in (0, \varepsilon_1^*]} V(\tilde{x}). \tag{3.4.13}
$$

The right-hand side is well-defined since $Q_i^{-1}(\varepsilon)$ is bounded by Lemma 2.3.4. Let $\varepsilon^* \in (0, \varepsilon_1^*]$ be such that $\tilde{x} \in L_V(c)$ implies $|F(\varepsilon) x|_\infty \leq \Delta$. The existence of such an ε^* is also due to Lemma 2.2.4. It then follows from (3.4.11) that any trajectory that starts from $\mathcal{W} \subset L_V(c)$ will remain inside $L_V(c)$ and hence (3.4.12) always holds for any $\varepsilon \in (0, \varepsilon^*]$. This completes the proof of the state feedback result.

Output Feedback Case:

From the proof of the state feedback case, it is straightforward to see that under the family of output feedback laws (3.4.4), the closed-loop system can be written as,

$$
\begin{cases}
\tilde{x}_1^+ = \tilde{J}_1(\varepsilon)\tilde{x}_1 + \sum_{j=2}^{l} S_1(\varepsilon)Q_1^{-1}(\varepsilon)B_{1j}F_j(\varepsilon)[Q_j(\varepsilon)S_j^{-1}(\varepsilon)\tilde{x}_j - e_j] \\
\qquad\quad -S_1(\varepsilon)Q_1^{-1}(\varepsilon)B_1 F_1(\varepsilon)e_1, \\
\tilde{x}_2^+ = \tilde{J}_2(\varepsilon)\tilde{x}_2 + \sum_{j=3}^{l} S_2(\varepsilon)Q_2^{-1}(\varepsilon)B_{2j}F_j(\varepsilon)[Q_j(\varepsilon)S_j^{-1}(\varepsilon)\tilde{x}_j - e_j] \\
\qquad\quad -S_2(\varepsilon)Q_2^{-1}(\varepsilon)B_2 F_2(\varepsilon)e_2, \\
\qquad\qquad\qquad\qquad\vdots \\
\tilde{x}_l^+ = \tilde{J}_l(\varepsilon)\tilde{x}_l - S_l(\varepsilon)Q_l^{-1}(\varepsilon)B_l F_l(\varepsilon)e_l, \\
x_0^+ = A_0 x_0 + \sum_{j=1}^{l} B_{0j}F_j(\varepsilon)[Q_j(\varepsilon)S_j^{-1}(\varepsilon)\tilde{x}_j - e_j], \\
e^+ = (A + LC)e,
\end{cases}
\tag{3.4.14}
$$

where

$$
e = \begin{bmatrix} e_1 \\ e_2 \\ \vdots \\ e_l \end{bmatrix} = x - \hat{x}.
$$

For this closed-loop system, let us consider the Lyapunov function

$$
V(\tilde{x}, e) = \sum_{i=1}^{l} \kappa^i \tilde{x}_i' \tilde{P}_i(\varepsilon)\tilde{x}_i + \tilde{x}_0' \tilde{P}_0 \tilde{x}_0 + \kappa^{l+1} e' P_e e,
\tag{3.4.15}
$$

where $\tilde{P}_0 > 0$ is the same as in (3.4.9) and P_e is such that $(A + LC)'P_e(A + LC) - P_e = -I$ and $\kappa > 0$ is constant whose value is to be determined later. Her again the existence of such a P_e is due to the fact that $A + LC$ is asymptotically stable.

The difference of V along the trajectories of the closed-loop system can be evaluated as

$$
\begin{aligned}
\Delta V = \sum_{i=1}^{l} \Bigg[&-\varepsilon\kappa^i \tilde{x}_i'\tilde{x}_i + 2 \sum_{j=i+1}^{l} \kappa^i \tilde{x}_i' \tilde{J}_i'(\varepsilon)\tilde{P}_i(\varepsilon)S_i(\varepsilon)Q_i^{-1}(\varepsilon) \\
&\times B_{ij}F_j(\varepsilon)[Q_j(\varepsilon)S_j^{-1}(\varepsilon)\tilde{x}_j - e_j] \\
&-2\kappa^i \tilde{x}_i' \tilde{J}_i'(\varepsilon)\tilde{P}_i'(\varepsilon)S_i(\varepsilon)Q_i^{-1}(\varepsilon)B_i F_i(\varepsilon)e_i \\
&+\kappa^i \left(\sum_{j=i+1}^{l} S_i(\varepsilon)Q_i^{-1}(\varepsilon)B_{ij}F_j(\varepsilon)[Q_j(\varepsilon)S_j^{-1}(\varepsilon)\tilde{x}_j - e_j] \right)' \tilde{P}_i(\varepsilon) \\
&\times \left(\sum_{j=i+1}^{l} S_i(\varepsilon)Q_i^{-1}(\varepsilon)B_{ij}F_j(\varepsilon)[Q_j(\varepsilon)S_j^{-1}(\varepsilon)\tilde{x}_j - e_j] \right) \Bigg]
\end{aligned}
$$

$$+\kappa^i e_i' F_i'(\varepsilon) B_i'(Q_i^{-1}(\varepsilon))' S_i'(\varepsilon) \tilde{P}_i(\varepsilon) S_i(\varepsilon) Q_i^{-1}(\varepsilon) B_i F_i(\varepsilon) e_i$$

$$-\tilde{x}_0' \tilde{x}_0 + 2 \sum_{j=1}^{n} \tilde{x}_0' A_0 \tilde{P}_0 B_{0j} F_j(\varepsilon) [Q_j(\varepsilon) S_j^{-1}(\varepsilon) - e_j \tilde{x}_j]$$

$$+ \left(\sum_{i=1}^{m} B_{0j} F_j(\varepsilon) [Q_j(\varepsilon) S_j^{-1}(\varepsilon) \tilde{x}_j - e_j] \right)' \tilde{P}_0$$

$$\times \left(\sum_{i=1}^{m} B_{0j} F_j(\varepsilon) [Q_j(\varepsilon) S_j^{-1}(\varepsilon) \tilde{x}_j - e_j] \right) - \kappa^{l+1} e'e. \tag{3.4.16}$$

In view of Lemmas 2.3.3 and 2.3.4, it is straightforward to verify that there exist a constant $\kappa > 0$ and an $\varepsilon_1^* \in (0,1]$ such that,

$$\Delta V \leq -\frac{\varepsilon}{2} \tilde{x}' \tilde{x} - \frac{1}{2} e'e, \quad \forall \varepsilon \in (0, \varepsilon_1^*], \tag{3.4.17}$$

which shows that the closed-loop system is exponentially stable for all $\varepsilon \in (0, \varepsilon_1^*]$. It remains to show that, for the given $\mathcal{W}$, there exists an $\varepsilon^* \in (0, \varepsilon_1^*]$ such that, for any $\varepsilon \in (0, \varepsilon^*]$,

$$|F(\varepsilon)\hat{x}(k)|_\infty = \max_j \left| F_j(\varepsilon) [Q_j(\varepsilon) S_j^{-1}(\varepsilon) \tilde{x}_j(k) - e_j(k)] \right| \leq \Delta,$$

$$\forall k \geq 0 \text{ and } \forall (x(0), \hat{x}(0)) \in \mathcal{W}. \tag{3.4.18}$$

To this end, let $c > 0$ be such that

$$c \geq \sup_{(x, \hat{x}) \in \mathcal{W}, \varepsilon \in (0, \varepsilon_1^*]} V(\tilde{x}, e). \tag{3.4.19}$$

The right-hand side is well-defined since $Q_i^{-1}(\varepsilon)$ is bounded by Lemma 2.3.4. Let $\varepsilon^* \in (0, \varepsilon_1^*]$ be such that $\tilde{x} \in L_V(c)$ implies $|F(\varepsilon)\hat{x}|_\infty \leq \Delta$. The existence of such an ε^* is due to Lemmas 2.3.1 and 2.3.4. It then follows from (3.4.17) that any trajectory that starts from $\mathcal{W} \subset L_V(c)$ will remain inside $L_V(c)$ and hence (3.4.18) always holds for any $\varepsilon \in (0, \varepsilon^*]$. This completes the proof of the output feedback result and of Theorem 3.4.1. ⊠

3.4.3. ARE Based Design

Design Algorithm:

Step 1 - State Feedback Design. For the matrix pair (A, B), construct a family of low gain state feedback laws as,

$$u = -(B' P(\varepsilon) B + I)^{-1} B' P(\varepsilon) Ax, \tag{3.4.20}$$

where $P(\varepsilon) > 0$ is the solution to the discrete-time ARE

$$P = A'PA - A'PB(B'PB + I)^{-1}B'PA + \varepsilon I = 0, \ \varepsilon \in (0, 1]. \quad (3.4.21)$$

The existence of such a $P(\varepsilon)$ is guaranteed by Assumption 3.4.1.

We recall that the above is the ARE based low gain feedback design of Section 2.3.2. Hence, Lemma 2.3.6 applies.

Step 2 - Output Feedback Design. Construct a family of low gain output feedback laws as,

$$\begin{cases} \hat{x}^+ = A\hat{x} + Bu - L(y - C\hat{x}), \\ u = -(B'P(\varepsilon)B + I)^{-1}B'P(\varepsilon)A\hat{x}, \end{cases} \quad (3.4.22)$$

where L is chosen such that $A+LC$ is asymptotically stable. The existence of such an L is due to Assumption 3.4.2.   Ⓐ

With these two families of feedback laws, we have the following results.

Theorem 3.4.2. *Consider system (3.4.1) with $\sigma \in S_2(\Delta, b)$, where $\Delta > 0$ is any arbitrarily small constant and b is any arbitrarily large constant. If Assumption 3.4.1 is satisfied, then the family of state feedback laws (3.4.20) solves Problem 3.4.1. If Assumptions 3.4.1 and 3.4.2 are satisfied, then the family of output feedback laws (3.4.22) solves Problem 3.4.2.*   Ⓣ

Proof of Theorem 3.4.2. We will separate the proof into state feedback and output feedback case. For each case, we will show that there exists an $\varepsilon^* \in (0, 1]$ such that for each $\varepsilon \in (0, \varepsilon^*]$, the closed-loop system is locally exponentially stable with $\mathcal{W}$ contained in its basin of attraction.

State Feedback Case:

With the given state feedback laws (3.4.20), the closed-loop system takes the form of

$$\begin{aligned} x^+ &= Ax + B\sigma(u) \\ &= \left(A - B(B'PB + I)^{-1}B'PA\right)x + B[\sigma(u) - u], \quad (3.4.23) \end{aligned}$$

where, for notational brevity, we have dropped the dependency on ε of the matrix $P(\varepsilon)$.

It follows from (2.2.51) that

$$\left(A - B(B'PB + I)^{-1}B'PA\right)'P\left(A - B(B'PB + I)^{-1}B'PA\right) - P = -\varepsilon I - Q_0, \quad (3.4.24)$$

where $Q_0 := A'PB(B'PB + I)^{-2}B'PA \geq 0$.

We now pick the Lyapunov function

$$V(x) = x'Px \tag{3.4.25}$$

and let c be a strictly positive real number such that

$$c \geq \sup_{x \in W, \varepsilon \in (0,1]} x'Px. \tag{3.4.26}$$

The right hand side is well defined since $\lim_{\varepsilon \to 0} P(\varepsilon) = 0$ by Lemma 2.3.6 and W is bounded. Let ε_1^* be such that for all $\varepsilon \in (0, \varepsilon_1^*]$, $x \in L_V(c)$ implies that $|(B'PB + I)^{-1}B'PAx|_\infty \leq \Delta$. Such an ε_1^* exists also because of Lemma 2.3.6 and the fact that $\lim_{\varepsilon \to 0} P(\varepsilon) = 0$.

The evaluation of the difference of V along the trajectories of the closed-loop system (3.4.23), using (3.4.24) and Remark 3.2.1, shows that for all $x \in L_V(c)$,

$$
\begin{aligned}
\Delta V &= -x'(\varepsilon I + Q_0)x + [\sigma(u) - u]'B'PB[\sigma(u) - u] \\
&\quad + 2x'\left(A - B(B'PB + I)^{-1}B'PA\right)'PB[\sigma(u) - u] \\
&\leq -\varepsilon x'x - u'u + [\sigma(u) - u]'B'PB[\sigma(u) - u] \\
&\quad + 2x'A'PB(B'PB + I)^{-1}[\sigma(u) - u] \\
&= -\varepsilon x'x - u'u + [\sigma(u) - u]'B'PB[\sigma(u) - u] - 2u'[\sigma(u) - u] \\
&\leq -\varepsilon x'x - u'u + (b - 1)^2 \lambda_{\max}(B'PB)u'u. \tag{3.4.27}
\end{aligned}
$$

Again recalling that $\lim_{\varepsilon \to 0} P = 0$, we easily see that there exists an $\varepsilon^* \in (0, \varepsilon_1^*]$ such that, for all $\varepsilon \in (0, \varepsilon^*]$, $(b - 1)^2 \lambda_{\max}(B'PB) - 1 \leq 0$. This shows that for any $\varepsilon \in (0, \varepsilon^*]$,

$$\Delta V \leq -\varepsilon x'x, \quad \forall x \in L_V(c), \tag{3.4.28}$$

which in turn shows that, for any $\varepsilon \in (0, \varepsilon^*]$, the equilibrium $x = 0$ of the closed-loop system is locally exponentially stable and its basin of attraction contains the set $L_V(c)$. This completes our proof of the state feedback case since $W \subset L_V(c)$.

Output Feedback Case:

With the output feedback laws (3.4.22), the closed-loop system takes the following form

$$
\begin{cases}
x^+ = Ax + B\sigma\left(-(B'PB + I)^{-1}B'PA\hat{x}\right), \\
\hat{x}^+ = \left(A + LC - B(B'PB + I)^{-1}B'PA\right)\hat{x} + LCx.
\end{cases} \tag{3.4.29}
$$

Letting $e = x - \hat{x}$, we rewrite the closed-loop system (3.4.29) as

$$
\begin{cases}
x^+ = \left(A - B(B'PB + I)^{-1}B'PA\right)x - Bu_e + B[\sigma(u) - u], \\
e^+ = (A + LC)e + B[\sigma(u) - u],
\end{cases} \tag{3.4.30}
$$

where we have defined as $u_e = -(B'PB + I)^{-1}B'PAe$. For later use, we also define $u_x = -(B'PB + I)^{-1}B'PAx$. Clearly, $u = u_x - u_e$.

Let $P_e > 0$ be the unique solution to the Lyapunov equation

$$(A + LC)'P_e(A + LC) - P_e = -I. \tag{3.4.31}$$

Such a unique solution exists since $A + LC$ is asymptotically stable. We then choose the Lyapunov function

$$V(x, e) = x'Px + \lambda_{\max}^{\frac{1}{2}}(P)e'P_e e. \tag{3.4.32}$$

Let c be a strictly positive real number such that

$$c \geq \sup_{(x,\hat{x}) \in \mathcal{W}, \varepsilon \in (0,1]} \left(x'Px + \lambda_{\max}(P)^{\frac{1}{2}}e'P_e e \right). \tag{3.4.33}$$

The right hand side is well defined since $\lim_{\varepsilon \to 0} P(\varepsilon) = 0$ by Lemma 2.3.6 and $\mathcal{W}$ is bounded. Let ε_1^* be such that for all $\varepsilon \in (0, \varepsilon_1^*]$, $(x, e) \in L_V(c)$ implies that $|(B'PB + I)^{-1}B'PA(x - e)|_\infty \leq \Delta$. Such an ε_1^* exists also because of Lemma 2.3.6.

The evaluation of the difference of V along the trajectories of the closed-loop system (3.4.30), using (3.4.24), (3.4.31) and Remark 3.2.1, shows that for all $(x, e) \in L_V(c)$,

$$\begin{aligned}
\Delta V = &-u_x'u_x + [\sigma(u) - u]'B'PB[\sigma(u) - u] \\
&-\varepsilon x'x + 2x'\left(A - B(B'PB + I)^{-1}B'PA\right)'PB[\sigma(u) - u] \\
&-2x'\left(A - B(B'PB + I)^{-1}B'PA\right)'PBu_e u_e + u_e'B'Bu_e \\
&-2u_e'B'B[\sigma(u) - u] - \lambda_{\max}^{\frac{1}{2}}(P)e'e + 2\lambda_{\max}^{\frac{1}{2}}(P)e'(A + LC)'P_e B[\sigma(u) - u] \\
&+\lambda_{\max}^{\frac{1}{2}}(P)[\sigma(u) - u]'B'P_e B[\sigma(u) - u] \\
\leq &-u_x'u_x + (b-1)^2\lambda_{\max}(B'PB)u'u - 2u_x'[\sigma(u) - u] + u_e'B'Bu_e \\
&-2u_e'B'B[\sigma(u) - u] - \lambda_{\max}^{\frac{1}{2}}(P)e'e \\
&-\varepsilon x'x + 2\lambda_{\max}^{\frac{1}{2}}(P)e'(A + LC)'P_e B[\sigma(u) - u)] \\
&+2u_x'u_e + \lambda_{\max}^{\frac{1}{2}}(P)[\sigma(u) - u]'B'P_e B[\sigma(u) - u] \\
\leq &-\varepsilon x'x - u_x'u_x + 2(b-1)^2\lambda_{\max}(B'PB)[u_x'u_x + u_e'u_e] \\
&+2(b-1)(1 + \lambda_{\max}(B'B))u_e'u_e + 4(b-1)^2(1 + \lambda_{\max}(B'B))^2 u_e'u_e \\
&+\frac{1}{4}u_e'u_e + 4u_e'u_e + \lambda_{\max}(B'B)u_e'u_e - \lambda_{\max}^{\frac{1}{2}}(P)e'e \\
&+\frac{1}{4}u_x'u_x + 2\alpha\lambda_{\max}^{\frac{3}{2}}(P)|A||B|e'e + 4\alpha^2\lambda_{\max}(P)e'e \\
&+\frac{1}{4}u_x'u_x + 2(b-1)^2\lambda_{\max}^{\frac{1}{2}}(P)\lambda_{\max}(B'P_e B)[u_x'u_x + u_e'u_e]
\end{aligned}$$

$$= -\varepsilon x'x - \lambda_{\max}^{\frac{1}{2}}(P)\left(1 - 4\alpha^2\lambda_{\max}^{\frac{1}{2}}(P) - 2\alpha\lambda_{\max}(P)|B||A|\right)e'e$$

$$-\left[\frac{1}{4} - 2(b-1)^2\lambda_{\max}(B'PB) - 2(b-1)^2\lambda_{\max}^{\frac{1}{2}}(P)\lambda_{\max}(B'P_eB)\right]u_x'u_x$$

$$+\left[2(b-1)^2\lambda(B'PB) + 4(b-1)^2(1 + \lambda_{\max}(BB'))^2 + 4\right.$$

$$2(b-1)(1 + \lambda_{\max}(B'B)) + \lambda_{\max}(B'B)$$

$$\left.+2(b-1)^2\lambda_{\max}^{\frac{1}{2}}(P)\lambda_{\max}(B'P_eB)\right]u_e'u_e, \tag{3.4.34}$$

where we have denoted $\alpha = (b-1)|(A + LC)'P_eB|$.

Recalling that $\lim_{\varepsilon \to 0} P = 0$ and α is independent of ε, we easily verify that there exists an $\varepsilon^* \in (0, \varepsilon_1^*]$ such that for all $\varepsilon \in (0, \varepsilon^*]$, the following hold,

$$\frac{1}{4} - 2(b-1)^2\lambda_{\max}(B'PB) - 2(b-1)^2\lambda_{\max}^{\frac{1}{2}}(P)\lambda_{\max}(B'P_eB) \geq 0,$$

$$4\alpha^2\lambda_{\max}^{\frac{1}{2}}(P) + 2\alpha\lambda_{\max}(P)|B||A| \leq \frac{1}{4},$$

and

$$\left[2(b-1)^2\lambda(B'PB) + 4(b-1)^2(1 + \lambda_{\max}(BB'))^2 + 2(b-1)(1 + \lambda_{\max}(B'B))\right.$$

$$\left.+ 4 + \lambda_{\max}(B'B) + 2(b-1)^2\lambda_{\max}^{\frac{1}{2}}(P)\lambda_{\max}(B'P_eB)\right]u_e'u_e \leq \frac{\lambda_{\max}^{\frac{1}{2}}(P)}{4}e'e.$$

Hence, it follows from (3.4.34) that for all $\varepsilon \in (0, \varepsilon^*]$,

$$\Delta V \leq \varepsilon x'x - \frac{\lambda_{\max}^{\frac{1}{2}}(P)}{2}e'e, \quad \forall(x, e) \in L_V(c), \tag{3.4.35}$$

which shows that, for any $\varepsilon \in (0, \varepsilon^*]$, the equilibrium $(x, e) = (0, 0)$ of the closed-loop system is locally exponentially stable and its basin of attraction contains the set $L_V(c)$. This completes our proof of Theorem 3.4.2 since $(x, \hat{x}) \in \mathcal{W}$ implies that $(x, e) \in L_V(c)$. $\boxtimes$

3.5. Concluding Remarks

In this chapter, we established in both continuous-time and discrete-time that, if a linear system is asymptotically null controllable with bounded controls, then, when subject to actuator magnitude saturation, it is semi-globally exponentially stabilizable by linear state feedback. If, in addition, the system is also detectable, then it is semi-globally exponentially stabilizable via linear output feedback. These results are shown by explicit construction of feedback laws. Both eigenstructure assignment and ARE based methods are utilized in the construction of feedback laws. We demonstrated that the ARE based design method, though requires solution of a family of parameterized algebraic Riccati equations, yields feedback laws that are more robust with respect to actuator nonlinearities.

Chapter 4

Robust Semi-Global Stabilization of Linear Systems with Saturating Actuators

4.1. Introduction

In Chapter 3, the notion of semi-global stabilization of linear systems with saturating actuators was introduced. The semi-global framework for stabilization requires feedback laws that yield a closed-loop system which has an asymptotically stable equilibrium whose basin of attraction includes an *a priori* given (arbitrarily large) bounded set. It was also shown in Chapter 3 that, for both continuous-time and discrete-time systems, under the condition of asymptotic null controllability with bounded controls, one can achieve semi-global stabilization of linear systems with saturating using *linear* feedback laws. Both the eigenstructure based and ARE based low gain feedback design techniques were utilized to construct semi-globally stabilizing controllers. These low gain control laws were constructed in such a way that the control input does not saturate for any *a priori* given (arbitrarily large) bounded set of initial conditions.

This chapter presents two new design techniques, low-and-high gain feedback design, and low gain based variable structure control design . Both methods are developed with the aim of fuller utilization of actuator capacity. The fuller utilization of actuator capacity in turn provides solutions to problems of semi-global asymptotic stabilization, robust semi-global stabilization for a class of input additive uncertainties and semi-global disturbance rejection. Both

65

low-and-high feedback design and the low gain based variable structure control design are sequential. First, a low gain control law is designed using the techniques of Chapter 2. Then, utilizing an appropriate Lyapunov function for the closed-loop system under this low gain control law, a variable structure control law or a high gain control law is constructed. A low-and-high gain feedback law is then obtained by simply adding the low gain and high gain feedback laws.

Both low gain and high gain controllers are equipped with tuning parameters. The roles of the low gain and that of the high gain controller are completely separated. The role of the low gain control law is to insure, independent from the high gain controller, (i) the asymptotic stability of the equilibrium of the closed-loop system and (ii) that the basin of attraction of the closed-loop system contains an *a priori* given bounded set. As seen in Chapter 3, the tuning parameter in the low gain controller can be tuned to increase the size of the basin of attraction of the equilibrium of the closed-loop system to include any *a priori* given (arbitrarily large) bounded set. On the other hand, the role of the high gain controller is to achieve performance beyond stabilization such as disturbance rejection, robustness and enhancing the utilization of the control capacity. Again, this performance is achieved by appropriate choice of the tuning parameter of the high gain controller.

The low-and-high gain design technique was basically conceived for semi-global control problems beyond stabilization and was related to the performance issues such as semi-global stabilization with enhanced utilization of the available control capacity and semi-global disturbance rejection. A conceptually similar design for local stabilization was proposed earlier in [20].

Similarly, the low gain based variable structure controllers are equipped with a low gain parameter, and a pair of variable structure control parameters. The role of the low gain parameter here is the same as that in the low-and-high gain feedback. The pair of variable structure control parameters play the same role as the high gain parameter in the low-and-high gain feedback does.

In the design of low gain feedback laws, one can utilize either the ARE based method or the eigenstructure assignment method. This leads to two types of low-and-high gain feedback laws and two types of low gain based variable structure control laws. While the ARE based feedback law is more compact and elegant, the eigenstrcuture assignment based feedback law seems to be numerically superior.

The remainder of this chapter is organized as follows. In Section 4.2, we define two new class of saturation functions, referred to as the class S_4 and class S_5 saturation functions. These two classes of saturation functions are more general than the other classes of saturation function we have defined in Chapter

3. In Section 4.3, we pose the problems to be solved by using either the low-and-high gain feedback laws or the low gain based variable structure control laws. In Sections 4.4-4.6 respectively, we present the ARE based low-and-high gain feedback design technique, the eigenstructure assignment based low-and-high gain design technique and the low gain based variable structure control design techniques, all of which lead to control laws that solve the problems posed in Section 4.3. Finally, in Section 4.7, we draw some concluding remarks regarding comparison between the three different designs presented in this chapter.

4.2. Definition of Saturations Functions

In this section we define another two classes of saturation functions, the class S_4 and class S_5 saturation functions. These two classes of saturation functions are more general than those defined in Chapter 3, in the sense that they also include the dead zone nonlinearities. As will become clear shortly, it is the two new design techniques to be presented in this chapter that give us the ability to cope with the generality of the class S_4 or class S_5 saturation functions.

Definition 4.2.1. *A function* $\sigma : \mathbf{R}^m \to \mathbf{R}^m$ *is called a class* S_5 *saturation function if*

1. *$\sigma(u)$ is decentralized, i.e., $\sigma(u) = [\sigma_1(u_1), \ldots, \sigma_m(u_m)]'$; and for each $i = 1$ to m,*

2. *σ_i is locally Lipschitz;*

3. *$s\sigma_i(s) \geq 0$ whenever $s \neq 0$;*

4. *$\liminf_{|s| \to \infty} |\sigma_i(s)| > 0$.* ◻

Remark 4.2.1. *1. In comparison with the three classes of saturation functions defined in Chapter 3, σ_i is not required to satisfy $s\sigma_i(s) > 0$, $\forall s \neq 0$ and it can include dead zone phenomenon in it.*

2. *Graphically, the class S_5 saturation function resides in the shaded area for some constants $\Delta > 0$, $b \geq 0$ and $k > 0$. Among these three constants, Δ represents the saturation level, b the dead zone break points and k the slope. Since our designs will require knowledge of (a lower bound on) k, we can, without loss of generality, assume that $k = 1$. Otherwise, if $k < 1$, we can redefine B to be kB and redefine σ to be $k^{-1}\sigma$.*

   Analytically,

$$s[\sigma_i(s + \text{sign}(s)\bar{b}) - \text{sat}_\Delta(s)] \geq 0, \quad \forall s \in \mathbf{R}, \forall \bar{b} \geq b. \tag{4.2.1}$$

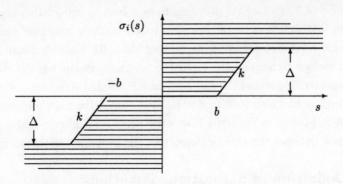

Figure 4.2.1: Qualitative description of the input output characteristics of class $\mathcal{S}_5$ saturation functions

3. Since σ_i is locally Lipschitz, there exists a function $\delta : \mathbb{R}^+ \to \mathbb{R}^+$ such that, for each i,

$$|\sigma_i(s+d) - \sigma_i(s)| \leq \delta(|d|)|d| \qquad \forall s : |s| \leq \Delta \text{ and } \forall d \in \mathbb{R}. \qquad (4.2.2)$$

ℝ

Remark 4.2.2. *It follows directly from Definition 4.2.1 that the functions $\sigma(t) = t$, $\arctan(t)$, $\tanh(t)$, the standard saturation function, the ideal dead zone characteristics, and standard saturation with ideal dead zone characteristics are all class $\mathcal{S}_5$ saturation functions.* ℝ

Definition 4.2.2. *The set of all saturation functions that satisfy the properties (4.2.1) and (4.2.2) with fixed constants Δ, b, $k = 1$ and a function δ is denoted by $\mathcal{S}_5(\Delta, b, \delta)$. The set of $\mathcal{S}_5(\Delta, b, \delta)$ functions which are linear with a unity slope for $b \leq |s| \leq b + \Delta$ and identically zero for $|s| < b$ is denoted as $\mathcal{S}_4(\Delta, b, \delta)$. That is, for $\sigma \in \mathcal{S}_4(\Delta, b, \delta)$,*

$$\sigma(s) \begin{cases} = 0 & \text{if } |s| < b, \\ = s - \text{sign}(s)b & \text{if } b \leq |s| \leq b + \Delta, \\ \geq \Delta & \text{if } s > b + \Delta, \\ \leq -\Delta & \text{if } s < -b - \Delta. \end{cases} \qquad (4.2.3)$$

Ⓓ

4.3. Problem Statement

Consider a class of nonlinear systems that are linear systems subject to actuator magnitude saturation and dead zone phenomenon and input additive uncertainties and disturbances. More specifically, the systems considered are in the following form,

$$\Sigma_0 : \quad \begin{cases} \dot{x} = Ax + B\sigma(u + g(x,t)), \\ y = Cx, \end{cases} \tag{4.3.1}$$

where $x \in \mathbb{R}^n$ is the state, $u \in \mathbb{R}^m$ is the control input, $y \in \mathbb{R}^p$ is the measurement output, $g : \mathbb{R}^n \times \mathbb{R}_+ \to \mathbb{R}^m$ represents disturbances (possibly time-varying) uncertainties, and $\sigma : \mathbb{R}^m \to \mathbb{R}^m$ is a class $\mathcal{S}_4$ or class $\mathcal{S}_5$ saturation function that characterizes the saturation and dead zone nonlinearities in the actuators.

Remark 4.3.1. *While the input additive disturbances are common, the input additive uncertainties is not. However, some other problems might be cast into one with input additive uncertainties. For example, in [52] where we design a stabilizing control laws for an F-16 fighter aircraft with rate saturating deflectors, the uncertainties in the actuator time constants are cast into input additive uncertainties $g(x,t)$.* ®

We will be interested in controllers, both state feedback and output feedback, which guarantee (possibly robust) semi-global stabilization and disturbance attenuation with respect to input additive uncertainties and disturbances.

The following assumption is clearly necessary for semi-global stabilization.

Assumption 4.3.1. *The pair (A, B) is asymptotically null controllable with bounded controls (ANCBC), i.e.,*

1. *The eigenvalues of A are all located in the closed left-half s-plane;*

2. *The pair (A, B) is stabilizable.* ⊞

Regarding the uncertain element g, we only require knowing an upper bound on the norm of g. More specifically, we make the following assumptions.

Assumption 4.3.2. *The uncertain element $g(x,t)$ is piecewise continuous in t, locally Lipschitz in x and its norm is bounded by a known function*

$$|g(x,t)| \le g_0(|x|) + D_0, \quad \forall (x,t) \in \mathbb{R}^n \times \mathbb{R}_+, \tag{4.3.2}$$

where D_0 is a known nonnegative constant, and the known function $g_0 : \mathbb{R}_+ \to \mathbb{R}_+$ is locally Lipschitz and satisfies $g_0(0) = 0$. ⊞

We will be interested in finding state feedback and output feedback controllers that achieve semi-global results independent of the precise $\sigma \in \mathcal{S}_4(\Delta, b, \delta)$ and independent of the precise g that satisfies Assumption 4.3.2. To state the problems we will solve, we will need to make some preliminary definitions.

Definition 4.3.1. The data $(\Delta, b, \delta, g_0, D_0, \mathcal{W}, \mathcal{W}_0)$ is said to be admissible for state feedback if Δ is a strictly positive real number, b is a nonnegative real number, $\delta : \mathbb{R}^+ \to \mathbb{R}^+$ is continuous, $g_0 : \mathbb{R}^+ \to \mathbb{R}^+$ is locally Lipschitz with $g_0(0) = 0$, D_0 is a nonnegative real number, $\mathcal{W}$ is a bounded subset of $\mathbb{R}^n$ and $\mathcal{W}_0$ is a subset of $\mathbb{R}^n$ which contains the origin as an interior point. $\boxed{\text{D}}$

The main state feedback problem we will consider is the following:

Problem 4.3.1. Given the data $(\Delta, b, \delta, g_0, D_0, \mathcal{W}, \mathcal{W}_0)$, admissible for state feedback, find a feedback gain matrix F such that, for all $\sigma \in \mathcal{S}_4(\Delta, b, \delta)$ [or $\sigma \in \mathcal{S}_5(\Delta, b, \delta)$] and all $g(x, t)$ satisfying Assumption 4.3.2 with (g_0, D_0), the closed-loop system with the control $u = Fx$ satisfies

1. if $D_0 = 0$ and $b = 0$, the point $x = 0$ is locally uniformly asymptotically stable and $\mathcal{W}$ is contained in its basin of attraction;

2. if $D_0 > 0$ or $b > 0$, every trajectory starting from $\mathcal{W}$ enters and remains in $\mathcal{W}_0$ after some finite time. $\boxed{\text{P}}$

Remark 4.3.2. Corresponding to specific values for (b, g_0, D_0), this problem is given special names. For the case when $g_0 \equiv 0$ and $D_0 = b = 0$, this is called the semi-global stabilization by state feedback problem. When $g_0 \not\equiv 0$ but $D_0 = b = 0$, this is called the robust semi-global stabilization by state feedback problem. When $g_0 \equiv 0$ but $D_0 > 0$, this is called the semi-global disturbance rejection by state feedback problem. When $g_0 \not\equiv 0$ and $D_0 > 0$, this is called the robust semi-global disturbance rejection by state feedback problem. Since the choice of F depends on (b, g_0, D_0), the solution to Problem 4.3.1 is automatically adapted to the appropriate special problem. $\boxed{\text{R}}$

Definition 4.3.2. The data $(\Delta, b, \delta, g_0, D_0, \mathcal{W}, \mathcal{W}_0)$ is said to be admissible for output feedback if Δ is a strictly positive real number, b is a nonnegative number, $\delta : \mathbb{R}^+ \to \mathbb{R}^+$ is continuous, $g_0 : \mathbb{R}^+ \to \mathbb{R}^+$ is locally Lipschitz with $g_0(0) = 0$, D_0 is a nonnegative real number, $\mathcal{W}$ is a bounded subset of $\mathbb{R}^{2n}$ and $\mathcal{W}_0$ is a subset of $\mathbb{R}^{2n}$ which contains the origin as an interior point. $\boxed{\text{D}}$

The main output feedback problem we will consider is the following:

Problem 4.3.2. *Given the data* $(\Delta, b, \delta, g_0, D_0, \mathcal{W}, \mathcal{W}_0)$, *admissible for output feedback, find matrices* (A_c, B_c, C_c) *such that, for all* $\sigma \in \mathcal{S}_4(\Delta, b, \delta)$ *[or* $\sigma \in \mathcal{S}_5(\Delta, b, \delta)]$ *which are uniformly bounded over* $\mathcal{S}_4(\Delta, b, \delta)$ *[or* $\mathcal{S}_5(\Delta, b, \delta)]$ *and all* $g(x, t)$ *satisfying Assumption 4.3.2 with* (g_0, D_0), *the closed-loop system with the compensator*

$$\begin{cases} \dot{x}_c = A_c x_c + B_c y, & x_c \in \mathbf{R}^n, \\ u = C_c x_c \end{cases}$$

satisfies

1. *if* $D_0 = 0$ *and* $b = 0$, *the point* $(x, x_c) = (0, 0)$ *is locally asymptotically stable and* $\mathcal{W}$ *is contained in its basin of attraction;*

2. *if* $D_0 > 0$ *or* $b > 0$, *every trajectory starting from* $\mathcal{W}$ *enters and remains in* $\mathcal{W}_0$ *after some finite time.* Ⓟ

Remark 4.3.3. *The comments of Remark 4.3.2 apply again, this time for output feedback.* Ⓡ

Remark 4.3.4. *The condition that* σ *be uniformly bounded over* $\mathcal{S}_4(\Delta, b, \delta)$ *[or* $\mathcal{S}_5(\Delta, b, \delta)]$ *is a condition that is used to guarantee that the compensator can be chosen to be linear. If this property does not hold, it turns out that the problem has a solution if one allows a nonlinear compensator. The approach is then to saturate the control term of the compensator outside of the domain of interest so that, effectively,* σ *is bounded. For a further discussion of this idea, see Remark 4.4.6.* Ⓡ

4.4. ARE Based Low-and-High Gain Design

In this section, we will construct a family of ARE based low-and-high gain state feedback laws and a family of ARE based low-and-high gain output feedback laws. These two families of controllers are shown to solve Problems 4.3.1 and 4.3.2, respectively. For clarity, we separate the state feedback and output feedback results into two subsections, Sections 4.4.1 and 4.4.2.

4.4.1. State Feedback Design

Our algorithm for the design of ARE based low-and-high gain state feedback laws is divided into three steps. Steps 1 and 2 deal respectively with the design of the low gain control and the high gain control. In Step 3, the low-and-high gain control is composed by adding together the low gain and the high gain control designed in Steps 1 and 2.

Step 1 - Low Gain Design. The low gain state feedback control law is formed as

$$u_{\text{L}} = F_{\text{L}}(\varepsilon)x, \tag{4.4.1}$$

where

$$F_{\text{L}}(\varepsilon) = -B'P(\varepsilon), \tag{4.4.2}$$

and where $P(\varepsilon)$ is the positive definite solution of the ARE,

$$A'P + PA - PBB'P + \varepsilon I = 0. \tag{4.4.3}$$

Recall that this is the same low gain design as given in Section 2.2.1 and ε is referred to as the low gain parameter.

Step 2 - High Gain Design. Form the high gain state feedback control law as,

$$u_{\text{H}} = F_{\text{H}}(\varepsilon, \rho)x, \tag{4.4.4}$$

where

$$F_{\text{H}}(\varepsilon, \rho) = -\rho B'P(\varepsilon), \quad \rho \geq 0. \tag{4.4.5}$$

We refer to the nonnegative parameter ρ as the high gain parameter.

Step 3: Low-and-High Gain Design. The family of parameterized low-and-high gain state feedback laws, denoted by $\Sigma_{\text{LH}}^{\text{S}}(\varepsilon, \rho)$, is simply formed by adding together the low gain control and the high gain control as designed in the previous steps. Namely,

$$\Sigma_{\text{LH}}^{\text{S}}(\varepsilon, \rho): \quad u = u_{\text{L}} + u_{\text{H}} = F_{\text{LH}}(\varepsilon, \rho)x, \tag{4.4.6}$$

where

$$F_{\text{LH}}(\varepsilon, \rho) = F_{\text{L}}(\varepsilon) + F_{\text{H}}(\varepsilon, \rho) = -(1 + \rho)B'P(\varepsilon). \tag{4.4.7}$$

Ⓐ

Remark 4.4.1. *The ARE based low-and-high gain state feedback design is actually an optimal design for the linear system (A, B) in the absence of actuator saturation with appropriately chosen $\bar{Q}$ and $\bar{R}$. More specifically, choosing $\bar{R} = I/(1 + \rho)$ and $\bar{Q} = Q + \rho PBB'P$ it is easy to verify that P is the solution to the new ARE,*

$$A'P + PA - PB\bar{R}^{-1}B'P + \bar{Q} = 0, \tag{4.4.8}$$

and hence $u = -(1 + \rho)B'Px = -\bar{R}^{-1}B'Px$ is an optimal control.   Ⓡ

The next theorem shows that the family of ARE based low-and-high gain state feedback laws solves Problem 4.3.1.

Theorem 4.4.1. *Let Assumption 4.3.1 hold. Let σ be a class S_5 saturation function. Given the data $(\Delta, b, \delta, g_0, D_0, \mathcal{W}, \mathcal{W}_0)$, admissible for state feedback, there exists an $\varepsilon^*(\Delta, \mathcal{W})$ and, for each $\varepsilon \in (0, \varepsilon^*]$, there exists $\rho^*(\varepsilon, \Delta, b, \delta, g_0, D_0, \mathcal{W}, \mathcal{W}_0)$ such that, for $\varepsilon \in (0, \varepsilon^*]$ and $\rho \geq \rho^*$, the low-and-high gain feedback gain matrix of (4.4.6) solves Problem 4.3.1. Moreover, if $D_0 = b = 0$ then ρ^* is independent of $\mathcal{W}_0$; if, in addition, $g_0 \equiv 0$ then $\rho^* \equiv 0$.* ⊤

Remark 4.4.2. *We note that the low-and-high gain control law which solves Problem 4.3.1 has infinite gain margin. That is, one can increase the gain of the controller arbitrarily and the resulting closed-loop system is guaranteed to satisfy the property required in the solution of Problem 4.3.1. This follows from the fact that the effect of increasing the controller gain is the same as the effect of increasing ρ.* ℝ

Remark 4.4.3. *The freedom in choosing the high gain parameter ρ arbitrarily large can be employed to achieve full utilization of the available control capacity. In particular, by increasing ρ, we can increase the utilization of the available control capacity. In fact, as $\rho \to \infty$, $\sigma(u)$ appears to be a bang-bang control.* ℝ

To prove the state feedback result, Theorem 4.4.1, we will need the following lemma.

Lemma 4.4.1. *Given $(\Delta, b, \delta, g_0, D_0)$, a subset of admissible data for state feedback, let $\varepsilon \in (0, 1]$ and c be a strictly positive real number such that, with $P(\varepsilon)$ satisfying (4.4.3) and using the notation $P := P(\varepsilon)$, we have*

$$|B'Px|_\infty \leq \Delta, \qquad \forall x \in \{x \in \mathbb{R}^n : x'Px \leq c\} . \tag{4.4.9}$$

Let m be the number of inputs in the system (4.3.1) and define

$$F = \sqrt{c\lambda_{\min}^{-1}(P)} , \tag{4.4.10}$$

$$M = \sup_{s \in (0,F]} \left\{ \frac{g_0(s)}{s} \right\}, \quad N = \max_{s \in [0,D_0+MF]} \delta(2s+b), \quad R = b^2\delta^2(b), \tag{4.4.11}$$

$$\rho_1^* := \rho_1^*(\varepsilon) = 30mM^2N\frac{\lambda_{\max}(P)}{\varepsilon\lambda_{\min}(P)}, \tag{4.4.12}$$

$$\rho_2^* := \rho_2^*(\varepsilon) = \left[30mN\left(D_0^2 + b^2\right) + 2R\right]\frac{\lambda_{\max}(P)}{\varepsilon}\frac{2}{c}, \tag{4.4.13}$$

and

$$\rho^* := \rho^*(\varepsilon) = \max\left\{\rho_1^*, \rho_2^*\right\}. \tag{4.4.14}$$

Assume $\rho \geq \rho^*$*. For the system (4.3.1) where* g *satisfies Assumption 4.3.2 with* (g_0, D_0) *and with the control (4.4.6), the function* $V(x) = x'Px$ *satisfies:*

1. *if* $\rho^* = 0$ *then*

$$x \in \{x \in \mathbb{R}^n : 0 < V(x) \leq c\} \Longrightarrow \dot{V} < 0; \tag{4.4.15}$$

2. *if* $\rho^* > 0$ *then*

$$x \in \left\{x \in \mathbb{R}^n : \frac{\rho_2^*}{\rho}\frac{c}{2} < V(x) \leq c\right\} \Longrightarrow \dot{V} < 0. \tag{4.4.16}$$

□

Proof of Lemma 4.4.1. Under the feedback law (4.4.6), the closed-loop system is given by,

$$\dot{x} = Ax + B\sigma((1 + \rho)u_{\text{L}} + g(x, t)). \tag{4.4.17}$$

Consider the function $V = x'Px$ and its derivative in the level set $L_V(c)$. Using (4.4.3) and the definition of u_{L} we have

$$\dot{V} = -x'(\varepsilon I + PBB'P)x - 2u_{\text{L}}'[\sigma((1 + \rho)u_{\text{L}} + g(x, t)) - u_{\text{L}}]. \tag{4.4.18}$$

Using u_{L}^i and g_i to denote the ith component of the vectors u_{L} and g, respectively, it follows from (4.2.1) and (4.4.9) that, for all $x \in L_V(c)$,

$$|\rho u_{\text{L}}^i| \geq |g_i| + b \Longrightarrow -2u_{\text{L}}^i[\sigma_i((1 + \rho)u_{\text{L}}^i + g_i) - u_{\text{L}}^i] \leq 0. \tag{4.4.19}$$

In particular, if $\rho^* = 0$ then $g \equiv 0$, $b = 0$, and point 1 of the lemma follows. In addition, for $\rho \geq \rho^* > 0$, it follows from (4.2.1) and (4.2.2) that

$$\begin{aligned}
|\rho u_{\text{L}}^i| \leq |g_i| + b &\Longrightarrow -2u_{\text{L}}^i[\sigma_i((1 + \rho)u_{\text{L}}^i + g_i) - u_{\text{L}}^i] \\
&= -2u_{\text{L}}^i[\sigma_i((1 + \rho)u_{\text{L}}^i + g_i) - \sigma_i(u_{\text{L}}^i) \\
&\quad -\sigma_i(u_{\text{L}}^i + \text{sign}(u_{\text{L}}^i)b) + \sigma_i(u_{\text{L}}^i) \\
&\quad +\sigma_i(u_{\text{L}}^i + \text{sign}(u_{\text{L}}^i)b) - \text{sat}_\Delta(u_{\text{L}}^i)] \\
&\leq \frac{2(|g_i| + b)[(2|g_i| + b)\delta(2|g_i| + b) + b\delta(b)]}{\rho} \\
&\leq \frac{5(|g_i| + b)^2\delta(2|g_i| + b) + b^2\delta^2(b)}{\rho} \\
&\leq \frac{15(D_0^2 + b^2 + M^2|x|^2)N + R}{\rho}.
\end{aligned} \tag{4.4.20}$$

Hence, we can conclude, for all $x \in L_V(c)$,

$$\dot{V} \leq -x'(\varepsilon I + PBB'P)x + \frac{1}{\rho}[15mD_0^2 N + 15mb^2 N + R + 15mM^2 N|x|^2]$$

$$\leq -\left[\frac{\varepsilon}{2}\lambda_{\max}^{-1}(P) - \frac{1}{\rho}15mM^2 N\lambda_{\min}^{-1}(P)\right]V$$

$$-\left[\frac{\varepsilon}{2}\lambda_{\max}^{-1}(P)V - \frac{1}{\rho}\left[15mN\left(D_0^2 + b^2\right) + R\right]\right]. \tag{4.4.21}$$

Since $\rho \geq \rho_1^*$, we get, for all $x \in L_V(c)$,

$$\dot{V} \leq -\frac{\varepsilon}{2}\lambda_{\max}^{-1}(P)\left(V - \frac{\rho_2^*}{\rho}\frac{c}{2}\right), \tag{4.4.22}$$

which is point 2 of the lemma. ⊠

We are now ready to prove Theorem 4.4.1.

Proof of Theorem 4.4.1. Let c be a strictly positive real number such that

$$c \geq \sup_{x \in \mathcal{W}, \varepsilon \in (0,1]} x'P(\varepsilon)x. \tag{4.4.23}$$

The right hand side is well defined since $\lim_{\varepsilon \to 0} P(\varepsilon) = 0$ by Lemma 2.2.6 and $\mathcal{W}$ is bounded. Let $\varepsilon^* \in (0,1]$ be such that (4.4.9) is satisfied for each $\varepsilon \in (0, \varepsilon^*]$. Such an ε^* exists also as a result of Lemma 2.2.6. Moreover, ε^* depends only on $\mathcal{W}$ and Δ. Fix $\varepsilon \in (0, \varepsilon^*]$.

Consider the case where $D_0 = b = 0$. Then ρ_2^* defined in (4.4.13) is equal to zero. So, if $\rho \geq \rho_1^*$, it follows from point 2 of Lemma 4.4.1 that the point $x = 0$ is locally asymptotically stable with its basin of attraction containing the set $\mathcal{W}$. Notice also that ρ^* is independent of $\mathcal{W}_0$. Moreover, if we also have $g_0 \equiv 0$ then $\rho_1^* = 0$.

Now consider the case where $D_0 > 0$ or $b > 0$. Let $\nu(\varepsilon)$ be a strictly positive real number such that, with $V = x'P(\varepsilon)x$,

$$L_V(\nu) \subset \mathcal{W}_0. \tag{4.4.24}$$

Such a strictly positive real number exists because $\mathcal{W}_0$ has the origin as an interior point and $P(\varepsilon) > 0$. It then follows from Lemma 4.4.1 that if we set

$$\rho^* = \max\left\{\rho_1^*, \rho_2^*, \frac{\rho_2^* c}{2\nu}\right\}, \tag{4.4.25}$$

then we get

$$x \in \{x \in \mathbb{R}^n : \nu < V(x) \leq c\} \implies \dot{V} < 0. \tag{4.4.26}$$

By the choices of c and ν, the solutions that start in $\mathcal{W}$ enter and remain in the set $\mathcal{W}_0$ after some finite time. ⊠

Remark 4.4.4. *As can be seen from the proof of Theorem 4.4.1, when there are open loop eigenvalues in the open right-half s-plane, we don't get that $P \to 0$. Instead, we can fix Q to be positive definite which fixes $P > 0$. Now, we can pick c such that (4.4.9) holds. Then, with any $\mathcal{W} \subset L_V(c)$ (instead of arbitrarily large), the same results hold. This remark also applies to the output feedback results.*   ⌐R⌐

4.4.2. Output Feedback Design

In this section, we construct a family of parameterized low-and-high gain output feedback control laws, and show that it actually solves Problem 4.3.2. The family of control laws we construct have observer-based structure and are constructed by utilizing the high gain observer as developed in [93] to implement the low-and-high gain state feedback laws constructed previously. In order to utilize the high gain observer, we make the following assumption,

Assumption 4.4.1. *The linear system represented by (A, B, C) is left invertible and of minimum-phase.*

This family of parameterized high gain observer based low-and-high gain output feedback control laws, denoted as $\Sigma_{\mathrm{LH}}^{\mathrm{o}}(\varepsilon, \rho, \ell)$, takes the form of,

$$\Sigma_{\mathrm{LH}}^{\mathrm{o}}(\varepsilon, \rho, \ell) : \quad \begin{cases} \dot{\hat{x}} = A\hat{x} + Bu - L(\ell)(y - C\hat{x}), \\ u = F_{\mathrm{LH}}(\varepsilon, \rho)\hat{x}, \end{cases} \tag{4.4.27}$$

where $L(\ell)$ is the high gain observer gain and ℓ is referred to as the *high gain observer parameter*. The high gain observer gain $L(\ell)$ is constructed in the following three steps.

Step 1: By Assumption 4.4.1, the linear system

$$\begin{cases} \dot{x} = Ax + Bu, \\ y = Cx \end{cases} \tag{4.4.28}$$

is left invertible. By Theorem A.1.1 (SCB), there exist a nonsingular state transformation and output transformation,

$$x = \Gamma_{\mathrm{s}}\bar{x}, \quad y = \Gamma_{\mathrm{o}}\bar{y},$$

such that

$$\bar{x} = [x_a', x_b', x_d']',$$
$$x_b = [x_{b1}', x_{b2}', \cdots, x_{bp-m}']', \ x_{bi} = [x_{bi1}, x_{bi2}, \cdots, x_{bir_i}]',$$
$$x_d = [x_1', x_2', \cdots, x_m']', \ x_i = [x_{i1}, x_{i2}, \cdots, x_{iq_i}]',$$
$$\bar{y} = [y_b', y_d']', \ y_b = [y_{b1}, y_{b2}, \cdots, y_{bp-m}]', \ y_d = [y_1, y_2, \cdots, y_m]',$$
$$u = [u_1, u_2, \cdots, u_m]',$$

$$\dot{x}_a = A_{aa}x_a + L_{ab}y_b + L_{ad}y_d, \qquad (4.4.29)$$

and for $i = 1$ to $p - m$,

$$\dot{x}_{bi} = A_{r_i}x_{bi} + L_{bib}y_b + L_{bid}y_d, \qquad (4.4.30)$$

$$y_{bi} = C_{r_i}x_{bi} = x_{bi1}, \qquad (4.4.31)$$

for $i = 1$ to m,

$$\dot{x}_i = A_{q_i}x_i + L_{id}y_d + B_{q_i}[u_i + E_{ia}x_a + E_{ib}x_b + E_{id}x_d], \quad (4.4.32)$$

$$y_i = C_{q_i}x_i = x_{i1}, \qquad (4.4.33)$$

where for an integer $r \geq 1$,

$$A_r = \begin{bmatrix} 0 & I_{r-1} \\ 0 & 0 \end{bmatrix}, \ B_r = \begin{bmatrix} 0 \\ 1 \end{bmatrix}, \ C_r = [1 \ \ 0].$$

Step 2: For $i = 1$ to $p - m$, choose $L_{bi} \in \mathbf{R}^{r_i \times 1}$ such that

$$\lambda(A_{r_i}^c) \in \mathbf{C}^-, \ \ A_{r_i}^c := A_{r_i} + L_{bi}C_{r_i}.$$

Note that the existence of such an L_{bi} is guaranteed by the special structure of the matrix pair (A_{r_i}, C_{r_i}).

Similarly, for $i = 1$ to m, choose $L_{di} \in \mathbf{R}^{q_i \times 1}$ such that

$$\lambda(A_{q_i}^c) \in \mathbf{C}^-, \ \ A_{q_i}^c := A_{q_i} + L_{di}C_{q_i}.$$

Again, the existence of such an L_{di} is guaranteed by the special structure of the matrix pair (A_{q_i}, C_{q_i}).

Step 3: For any $\ell \in (0, 1]$, define a matrix $L(\ell) \in \mathbf{R}^{n \times p}$ as

$$L(\ell) = \Gamma_s \begin{bmatrix} -L_{ab} & -L_{ad} \\ -L_{bb} + L_b(\ell) & -L_{bd} \\ 0 & -L_{dd} + L_d(\ell) \end{bmatrix} \Gamma_o^{-1}, \qquad (4.4.34)$$

where

$$L_{bb} = \begin{bmatrix} L_{b1b} \\ L_{b2b} \\ \vdots \\ L_{bp-mb} \end{bmatrix}, \ \ L_{bd} = \begin{bmatrix} L_{b1d} \\ L_{b2d} \\ \vdots \\ L_{bmd} \end{bmatrix}, \ \ L_{dd} = \begin{bmatrix} L_{1d} \\ L_{2d} \\ \vdots \\ L_{md} \end{bmatrix},$$

$$L_b(\ell) = \begin{bmatrix} S_{r_1}(\ell)L_{b1} & 0 & \cdots & 0 \\ 0 & S_{r_2}(\ell)L_{b2} & \cdots & 0 \\ \vdots & \vdots & \ddots & \vdots \\ 0 & 0 & \cdots & S_{r_{p-m}}(\ell)L_{bp-m} \end{bmatrix},$$

$$L_d(\ell) = \begin{bmatrix} S_{q_1}(\ell)L_{d1} & 0 & \cdots & 0 \\ 0 & S_{q_2}(\ell)L_{d2} & \cdots & 0 \\ \vdots & \vdots & \ddots & \vdots \\ 0 & 0 & \cdots & S_{q_m}(\ell)L_{dm} \end{bmatrix},$$

and where for any integer $r \geq 1$,

$$S_r(\ell) = \begin{bmatrix} \ell & 0 & \cdots & 0 \\ 0 & \ell^2 & \cdots & 0 \\ \vdots & \vdots & \ddots & \vdots \\ 0 & 0 & \cdots & \ell^r \end{bmatrix}.$$

Ⓐ

The next theorem then shows that the family of ARE based low-and-high gain output feedback laws solves Problem 4.3.2.

Theorem 4.4.2. *Let Assumptions 4.3.1 and 4.4.1 hold. Let σ be a class S_5 saturation function. Given the data $(\Delta, b, \delta, g_0, D_0, \mathcal{W}, \mathcal{W}_0)$, admissible for the output feedback problem, there exists $\varepsilon^*(\Delta, \mathcal{W})$, for each $\varepsilon \in (0, \varepsilon^*]$ there exists $\rho^*(\varepsilon, \Delta, b, \delta, g_0, D_0, \mathcal{W}, \mathcal{W}_0)$, and for each $\varepsilon \in (0, \varepsilon^*]$, $\rho \geq \rho^*$ there exists $\ell^*(\varepsilon, \rho) > 0$, such that, for $\ell \geq \ell^*(\varepsilon, \rho)$, $\rho \geq \rho^*$, $\varepsilon \in (0, \varepsilon^*]$, the matrices of the high gain observer based low-and-high gain output feedback control law as given by (4.4.27) solve Problem 4.3.2. Moreover, if $D_0 = b = 0$ then ρ^* is independent of $\mathcal{W}_0$; if, in addition, $g_0 \equiv 0$ then $\rho^* \equiv 0$.*   Ⓣ

Remark 4.4.5. *This result is obtained by utilizing high gain observers, which motivated Assumption 4.4.1. High gain observers are not needed when $g \equiv 0$ and $b = 0$, and ρ is chosen equal to zero (see Chapter 3).*   Ⓡ

To prove the output feedback result, Theorem 4.4.2, we will need a lemma. Consider a system of the form

$$\dot{x} = Ax + B[\sigma(u + g(x + Te, t)) + Ee], \tag{4.4.35}$$

$$\dot{e} = A_o e, \tag{4.4.36}$$

where $x \in \mathbb{R}^n$, $e \in \mathbb{R}^m$. Assume A_o is asymptotically stable and let P_o be the positive definite solution to the Lyapunov equation

$$A_o'P_o + P_oA_o = -I. \tag{4.4.37}$$

Also let

$$\tau = \sqrt{\lambda_{\max}(E'E)}, \quad \kappa = \sqrt{\lambda_{\max}(T'T)}. \tag{4.4.38}$$

Lemma 4.4.2. *Given* $(\Delta, b, \delta, g_0, D_0)$, *a subset of admissible data for output feedback, let* $\varepsilon \in (0, 1]$ *and* c *be a strictly positive real number such that, with* $P(\varepsilon)$ *satisfying (4.4.3) and using the notation* $P := P(\varepsilon)$, *we have*

$$|B'Px|_\infty \le \Delta, \quad \forall x \in \{x \in \mathbb{R}^n : x'Px \le c+1\}. \tag{4.4.39}$$

Let m *be the number of inputs in the system (4.4.35)-(4.4.36) and define*

$$\gamma = \min \left\{ \frac{\varepsilon}{\lambda_{\max}(P)}, \frac{1}{(\tau^2 + 1)\lambda_{\max}(P_o)} \right\}, \tag{4.4.40}$$

$$F = \sqrt{c+1} \left(\sqrt{\lambda_{min}^{-1}(P)} + \kappa \sqrt{[(\tau^2 + 1)\lambda_{min}(P_o)]^{-1}} \right), \tag{4.4.41}$$

$$M = \sup_{s \in (0, F]} \left\{ \frac{g_0(s)}{s} \right\}, \quad N = \max_{s \in [0, D_0 + MF]} \delta(2s + b), \quad R = b^2 \delta^2(b), \tag{4.4.42}$$

$$\rho_1^* := \rho_1^*(\varepsilon) = 60mM^2N \frac{\lambda_{max}(P)}{\varepsilon \lambda_{min}(P)}, \tag{4.4.43}$$

$$\rho_2^* := \rho_2^*(\varepsilon) = 60mM^2N\kappa^2, \quad \rho_3^* := \rho_3^*(\varepsilon) = \frac{60m\left(D_0^2 + b^2\right)N + 4R}{\gamma(c+1)}, \tag{4.4.44}$$

and

$$\rho^* := \rho^*(\varepsilon) = \max\{\rho_1^*, \rho_2^*, \rho_3^*\}. \tag{4.4.45}$$

Assume $\rho \ge \rho^*$. *For the system (4.4.35)-(4.4.36) where* g *satisfies Assumption 4.3.2 with* (g_0, D_0) *and with the control (4.4.6), there exists a continuous function* $\psi : \mathbb{R}^n \times \mathbb{R}^m$ *such that the function*

$$V(x, e) = x'Px + (\tau^2 + 1)e'P_o e \tag{4.4.46}$$

satisfies $\dot{V} \le -\psi(x, e)$ *and*

1. *if* $\rho^* = 0$ *then*

$$(x, e) \in L_V(c+1) \implies \psi(x, e) \ge 0.5\gamma V; \tag{4.4.47}$$

2. *if* $\rho^* > 0$ *then*

$$(x, e) \in L_V(c+1) \implies \psi(x, e) \ge \frac{1}{2}\gamma \left(V - \frac{\rho_3^*}{\rho} \frac{c+1}{2} \right). \tag{4.4.48}$$

$\boxed{\text{L}}$

Proof of Lemma 4.4.2. Under control (4.4.6), the closed-loop system can be written as,

$$\dot{z} = Ax + B[\sigma((1 + \rho)u_{\mathrm{L}} + g(x + Te, t) + Ee], \qquad (4.4.49)$$

$$\dot{e} = A_o e. \qquad (4.4.50)$$

Consider the function V defined in (4.4.46) and its derivative in the set $L_V(c+1)$. Using (4.4.3) and the definition of u_{L} we have

$$\dot{V} = -x'(\varepsilon I + PBB'P)x - 2u'_{\mathrm{L}}[\sigma((1 + \rho)u_{\mathrm{L}} + g(x + Te, t)) - u_{\mathrm{L}}]$$

$$+2u'_{\mathrm{L}}Ee - (\tau^2 + 1)e'e$$

$$\leq -\varepsilon x'x - 2u'_{\mathrm{L}}[\sigma((1 + \rho)u_{\mathrm{L}} + g(x + Te, t)) - u_{\mathrm{L}}] - e'e. \qquad (4.4.51)$$

Now, using u_{L}^i and g_i to denote the ith components of the vectors u_{L} and g, respectively, it follows from (4.2.1) and (4.4.39) that, for all $(z, e) \in L_V(c+1)$,

$$|\rho u_{\mathrm{L}}^i| \geq |g_i| + b \implies -2u_{\mathrm{L}}^i[\sigma_i((1 + \rho)u_{\mathrm{L}}^i + g_i) - u_{\mathrm{L}}^i] \leq 0. \qquad (4.4.52)$$

In particular, if $\rho^* = 0$ then $g \equiv 0$, $b = 0$, and point 1 of the lemma follows. In addition, for $\rho \geq \rho^* > 0$, it follows from (4.2.1) and (4.2.2) that

$$|\rho u_{\mathrm{L}}^i| \leq |g_i| + b \implies -2u_{\mathrm{L}}^i[\sigma_i((1 + \rho)u_{\mathrm{L}}^i + g_i) - u_{\mathrm{L}}^i]$$

$$= -2u_{\mathrm{L}}^i[\sigma_i((1 + \rho)u_{\mathrm{L}}^i + g_i) - \sigma_i(u_{\mathrm{L}}^i)$$

$$-\sigma_i(u_{\mathrm{L}}^i + \mathrm{sign}(u_{\mathrm{L}}^i)b) + \sigma_i(u_{\mathrm{L}}^i)$$

$$\sigma_i(u_{\mathrm{L}}^i + \mathrm{sign}(u_{\mathrm{L}}^i)b) - \mathrm{sat}_\Delta(u_{\mathrm{L}}^i)]$$

$$\leq \frac{2(|g_i| + b)[(2|g_i| + b)\delta(2|g_i| + b) + b\delta(b)]}{\rho}$$

$$\leq \frac{15(D^2 + b^2 + M^2|x + Te|^2)N + R}{\rho}. \qquad (4.4.53)$$

Hence, we can conclude, for all $(x, e) \in L_V(c+1)$,

$$\dot{V} \leq -\varepsilon x'x - e'e + \frac{1}{\rho}\left[15m\left(D_0^2 + b^2\right)N + R + 30mM^2N\left(|x|^2 + \kappa^2|e|^2\right)\right]$$

$$\leq -\left[\frac{\varepsilon}{2}\lambda_{\max}^{-1}(P) - \frac{1}{\rho}30mM^2N\lambda_{\min}^{-1}(P)\right]V - \left[\frac{1}{2} - \frac{1}{\rho}30mM^2N\kappa^2\right]|e|^2$$

$$-\frac{1}{2}|e|^2 - \left[\frac{\varepsilon}{2}\lambda_{\max}^{-1}(P)V - \frac{1}{\rho}\left[15m\left(D_0^2 + b^2\right)N + R\right]\right]. \qquad (4.4.54)$$

Since $\rho \geq \max\{\rho_1^*, \rho_2^*\}$, we get, for all $(x, e) \in L_V(c+1)$,

$$\dot{V} \leq -\frac{1}{2}|e|^2 - \left[\frac{\varepsilon}{2}\lambda_{\max}^{-1}(P)V - \frac{1}{\rho}\left[15m\left(D_0^2 + b^2\right)N + R\right]\right]$$

$$\leq -\frac{1}{2}\gamma V + \frac{1}{\rho}\left[15m\left(D_0^2 + b^2\right)N + R\right]$$

$$\leq -\frac{1}{2}\gamma\left(V - \frac{\rho_3^*}{\rho}\frac{c+1}{2}\right). \tag{4.4.55}$$

This completes the proof of Lemma 4.4.2.   ⊠

We are now ready to prove Theorem 4.4.2.

Proof of Theorem 4.4.2. For the system (4.3.1) under the family of low-and-high gain output feedback laws $\Sigma_{\mathrm{LH}}^{\mathrm{o}}(\varepsilon, \rho, \ell)$ as given by (4.4.27), the closed-loop system takes the form of

$$\dot{x} = Ax + B\sigma(u + g(x, t, d(t))), \tag{4.4.56}$$

$$\dot{\hat{x}} = A\hat{x} + Bu + L(\ell)(y - C\hat{x}), \tag{4.4.57}$$

$$u = F_{\mathrm{LH}}(\varepsilon, \rho)\hat{x}. \tag{4.4.58}$$

Recall that Γ_{s} and Γ_{o} are the state and output transformation that take the system into its SCB form. Partition the state $\bar{x} = \Gamma_{\mathrm{s}}^{-1}x$ and $\bar{\hat{x}} = \Gamma_{\mathrm{s}}^{-1}\hat{x}$ as

$$\bar{x} = [x_a', x_b', x_d']', \quad \bar{\hat{x}} = [\hat{x}_a', \hat{x}_b', \hat{x}_d']',$$

where $x_a, \hat{x}_a \in \mathbb{R}^{n_a}$, $x_b, \hat{x}_b \in \mathbb{R}^{n_b}$, $x_d, \hat{x}_d \in \mathbb{R}^{n_d}$. We then perform a state transformation as follows,

$$\tilde{x} = \Gamma_{\mathrm{s}}[\hat{x}_a', x_b', x_d']', \quad \tilde{e} = [\tilde{e}_a', \tilde{e}_b', \tilde{e}_d']',$$

where $\tilde{e}_a = x_a - \hat{x}_a$, $\tilde{e}_b = S_b(\ell)(x_b - \hat{x}_b)$, $\tilde{e}_d = S_d(\ell)(x_d - \hat{x}_d)$ and where

$$S_b(\ell) = \begin{bmatrix} \ell^{r_1}S_{r_1}^{-1}(\ell) & 0 & \cdots & 0 \\ 0 & \ell^{r_2}S_{r_2}^{-1}(\ell) & \cdots & 0 \\ \vdots & \vdots & \ddots & \vdots \\ 0 & 0 & \cdots & \ell^{r_{p-m}}S_{r_{p-m}}^{-1}(\ell) \end{bmatrix}$$

and

$$S_d(\ell) = \begin{bmatrix} \ell^{q_1}S_{q_1}^{-1}(\ell) & 0 & \cdots & 0 \\ 0 & \ell^{q_2}S_{q_2}^{-1}(\ell) & \cdots & 0 \\ \vdots & \vdots & \ddots & \vdots \\ 0 & 0 & \cdots & \ell^{q_m}S_{q_m}^{-1}(\ell) \end{bmatrix}.$$

Denoting $\tilde{e}_{bd} = [\tilde{e}_b', \tilde{e}_d']'$, we write the closed-loop system (4.4.56)-(4.4.58) in the new state $\tilde{x}$, $\tilde{e}_a$ and $\tilde{e}_{bd}$ as,

$$\dot{\tilde{x}} = A\tilde{x} + B[\sigma(u + g(\tilde{x} + \Gamma_{\mathrm{sa}}\tilde{e}_a), t, d(t))) + E_a\tilde{e}_a], \tag{4.4.59}$$

$$\dot{\tilde{e}}_a = A_{aa}\tilde{e}_a, \tag{4.4.60}$$

$$\dot{\tilde{e}}_{bd} = \ell A_{bd}\tilde{e}_{bd} + B_{bd}[\sigma(u + g(\tilde{x} + \Gamma_{sa}\tilde{e}_a, t, d(t))) - u$$

$$+ E_{bd}S_{bd}^{-1}(\ell)\tilde{e}_{bd} + E_a\tilde{e}_a], \tag{4.4.61}$$

$$u = F_{\mathrm{LH}}(\varepsilon, \rho)[\tilde{x} - \Gamma_{sbd}S_{bd}^{-1}(\ell)\tilde{e}_{bd}], \tag{4.4.62}$$

where

$$A_{bd} = \mathrm{blkdiag}\left\{A_{r_1}^c, A_{r_2}^c, \cdots, A_{r_{p-m}}^c, A_{q_1}^c, A_{q_2}^c, \cdots, A_{q_m}^c\right\},$$

$$B_{bd} = [0, \mathrm{blkdiag}\{B_{q_1}, B_{q_2}, \cdots, B_{q_m}\}']',$$

$$E_{bd} = \begin{bmatrix} E_{1b} & E_{1d} \\ E_{2b} & E_{2d} \\ \vdots & \vdots \\ E_{mb} & E_{md} \end{bmatrix},$$

$$S_{bd}(\ell) = \mathrm{blkdiag}\{S_b(\ell), S_d(\ell)\},$$

and Γ_{sbd} is defined through the following partitioning,

$$\Gamma = [\Gamma_{sa}, \Gamma_{sbd}], \quad \Gamma_{sa} \in \mathbb{R}^{n \times n_a}, \Gamma_{sbd} \in \mathbb{R}^{n_a \times (n_b + n_d)}.$$

We note here that A_{bd} is asymptotically stable since $A_{r_i}^c$'s and $A_{q_i}^c$'s are all asymptotically stable.

This system is in the form (A.2.1)-(A.2.2) of Lemma A.2.1 with $(\tilde{x}, \tilde{e}_a) = z$ and $\tilde{e}_{bd} = e$. To insure the assumptions of Lemma A.2.1, we will apply Lemma 4.4.2. To that end, we set $\tilde{e}_{bd} = 0$ in the closed-loop system equations. The equations (4.4.59)-(4.4.60) then become

$$\dot{\tilde{x}} = A\tilde{x} + B[\sigma(u + g(\tilde{x} + \Gamma_{sa}\tilde{e}_a, t, d(t))) + E_a\tilde{e}_a], \tag{4.4.63}$$

$$\dot{\tilde{e}}_a = A_{aa}\tilde{e}_a, \tag{4.4.64}$$

$$u = F_{\mathrm{LH}}(\varepsilon, \rho)\tilde{x}, \tag{4.4.65}$$

which is in the form of (4.4.35)-(4.4.36). By Theorem A.1.1 and Assumption 4.4.1, A_{aa} is asymptotically stable. Let $P_a > 0$ be such that

$$A_{aa}'P_a + P_aA_{aa} = -I. \tag{4.4.66}$$

Following Lemma 4.4.2, we define

$$V_1(\tilde{x}, \tilde{e}_a) = \tilde{x}'P\tilde{x} + (\tau^2 + 1)\tilde{e}_a'P_a\tilde{e}_a, \tag{4.4.67}$$

where $\tau = \sqrt{\lambda_{\max}(E_a'E_a)}$. Let the real number $c_1 \geq 1$ be such that

$$c_1 \geq \sup_{(x,\hat{x}) \in \mathcal{W}, \varepsilon \in (0,1]} V_1(\tilde{x}, \tilde{e}_a). \tag{4.4.68}$$

Such a c_1 exists since $\tilde{x}$ and $\tilde{e}_a$ are both independent of ℓ, $\lim_{\varepsilon \to 0} P(\varepsilon) = 0$ and the set $\mathcal{W}$ is bounded. Let $\varepsilon^* \in (0, 1]$ be such that (4.4.9) is satisfied for each $\varepsilon \in (0, \varepsilon^*]$. Such an ε^* exists as a result of Lemma 2.3.6. Moreover, ε^* depends only on $\mathcal{W}$ and Δ. Fix $\epsilon \in (0, \varepsilon^*]$.

Consider the case where $D_0 = b = 0$. Then ρ_3^* defined in (4.4.44) is equal to zero. So, if $\rho \geq \max\{\rho_1^*, \rho_2^*\}$, it follows from point 2 of Lemma 4.4.2 that $V_1 \leq -\psi_1(\tilde{x}, \tilde{e}_a)$ where

$$(\tilde{x}, \tilde{e}_a) \in \{(\tilde{x}, \tilde{e}) \in \mathbb{R}^n \times \mathbb{R}^m : 0 < V_1(\tilde{x}, \tilde{e}_a) \leq c_1 + 1\} \implies -\psi_1(\tilde{x}, \tilde{e}_a) < 0 .$$
(4.4.69)

Notice that $\max\{\rho_1^*, \rho_2^*\}$ is independent of $\mathcal{W}_0$. Moreover, $\max\{\rho_1^*, \rho_2^*\} = 0$ when we also have $g_0 \equiv 0$.

Now consider the case where $D_0 > 0$ or $b > 0$. Let P_{bd} satisfy the Lyapunov equation

$$A_{bd}' P_{bd} + P_{bd} A_{bd} = -I,$$
(4.4.70)

and let

$$V_3 = \tilde{e}_{bd}' P_{bd} e_{bd}.$$
(4.4.71)

Observe that, from the definition of $S_r(\ell)$, if we restrict our attention to the case where $\ell \geq 1$ we have that there exists $k > 0$ such that, for each $r \geq 0$,

$$|(\tilde{x}', \tilde{e}_a', e_{bd}')'| \leq r \implies |(x', \hat{x}')'| \leq kr.$$
(4.4.72)

Now, since we have not yet specified ℓ, the level sets of V_3, expressed in the original coordinates, will depend on ℓ. Nevertheless, we can pick $\nu < 1$, a strictly positive real number such that, for all $\ell \geq 1$,

$$L_{V_1}(\nu) \times L_{V_3}(\exp(\nu) - 1) \subset \mathcal{W}_0.$$
(4.4.73)

Such a strictly positive real number exists because $\mathcal{W}_0$ has the origin as an interior point, $P(\epsilon)$, P_a and P_{bd} are all positive definite and (4.4.72) holds. It then follows from Lemma 4.4.2 that if we set

$$\rho^* = \max\left\{\rho_1^*, \rho_2^*, \rho_3^*, \frac{2\rho_3^*(c_1 + 1)}{\nu}\right\},$$
(4.4.74)

then we get $\dot{V}_1 \leq -\psi_1(\tilde{x}, \tilde{e}_a)$ where

$$(\tilde{x}, \tilde{e}_a) \in \{(\tilde{x}, \tilde{e}_a) \in \mathbb{R}^n \times \mathbb{R}^m : \frac{\nu}{4} < V_1(\tilde{x}, \tilde{e}_a) \leq c_1 + 1\} \implies -\psi_1(\tilde{x}, \tilde{e}_a) < 0.$$
(4.4.75)

Henceforth $\rho \geq \rho^*(\epsilon)$ will be fixed. To this point we have that the first assumption of Lemma A.2.1 is satisfied with $\mathcal{W}_1 = \mathbb{R}^{n+n_a}$, V_1, ψ_1, $\nu_1 = \frac{\nu}{2}$ and

c_1. For the case where $D_0 > 0$ or $b > 0$, ν has been specified. For the case where $D_0 = b = 0$, ν is still arbitrary.

Concerning the conditions (A.2.3), they follow readily from a comparison of (4.4.59)-(4.4.62) with (A.2.1)-(A.2.2) and the fact that σ is locally Lipschitz and uniformly bounded.

Remark 4.4.6. *If σ were not uniformly bounded, we could redefine u so that, with $\tilde{e}_{bd} = 0$, it saturated outside of the set $L_{V_1}(c_1 + 1)$. This would not change any of the properties deduced thus far and would induce the properties in (A.2.3). Of course, this action would make the compensator nonlinear, and the location of the saturation would be a function of (a subset of) the admissible data. This idea was first introduced in [15] and later exploited for very general nonlinear output feedback problems in [114,115].* Ⓡ

Now we choose $c_2(\ell) = \ln(1 + \lambda_{\max}(P_{bd}) R_0^2 \ell^{2(n_b + n_d)})$, where R_0 is such that $(x, \hat{x}) \in \mathcal{W}$ implies $|x_b - \hat{x}_b| \le R_0/2$ and $|x_d - \hat{x}_d| \le R_0/2$. The function $c_2(\ell)$ is obviously of class $\mathcal{K}_\infty$. Furthermore, it satisfies $\ln(1 + \tilde{e}_{bd}(0)' P_{bd} \tilde{e}_{bd}(0)) \le c_2(\ell)$ and

$$\lim_{\ell \to \infty} \frac{\ell}{c_2^4(\ell)} = \infty.$$

We then define the Lyapunov function

$$V_2(\tilde{x}, \tilde{e}_a, \tilde{e}_{bd}) = \frac{c_1 V_1(\tilde{x}, \tilde{e}_a, \tilde{e}_{bd})}{c_1 + 1 - V_1(\tilde{x}, \tilde{e}_a, \tilde{e}_{bd})} + \frac{c_2(\ell) \ln(1 + \tilde{e}_{bd}' P_{bd} \tilde{e}_{bd})}{c_2(\ell) + 1 - \ln(1 + \tilde{e}_{bd}' P_{bd} \tilde{e}_{bd})} \quad (4.4.76)$$

and the set

$$\mathcal{W}_2 = \{(\tilde{x}, \tilde{e}_a) : V_1(\tilde{x}, \tilde{e}_a) < c_1 + 1\} \times \{\tilde{e}_{bd} : \ln(1 + \tilde{e}_{bd}' P_{bd} \tilde{e}_{bd}) < c_2(\ell) + 1\}. \quad (4.4.77)$$

It then follows from Lemma A.2.1 that for all $\ell > 0$, $V_2 : \mathcal{W}_2 \to \mathbb{R}_+$ is positive definite on $\mathcal{W}_2 \backslash \{0\}$ and proper on $\mathcal{W}_2$. Furthermore, there exists an $\ell_1^*(\varepsilon, \rho, \nu) \ge 1$, such that, for all $\ell \ge \ell_1^*(\varepsilon, \rho, \nu)$,

$$\dot{V}_2 \le -\psi_2(\tilde{x}, \tilde{e}_a, \tilde{e}_{bd}), \quad (4.4.78)$$

where $\psi_2(\tilde{x}, \tilde{e}_a, \tilde{e}_{bd})$ is continuous on $\mathcal{W}_2$ and is positive definite on $\mathcal{W}_3 = \{(\tilde{x}, \tilde{e}_a, \tilde{e}_{bd}), \frac{\nu}{2} \le V_2(\tilde{x}, \tilde{e}_a, \tilde{e}_{bd}) \le c_1^2 + c_2^2(\ell) + 1\}$.

Finally,

$$(x, \hat{x}) \in \mathcal{W} \implies V_2 \le c_1^2 + c_2^2(\ell), \quad (4.4.79)$$

and

$$V_2 \le \frac{\nu}{2} \implies \left\{ \begin{array}{ccc} V_1 & \le & \nu \\ V_3 & \le & \exp(\nu) - 1 \end{array} \right. \implies (x, \hat{x}) \in \mathcal{W}_0. \quad (4.4.80)$$

This establishes the result for the case where $D_0 > 0$ or $b > 0$.

For the case where $D_0 = b = 0$ recall that ν is arbitrary. So the result follows if there exists a neighborhood $\mathcal{A}$ of the origin and a positive real number ℓ_2^* such that, for all $\ell \geq \ell_2^*$, the origin of (4.4.59)-(4.4.62) is uniformly locally asymptotically stable with basin of attraction containing $\mathcal{A}$. But this is just a standard singular perturbation result since the origin of the $(\tilde{x}, \tilde{e}_a)$ subsystem is locally exponentially stable. For example, one could just follow the calculations in the proof of [29, Theorem 9.3] using the Lyapunov function candidate $V_1 + V_3$. The function V_1 has the appropriate properties since point 1 of Lemma 4.4.2 holds.   ⊠

4.5. Eigenstructure Assignment Based Low-and-High Design

In this section, we will construct a family of eigenstructure assignment based low-and-high gain state feedback laws and a family of eigenstructure assignment based low-and-high gain output feedback laws. These two families of controllers are shown to solve Problems 4.3.1 and 4.3.2, respectively. For clarity, we separate the state feedback and output feedback results into two subsections, Sections 4.5.1 and 4.5.2.

4.5.1. State Feedback Design

Our algorithm for the design of eigenstructure assignment based low-and-high gain state feedback laws is divided into three steps. Steps 1 and 2 deal respectively with the design of the low gain control and the high gain control. In Step 3, the low-and-high gain control is composed by adding together the low gain and the high gain controls designed in Steps 1 and 2.

Step 1 - Low Gain Design.

>   Step 1.1. Find nonsingular transformation matrices T_s and T_i such that the pair (A, B) is transformed into the following block diagonal control canonical form,

$$
T_s^{-1} A T_s = \begin{bmatrix} A_1 & 0 & \cdots & 0 & 0 \\ 0 & A_2 & \cdots & 0 & 0 \\ \vdots & \vdots & \ddots & \vdots & \vdots \\ 0 & 0 & \cdots & A_l & 0 \\ 0 & 0 & 0 & 0 & A_0 \end{bmatrix}, \qquad (4.5.1)
$$

$$T_s^{-1}BT_I = \begin{bmatrix} B_1 & B_{12} & \cdots & B_{1l} & * \\ 0 & B_2 & \cdots & B_{2l} & * \\ \vdots & \vdots & \ddots & \vdots & \vdots \\ 0 & 0 & \cdots & B_l & * \\ B_{01} & B_{02} & \cdots & B_{0l} & * \end{bmatrix}, \qquad (4.5.2)$$

where, A_0 contains all the open left-half plane eigenvalues of A, for each $i = 1$ to l, all eigenvalues of A_i are on the $j\omega$ axis and hence (A_i, B_i) is controllable as given by,

$$A_i = \begin{bmatrix} 0 & 1 & 0 & \cdots & 0 \\ 0 & 0 & 1 & \cdots & 0 \\ \vdots & \vdots & \vdots & \ddots & \vdots \\ 0 & 0 & 0 & \cdots & 1 \\ -a_{n_i}^i & -a_{n_i-1}^i & -a_{n_i-2}^i & \cdots & -a_1^i \end{bmatrix}, \; B_i = \begin{bmatrix} 0 \\ 0 \\ \vdots \\ 0 \\ 1 \end{bmatrix},$$

and finally, $*$'s represent submatrices of less interest.

We note that the existence of the above canonical form was shown in [126]. The software realization can be found in [9].

Step 1.2. For each (A_i, B_i), let $F_i(\varepsilon) \in \mathbb{R}^{1 \times n_i}$ be the state feedback gain such that

$$\lambda(A_i + B_i F_i(\varepsilon)) = -\varepsilon + \lambda(A_i) \in \mathbb{C}^-, \; \varepsilon \in (0, 1]. \qquad (4.5.3)$$

Note that $F_i(\varepsilon)$ is unique.

Step 1.3. Construct a family of low gain state feedback laws as,

$$u = F(\varepsilon)x, \qquad (4.5.4)$$

where the low gain matrix $F_L(\varepsilon)$ is given by

$$F_L(\varepsilon) = T_I \begin{bmatrix} F_1(\varepsilon) & 0 & \cdots & 0 & 0 & 0 \\ 0 & F_2(\varepsilon) & \cdots & 0 & 0 & 0 \\ \vdots & \vdots & \ddots & \vdots & \vdots & \vdots \\ 0 & 0 & \cdots & F_{l-1}(\varepsilon) & 0 & 0 \\ 0 & 0 & \cdots & 0 & F_l(\varepsilon) & 0 \\ 0 & 0 & \cdots & 0 & 0 & 0 \end{bmatrix} T_s^{-1}. \qquad (4.5.5)$$

We recall that this is the same low gain design as given in Section 2.2.1 and ε is referred to as the low gain parameter.

Step 2 - High Gain Design.

Step 2.1. Now for each $i = 1$ to l, let $Q_i(\varepsilon)$, $S_i(\varepsilon)$, $J_i(\varepsilon)$ and $\tilde{P}_i$ be as defined in Lemmas 2.2.2-2.2.4 for the triple $(A_i, B_i, F_i(\varepsilon))$. Let

$$P_i(\varepsilon) = (Q_i^{-1}(\varepsilon))' S_i(\varepsilon) \tilde{P}_i S_i(\varepsilon) Q_i^{-1}(\varepsilon). \tag{4.5.6}$$

Also let $P_0 > 0$ be such that $A_0' P_0 + P_0 A_0 = -I$. Such a P_0 exists since A_0 is asymptotically stable.

Step 2.2. Form a positive definite matrix

$$P(\varepsilon) = (T_s^{-1})' \begin{bmatrix} \kappa P_1(\varepsilon) & 0 & \cdots & 0 & 0 \\ 0 & \kappa^2 P_2(\varepsilon) & \cdots & 0 & 0 \\ \vdots & \vdots & \ddots & \vdots & \vdots \\ 0 & 0 & \cdots & \kappa^l P_l(\varepsilon) & 0 \\ 0 & 0 & \cdots & 0 & P_0 \end{bmatrix} T_s^{-1}, \tag{4.5.7}$$

where $\kappa > 0$ is a constant. It follows from the proof of Theorem 3.3.1 (see (3.3.11)) that there exist a constant $\kappa > 0$ and an $\varepsilon_1^* \in (0, 1]$ such that for all $\varepsilon \in (0, \varepsilon_1^*]$,

$$(A + BF_{\mathrm{L}}(\varepsilon))' P(\varepsilon) + P(\varepsilon)(A + BF_{\mathrm{L}}(\varepsilon)) \leq -Q(\varepsilon), \tag{4.5.8}$$

where

$$Q(\varepsilon) = \frac{\varepsilon}{2} \begin{bmatrix} (Q_1^{-1}(\varepsilon))' S_1^2(\varepsilon) Q_1^{-1}(\varepsilon) & 0 & \cdots \\ 0 & Q_2^{-1}(\varepsilon)' S_2^2(\varepsilon) Q_2^{-1}(\varepsilon) & \cdots \\ \vdots & \vdots & \ddots \\ 0 & 0 & \cdots \\ 0 & 0 & \cdots \end{bmatrix}$$

$$\begin{matrix} & 0 & 0 \\ & 0 & 0 \\ & \vdots & \vdots \\ (Q_l^{-1}(\varepsilon))' S_l^2(\varepsilon) Q_l^{-1}(\varepsilon) & 0 \\ 0 & I \end{matrix} . \tag{4.5.9}$$

Step 2.3. Form the high gain state feedback law as

$$u_{\mathrm{H}} = F_{\mathrm{H}}(\varepsilon, \rho) x, \tag{4.5.10}$$

where

$$F_{\mathrm{H}}(\varepsilon, \rho) = -\rho B' P(\varepsilon) x, \quad \rho \geq 0. \tag{4.5.11}$$

Step 3. The family of parameterized low-and-high gain state feedback laws, denoted by $\Sigma_{\mathrm{LH}}^{\mathrm{s}}(\varepsilon, \rho)$, is simply formed by adding together the low gain control and the high gain control as designed in the previous steps. Namely,

$$\Sigma_{\mathrm{LH}}^{\mathrm{s}}(\varepsilon, \rho): \quad u = u_{\mathrm{L}} + u_{\mathrm{H}} = F_{\mathrm{LH}}(\varepsilon, \rho) x, \tag{4.5.12}$$

where

$$F_{\text{LH}}(\varepsilon, \rho) = F_{\text{L}}(\varepsilon) + F_{\text{H}}(\varepsilon, \rho). \tag{4.5.13}$$

𝔸

The next theorem shows that the above family of eigenstructure assignment based low-and-high gain state feedback laws solves Problem 4.3.1.

Theorem 4.5.1. *Let Assumption 4.3.1 hold. Let σ be a class S_4 saturation function. Given the data $(\Delta, b, \delta, g_0, D_0, \mathcal{W}, \mathcal{W}_0)$, admissible for state feedback, there exists an $\varepsilon^*(\Delta, \mathcal{W})$ and, for each $\varepsilon \in (0, \varepsilon^*]$, there exists $\rho^*(\varepsilon, \Delta, g_0, D_0, \mathcal{W}, \mathcal{W}_0)$ such that, for $\varepsilon \in (0, \varepsilon^*]$ and $\rho \geq \rho^*$, the low-and-high gain feedback matrix of (4.5.12) solves Problem 4.3.1. Moreover, if $D_0 = b = 0$ then ρ^* is independent of $\mathcal{W}_0$; if, in addition, $g_0 \equiv 0$ then $\rho^* \equiv 0$.*

𝕋

Remark 4.5.1. *The freedom in choosing the high gain parameter ρ arbitrarily large can be employed to achieve full utilization of the available control capacity. In particular, by increasing ρ, we can increase the utilization of the available control capacity. In fact, as $\rho \to \infty$, $\sigma(u)$ appears to be a bang-bang control.*

ℝ

To prove the state feedback result, Theorem.4.5.1, we will need the following lemma, which is similar to Lemma 4.4.1.

Lemma 4.5.1. *Given $(\Delta, b, \delta, g_0, D_0)$, a subset of admissible data for state feedback. For $F_{\text{L}}(\varepsilon)$, $P(\varepsilon)$, ε_1^* and $Q(\varepsilon)$ as given in Steps 1 and 2 of the above design procedure, let $\varepsilon \in (0, \varepsilon_2^*]$ and c be a strictly positive real number such that, using the notation $P := P(\varepsilon)$, $F_{\text{L}} := F_{\text{L}}(\varepsilon)$ and $Q := Q(\varepsilon)$, we have*

$$|F_{\text{L}}x|_\infty \leq \Delta, \qquad \forall x \in \{x \in \mathbb{R}^n : x'Px \leq c\}. \tag{4.5.14}$$

Let m be the number of inputs in the system (4.3.1) and define

$$F = \sqrt{c\lambda_{\min}^{-1}(P)}, \tag{4.5.15}$$

$$M = \sup_{s \in (0, F]} \left\{ \frac{g_0(s)}{s} \right\}, \quad N = \max_{s \in [0, D_0 + MF]} \delta(2s + b), \quad R = b^2 \delta^2(b), \tag{4.5.16}$$

$$\rho_1^* := \rho_1^*(\varepsilon) = 30mM^2N \frac{\lambda_{\max}(P)}{\lambda_{\min}(P)\lambda_{\min}(Q)}, \tag{4.5.17}$$

$$\rho_2^* := \rho_2^*(\varepsilon) = \left[30mN\left(D_0^2 + b^2\right) + 2R\right] \frac{\lambda_{\max}(P)}{\lambda_{\min}(Q)} \frac{2}{c}, \tag{4.5.18}$$

and

$$\rho^* := \rho^*(\varepsilon) = \max\{\rho_1^*, \rho_2^*\}. \tag{4.5.19}$$

Assume $\rho \geq \rho^*$. For the system (4.3.1) where g satisfies Assumption 4.3.1 with (g_0, D_0) and with the control law $\Sigma_{\text{LH}}^{\text{S}}(\varepsilon, \rho)$ as given by (4.5.12), the function $V(x) = x'Px$ satisfies:

1. if $\rho^* = 0$ then

$$x \in \{x \in \mathbb{R}^n : 0 < V(x) \leq c\} \Longrightarrow \dot{V} < 0; \tag{4.5.20}$$

2. if $\rho^* > 0$ then

$$x \in \left\{x \in \mathbb{R}^n : \frac{\rho_2^*}{\rho}\frac{c}{2} < V(x) \leq c\right\} \Longrightarrow \dot{V} < 0. \tag{4.5.21}$$

$\boxed{\text{L}}$

Proof of Lemma 4.5.1. The proof of this lemma follows the same line as the proof of Lemma 4.4.1. The closed-loop system consisting of (4.3.1) and the control law $\Sigma_{\text{LH}}^{\text{S}}(\varepsilon, \rho)$ can be written as

$$\dot{x} = (A + BF_{\text{L}})x + B[-F_{\text{L}}x + \sigma(F_{\text{L}}x - \rho B'Px + g(x, t))]. \tag{4.5.22}$$

Consider the function $V = x'Px$ and its derivative in the set $L_V(c)$. Using (4.5.8), we have

$$\dot{V} \leq -x'Qx + 2x'PB[-F_{\text{L}}x + \sigma(F_{\text{L}}x - \rho B'Px + g(x, t))]$$
$$= -x'Qx - \sum_{i=1}^{m} 2v_i[-\mu_i + \sigma_i(\mu_i + \rho v_i + g_i)], \tag{4.5.23}$$

where we have used μ_i, v_i and g_i to denote respectively the ith elements of the vectors $F_{\text{L}}x$, $-B'Px$ and $g(x, t)$.

It then follows from (4.5.14) and the definition of σ that, for all $x \in L_V(c)$,

$$\rho v_i \geq |g_i| + b \Longrightarrow v_i \geq 0, \ \sigma_i(\mu_i + \rho v_i + g_i) \geq \mu_i$$
$$\Longrightarrow -v_i[-\mu_i + \sigma_i(\mu_i + \rho v_i + g_i)] \leq 0, \tag{4.5.24}$$

$$\rho v_i \leq -|g_i| - b \Longrightarrow v_i \leq 0, \ \sigma_i(\mu_i + \rho v_i + g_i) \leq \mu_i$$
$$\Longrightarrow -v_i[-\mu_i + \sigma_i(\mu_i + \rho v_i + g_i)] \leq 0. \tag{4.5.25}$$

In particular, if $\rho^* = 0$ then $g \equiv 0$, $b = 0$, and point 1 of the lemma follows. In addition, for $\rho \geq \rho^* > 0$, it follows from (4.2.2) that

$$
\begin{aligned}
|\rho v_i| \leq |g_i| + b \Longrightarrow |v_i| &\leq \frac{|g_i| + b}{\rho} \\
&\Longrightarrow -2v_i[-\mu_i + \sigma_i(\mu_i + \rho v_i + g_i)] \\
&= -2v_i[\sigma_i(\mu_i + \rho v_i + g_i) - \sigma_i(\mu_i) \\
&\quad -\sigma_i(\mu_i + \mathrm{sign}(\mu_i)b) + \sigma_i(\mu_i)] \\
&\leq \frac{2(|g_i| + b)[(2|g_i| + b)\delta(2|g_i| + b) + b\delta(b)]}{\rho} \\
&\leq \frac{15(D_0^2 + b^2 + M^2|x|^2)N + R}{\rho}.
\end{aligned}
\tag{4.5.26}
$$

Hence, we can conclude that, for all $x \in L_V(c)$,

$$
\begin{aligned}
\dot{V} &\leq -x'Qx + \frac{1}{\rho}[15m(D_0^2 + b^2)N + R + 15mM^2N|x|^2] \\
&\leq -\left[\frac{1}{2}\lambda_{\min}(Q)\lambda_{\max}^{-1}(P) - \frac{1}{\rho}15mM^2N\lambda_{\min}^{-1}(P)\right]V \\
&\quad -\left[\frac{1}{2}\lambda_{\min}(Q)\lambda_{\max}^{-1}(P)V - \frac{1}{\rho}\left[15m\left(D_0^2 + b^2\right)N + R\right]\right].
\end{aligned}
\tag{4.5.27}
$$

Since $\rho \geq \rho_1^*$, we get, for all $x \in L_V(c)$,

$$
\dot{V} \leq -\frac{1}{2}\lambda_{\min}(Q)\lambda_{\max}^{-1}(P)\left(V - \frac{\rho_2^*}{\rho}\frac{c}{2}\right),
\tag{4.5.28}
$$

from which we get point 2 of the lemma.   ☒

We are now ready to prove Theorem 4.5.1.

Proof of Theorem 4.5.1. Let c be a strictly positive real number such that

$$
c \geq \sup_{x \in \mathcal{W}, \varepsilon \in (0, \varepsilon_1^*]} x'P(\varepsilon)x.
\tag{4.5.29}
$$

The right hand side is well defined since $\mathcal{W}$ is bounded and $P(\varepsilon)$ is also bounded by its definition (4.5.7) and property (2.2.18) of $Q_i(\varepsilon)$ used in its definition. Let $\varepsilon^* \in (0, \varepsilon_1^*]$ be such that (4.5.14) is satisfied for each $\varepsilon \in (0, \varepsilon^*]$. Such an ε^* exists as a result of the fact

$$
\lim_{\varepsilon \to 0} |F_{\mathrm{L}}(\varepsilon)P^{-\frac{1}{2}}(\varepsilon)| = 0,
\tag{4.5.30}
$$

which is due to Lemma 2.2.4. Moreover, ε^* depends only on $\mathcal{W}$ and Δ. Fix $\varepsilon \in (0, \varepsilon^*]$.

Consider the case where $D_0 = b = 0$. Then ρ_2^* defined in (4.5.18) is equal to zero. So, if $\rho \geq \rho_1^*$, it follows from point 2 of Lemma 4.5.1 that the point $x = 0$ is locally asymptotically stable with basin of attraction containing the set $\mathcal{W}$. Notice also that ρ^* is independent of $\mathcal{W}_0$. Moreover, if we also have $g_0 \equiv 0$ then $\rho_1^* = 0$.

Now consider the case where $D_0 > 0$ or $b > 0$. Let $\nu(\varepsilon)$ be a strictly positive real number such that, with $V = x'P(\varepsilon)x$,

$$L_V(\nu) \subset \mathcal{W}_0. \tag{4.5.31}$$

Such a strictly positive real number exists because $\mathcal{W}_0$ has the origin as an interior point and $P(\varepsilon) > 0$. It then follows from Lemma 4.5.1 that if we set

$$\rho^* = \max\left\{\rho_1^*, \rho_2^*, \frac{\rho_2^* c}{2\nu}\right\}, \tag{4.5.32}$$

then we get

$$x \in \{x \in \mathbb{R}^n : \nu < V(x) \leq c\} \Longrightarrow \dot{V} < 0. \tag{4.5.33}$$

By the choices of c and ν, the solutions which start in $\mathcal{W}$ enter and remain in the set $\mathcal{W}_0$ after some finite time. ⊠

4.5.2. Output Feedback Design

In this section, we construct a family of parameterized low-and-high gain output feedback control laws, and show that it actually solves Problem 4.3.2. The family of control laws we construct have observer-based structure and are constructed by utilizing the high gain observer as developed in [93] to implement the eigenstructure based low-and-high gain state feedback laws constructed previously. In order to utilize the high gain observer, we make Assumption 4.4.1, as we did in Section 4.4.2.

This family of parameterized high gain observer based low-and-high gain output feedback control laws, denoted as $\Sigma_{\mathrm{LH}}^{\mathrm{o}}(\varepsilon, \rho, \ell)$, takes the form of,

$$\Sigma_{\mathrm{LH}}^{\mathrm{o}}(\varepsilon, \rho, \ell): \quad \begin{cases} \dot{\hat{x}} = A\hat{x} + Bu - L(\ell)(y - C\hat{x}), \\ u = F_{\mathrm{LH}}(\varepsilon, \rho)\hat{x}, \end{cases} \tag{4.5.34}$$

where $L(\ell)$ is the high gain observer gain and ℓ is referred to as the *high gain observer parameter*. The high gain observer gain $L(\ell)$ is as constructed in Section 4.4.2.

The next theorem then shows that the above family of eigenstructure assignment based low-and-high gain output feedback laws solves Problem 4.3.2.

Theorem 4.5.2. *Let Assumptions 4.3.1 and 4.4.1 hold. Let σ be a class $\mathcal{S}_4$ saturation function. Given the data $(\Delta, b, \delta, g_0, D_0, \mathcal{W}, \mathcal{W}_0)$, admissible for the output feedback problem, there exists $\varepsilon^*(\Delta, \mathcal{W})$, for each $\varepsilon \in (0, \varepsilon^*]$ there exists $\rho^*(\varepsilon, \Delta, b, \delta, g_0, D_0, \mathcal{W}, \mathcal{W}_0)$, and for each $\varepsilon \in (0, \varepsilon^*]$, $\rho \geq \rho^*$ there exists $\ell^*(\varepsilon, \rho) > 0$, such that, for $\ell \geq \ell^*(\varepsilon, \rho)$, $\rho \geq \rho^*$, $\varepsilon \in (0, \varepsilon^*]$, the matrices of the high gain observer based low-and-high gain output feedback control law as given by (4.5.34) solve Problem 4.3.2. Moreover, if $D_0 = b = 0$ then ρ^* is independent of $\mathcal{W}_0$; if, in addition, $g_0 \equiv 0$ then $\rho^* \equiv 0$.*   ⊤

Remark 4.5.2. *This result is obtained by utilizing high gain observers, which motivated Assumption 4.4.1. High gain observers are not needed when $g \equiv 0$ and $b = 0$, and ρ is chosen equal to zero (see Chapter 3).*   Ⓡ

To prove the output feedback result, Theorem 4.5.2, we will need a lemma. Consider a system of the form

$$\dot{x} = Ax + B[\sigma(u + g(x + Te, t)) + Ee], \tag{4.5.35}$$

$$\dot{e} = A_o e, \tag{4.5.36}$$

where $x \in \mathbb{R}^n$, $e \in \mathbb{R}^m$. Assume A_o is asymptotically stable and let P_o be the positive definite solution to the Lyapunov equation

$$A_o' P_o + P_o A_o = -I. \tag{4.5.37}$$

Also let

$$\tau = \sqrt{\lambda_{max}(E'E)}, \quad \kappa = \sqrt{\lambda_{max}(T'T)}. \tag{4.5.38}$$

Lemma 4.5.2. *Given $(\Delta, b, \delta, g_0, D_0)$, a subset of admissible data for output feedback. For $F_L(\varepsilon)$, $P(\varepsilon)$, ε_1^* and $Q(\varepsilon)$ as given in Steps 2 and 3 of the eigenstructure assignment based low-and-high gain feedback design of Section 4.5.1. Let $\varepsilon \in (0, \varepsilon_1^*]$ and c be a strictly positive real number such that, using the notation $P := P(\varepsilon)$, $F_L := F_L(\varepsilon)$ and $Q := Q(\varepsilon)$, we have*

$$\left| F_L x + \frac{1}{2} B' P x \right|_\infty \leq \Delta, \quad \forall x \in \{ x \in \mathbb{R}^n : x' P x \leq c + 1 \}. \tag{4.5.39}$$

Let m be the number of inputs of the system (4.5.35)-(4.5.36) and define

$$\gamma = \min \left\{ \frac{\lambda_{min}(Q)}{\lambda_{max}(P)}, \frac{1}{(\tau^2 + 1)\lambda_{max}(P_o)} \right\}, \tag{4.5.40}$$

$$F = \sqrt{c+1} \left(\sqrt{\lambda_{min}^{-1}(P)} + \kappa \sqrt{[(\tau^2 + 1)\lambda_{min}(P_o)]^{-1}} \right), \tag{4.5.41}$$

$$M = \sup_{s\in(0,F]} \left\{ \frac{g_0(s)}{s} \right\}, \ N = \max_{s\in[0,D_0+MF]} \delta(2s+b), \ R = b^2\delta^2(b), \quad (4.5.42)$$

$$\rho_1^* := \rho_1^*(\varepsilon) = 60mM^2N\frac{\lambda_{\max}(P)}{\lambda_{\min}(P)\lambda_{\min}(R)}, \quad (4.5.43)$$

$$\rho_2^* := \rho_2^*(\varepsilon) = 60mM^2N\kappa^2, \quad \rho_3^* := \rho_3^*(\varepsilon) = \frac{60m(D_0^2+b^2)N+4R}{\gamma(c+1)}, \quad (4.5.44)$$

and

$$\rho^* := \rho^*(\varepsilon) = \max\{\rho_1^*, \rho_2^*, \rho_3^*\}. \quad (4.5.45)$$

Assume $\rho \geq \rho^*$. For the system (4.5.35)-(4.5.36) where g satisfies Assumption 4.3.2 with (g_0, D_0) and with the control law $\Sigma_{\mathrm{LH}}^{\mathrm{s}}(\varepsilon, \rho+1/2)$ as given by (4.5.12), there exists a continuous function $\psi : \mathbb{R}^n \times \mathbb{R}^m$ such that the function

$$V(x,e) = x'Px + (\tau^2+1)e'P_oe \quad (4.5.46)$$

satisfies $\dot{V} \leq -\psi(x,e)$ and

1. if $\rho^* = 0$ then

$$(x,e) \in L_V(c+1) \implies \psi(x,e) \geq \frac{1}{2}\gamma V; \quad (4.5.47)$$

2. if $\rho^* > 0$ then

$$(x,e) \in L_V(c+1) \implies \psi(x,e) \geq \frac{1}{2}\gamma(V - \frac{\rho_3^*}{\rho}\frac{c+1}{2}). \quad (4.5.48)$$

□

Proof of Lemma 4.5.2. The proof of this lemma follows the same line as the proof of Lemma 4.4.2. The closed-loop system consisting of (4.5.35)-(4.5.36) and the control law $\Sigma_{\mathrm{LH}}^{\mathrm{s}}(\varepsilon, \rho+1/2)$ can be written as,

$$\dot{x} = (A+BF_{\mathrm{L}})x + B[-F_{\mathrm{L}}x$$
$$+\sigma(F_{\mathrm{L}}x - (\rho+1/2)B'Px + g(x+Te,t)) + Ee], \quad (4.5.49)$$
$$\dot{e} = A_oe. \quad (4.5.50)$$

Consider the function V defined in (4.5.46) and its derivative in the set $L_V(c+1)$. Using (4.5.8), we have

$$\dot{V} \leq -x'Qx + 2x'PB[-F_{\mathrm{L}}x + \sigma(F_{\mathrm{L}}x - (\rho+1/2)B'Px + g(x+Te,t))]$$
$$+2x'PBEe - (\tau^2+1)e'e$$
$$\leq -x'Qx - e'e$$
$$+2x'PB[-F_{\mathrm{L}}x + \frac{1}{2}B'Px + \sigma(F_{\mathrm{L}}x - (\rho+1/2)B'Px + g(x+Te,t))]$$
$$= -x'Qx - e'e - \sum_{i=1}^{m} 2v_i[-\mu_i + \sigma_i(\mu_i + \rho v_i + g_i)], \quad (4.5.51)$$

where we have used μ_i, v_i and g_i to denote respectively the ith element of the vectors $F_L x - \frac{1}{2} B' P x$, $-B' P x$ and $g(x + Te, t)$.

Now, using (4.5.39) and recalling the definition of σ, we have that, for all $(x, e) \in L_V(c + 1)$,

$$\rho v_i \geq |g_i| + b \Longrightarrow v_i \geq 0, \ \sigma_i(\mu_i + \rho v_i + g_i) \geq \mu_i$$
$$\Longrightarrow -v_i[-\mu_i + \sigma_i(\mu_i + \rho v_i + g_i)] \leq 0, \tag{4.5.52}$$

$$\rho v_i \leq -|g_i| - b \Longrightarrow v_i \leq 0, \ \sigma_i(\mu_i + \rho v_i + g_i) \leq \mu_i$$
$$\Longrightarrow -v_i[-\mu_i + \sigma_i(\mu_i + \rho v_i + g_i)] \leq 0. \tag{4.5.53}$$

In particular, if $\rho^* = 0$ then $g \equiv 0$ and point 1 of the lemma follows. In addition, for $\rho \geq \rho^* > 0$, it follows from (4.2.2) that

$$
\begin{aligned}
|\rho v_i| \leq |g_i| + b \Longrightarrow |v_i| &\leq \frac{|g_i| + b}{\rho} \\
&\Longrightarrow -2v_i[-\mu_i + \sigma_i(\mu_i + \rho v_i + g_i)] \\
&= -2v_i[\sigma_i(\mu_i + \rho v_i + g_i) - \sigma_i(\mu_i) \\
&\quad -\sigma(\mu_i + \text{sign}(\mu_i)b) + \sigma_i(\mu_i)] \\
&\leq \frac{2(|g_i| + b)[(2|g_i| + b)\delta(2|g_i| + b) + b\delta(b)]}{\rho} \\
&\leq \frac{15(D_0^2 + b^2 + M^2|x + Te|^2)N + R}{\rho}. \tag{4.5.54}
\end{aligned}
$$

Hence, we can conclude that, for all $(x, e) \in L_V(c + 1)$,

$$
\begin{aligned}
\dot{V} \leq &-x'Qx - e'e + \frac{1}{\rho}[15m(D_0^2 + b^2)N + R] + \frac{1}{\rho}30mM^2N(|x|^2 + \kappa^2|e|^2) \\
\leq &-\left[\frac{1}{2}\lambda_{\min}(Q)\lambda_{\max}^{-1}(P) - \frac{1}{\rho}30mM^2N\lambda_{\min}^{-1}(P)\right]V \\
&-\left[\frac{1}{2} - \frac{1}{\rho}30mM^2N\kappa^2\right]|e|^2 \\
&-\frac{1}{2}|e|^2 - \frac{1}{2}\lambda_{\min}(Q)\lambda_{\max}^{-1}(P)V + \frac{1}{\rho}[15m(D_0^2 + b^2)N + R]. \tag{4.5.55}
\end{aligned}
$$

Since $\rho \geq \max\{\rho_1^*, \rho_2^*\}$, we get, for all $(x, e) \in L_V(c + 1)$,

$$
\begin{aligned}
\dot{V} \leq &-\frac{1}{2}|e|^2 - \frac{1}{2}\lambda_{\min}(Q)\lambda_{\max}^{-1}(P)V + \frac{1}{\rho}[15m(D_0^2 + b^2)N + R] \\
\leq &-\frac{1}{2}\gamma V + \frac{1}{\rho}[15m(D_0^2 + b^2)N + R] \\
\leq &-\frac{1}{2}\gamma\left(V - \frac{\rho_3^*}{\rho}\frac{c^2 + 1}{2}\right), \tag{4.5.56}
\end{aligned}
$$

which implies point 2 of the lemma. ⊠

We are now ready to prove Theorem 4.5.2.

Proof of Theorem 4.5.2. The proof involves the application of Lemmas 4.5.2 and A.2.1. The procedure is exactly the same as that of proof of Theorem 4.4.2 and is omitted here. ⊠

4.6. Low Gain Based Variable Structure Control Design

In this section, we construct a family of low gain based variable structure state feedback control laws and a family of low gain based variable structure output feedback control laws. These families of feedback laws are then shown to solve Problems 4.3.1 and 4.3.2. Again we will separate our presentation into two subsections, one for state feedback and the other for output feedback.

4.6.1. State Feedback Design

The family of low gain based variable structure state feedback control laws is given by,

$$u = [u_1, u_2, \cdots, u_m]', \tag{4.6.1}$$

where

$$u_i = \begin{cases} -\rho \dfrac{B_i'P(\varepsilon)x}{|B_i'P(\varepsilon)x|} & \text{if } |B_i'P(\varepsilon)x| > \mu, \\[2mm] -\rho \dfrac{B_i'P(\varepsilon)x}{\mu} & \text{if } |B_i'P(\varepsilon)x| \le \mu, \end{cases} \tag{4.6.2}$$

and where B_i is the ith column of matrix B, $\varepsilon \in (0, 1]$, $\rho > 0$, $\mu \in (0, 1]$, and $P(\varepsilon) > 0$ is the unique solution to the ARE (4.4.3).

With this family of state feedback laws, we have the following theorem concerning the solution of Problem 4.3.1.

Theorem 4.6.1. *Consider system (4.3.1). Let Assumption 4.3.1 hold. Given the set of data* $(\Delta, b, \delta_0, g_0, D_0, \mathcal{W}, \mathcal{W}_0)$, *admissible for state feedback, there exists an* $\varepsilon^* \in (0, 1]$, *for each* $\varepsilon \in (0, \varepsilon^*]$ *there exists a* $\rho^*(\varepsilon) > 0$, *and for each* $\rho \ge \rho^*(\varepsilon)$, $\varepsilon \in (0, \varepsilon^*]$, *there exists a* $\mu^*(\varepsilon, \rho) > 0$, *such that, for all* $\mu \in (0, \mu^*(\varepsilon, \rho)]$, $\rho \ge \rho_0^*(\varepsilon)$, $\varepsilon \in (0, \varepsilon^*]$, *the closed-loop system consisting of (4.3.1) and (4.6.1) has the following property: for any* $\sigma \in \mathcal{S}_5(\Delta, b, \delta)$ *and for all* $g(x, t)$ *satisfying Assumption 4.3.2, every trajectory that starts from* $\mathcal{W}$ *enters and remains in* $\mathcal{W}_0$ *after some finite time.* Ⓣ

Remark 4.6.1. *This family of control laws can be referred to as soft switching control laws since they are globally Lipschitz and reduces to the switching*

*control laws as $\mu \to 0$. Theorem 4.6.1 shows that point 2 of Problem 4.3.1
is obtained. As will shortly become clear, as $\mathcal{W}_0$ decreases to $\{0\}$, μ is required
to decreases to 0, and in the case that $\mu = 0$, while infinitely fast switching is
required, it achieves point 1 of Problem 4.3.1 even for $D > 0$ and/or $b > 0$.* ▣

Proof of Theorem 4.6.1. We pick the Lyapunov function $V = x'P(\varepsilon)x$ and
let $c > 0$ be such that

$$c \geq \sup_{x \in W, \varepsilon \in (0,1]} x'P(\varepsilon)x. \tag{4.6.3}$$

Such a c exists since $\lim_{\varepsilon \to 0} P(\varepsilon) = 0$ by Lemma 2.2.6 and $\mathcal{W}$ is bounded.
Let $\varepsilon^* \in (0,1]$ be such that, for each $\varepsilon \in (0, \varepsilon^*]$, $x \in L_V(c)$ implies that
$|B'P(\varepsilon)x|_\infty \leq \Delta$. The existence of such an ε^* is again due to the fact that
$\lim_{\varepsilon \to 0} P(\varepsilon) = 0$. For each $\varepsilon \in (0, \varepsilon^*]$, pick the $\rho^*(\varepsilon) > 0$ such that

$$\rho^*(\varepsilon) > \sup_{x \in L_V(c)} g_0(|x|) + D_0 + \Delta + b. \tag{4.6.4}$$

Using (4.4.3), the evaluation of $\dot{V}$ along the trajectories of the closed-loop
system comprising (4.3.1) and (4.6.1) yields that, for $x \in L_V(c)$ and $\rho \geq \rho^*(\varepsilon)$,
$\varepsilon \in (0, \varepsilon^*]$,

$$\dot{V} = -x'(\varepsilon I + P(\varepsilon)BB'P(\varepsilon))x + 2x'P(\varepsilon)B[\sigma(u + g(x,t)) + B'P(\varepsilon)x]$$

$$\leq -\varepsilon x'x + 2\sum_{i=1}^{m} B_i'P(\varepsilon)x[\sigma_i(u_i + g_i(x)) + B_i'P(\varepsilon)x], \tag{4.6.5}$$

where $g_i(x)$ is the ith element of $g(x)$.

Hence, by property (4.2.2) of σ,

$$\text{if } |B_i'P(\varepsilon)x| > \mu \Longrightarrow B_i'P(\varepsilon)x[\sigma_i(u_i + g_i(x)) + B_i'P(\varepsilon)x] \leq 0, \text{ and,}$$

$$\text{if } |B_i'P(\varepsilon)x| \leq \mu \Longrightarrow B_i'P(\varepsilon)x[\sigma_i(u_i + g_i(x)) + B_i'P(\varepsilon)x] \leq \mu[(\rho + \Delta)\delta(\rho + \Delta) + \Delta].$$

In conclusion, we have that, for $x \in L_V(c)$ and $\rho \geq \rho^*(\varepsilon)$, $\varepsilon \in (0, \varepsilon^*]$,

$$\dot{V} \leq -\varepsilon x'x + 2m\mu[(\rho + \Delta)\delta(\rho + \Delta) + \Delta]. \tag{4.6.6}$$

Now let $c_0(\varepsilon) > 0$ be such that $L_V(c_0(\varepsilon)) \subset \mathcal{W}_0$. Such a $c_0(\varepsilon)$ exists since
$\mathcal{W}_0$ contains the origin as an interior point. Also let,

$$\mu^*(\varepsilon, \rho) = \frac{\varepsilon c_0(\varepsilon)}{3m\lambda_{\max}(P(\varepsilon))[(\rho + \Delta)\delta(\rho + \Delta) + \Delta]}. \tag{4.6.7}$$

We then have that, for all $\mu \in (0, \mu^*(\varepsilon, \rho)]$, $\rho \geq \rho^*(\varepsilon)$, $\varepsilon \in (0, \varepsilon^*]$,

$$\dot{V} < 0, \quad \forall x \in L_V(c) \setminus L_V^o(c_0(\varepsilon)), \tag{4.6.8}$$

which implies that $L_V(c)$ is an invariant set and every trajectory that starts
from $L_V(c) \supset \mathcal{W}$ will enter and remain in $L_V(c_0(\varepsilon)) \subset \mathcal{X}_0$ after some finite
time. This completes the proof. ▣

4.6.2. Output Feedback Design

In this section, we construct a family of parameterized low gain based variable structure output feedback control laws, and show that it solves Problem 4.3.2. The family of control laws we construct have observer-based structure and are constructed by utilizing the high gain observer as developed in [93] to implement the low gain based variable structure state feedback control laws constructed previously. In order to utilize the high gain observer, we make Assumption 4.4.1, as we did in Sections 4.4.2 and 4.5.2.

This family of parameterized high gain observer based variable structure output feedback control laws takes the form of

$$u = [u_1, u_2, \cdots, u_m]', \tag{4.6.9}$$

where

$$\dot{\hat{x}} = A\hat{x} + Bu - L(\ell)(y - C\hat{x}), \tag{4.6.10}$$

$$u_i = \begin{cases} -\rho \dfrac{B_i' P(\varepsilon)\hat{x}}{|B_i' P(\varepsilon)\hat{x}|} & \text{if } |B_i' P(\varepsilon)\hat{x}| > \mu, \\[4mm] -\rho \dfrac{B_i' P(\varepsilon)\hat{x}}{\mu} & \text{if } |B_i' P(\varepsilon)\hat{x}| \le \mu, \end{cases} \tag{4.6.11}$$

and where B_i is the ith column of matrix B, $\varepsilon \in (0, 1]$, $\rho > 0$, $\mu \in (0, 1]$, and $P(\varepsilon) > 0$ is the unique solution to the ARE (4.4.3), and $L(\ell)$ is the high gain observer gain and ℓ is referred to as the *high gain observer parameter*. The high gain observer gain $L(\ell)$ is as constructed in Section 4.4.2.

With this family of output feedback laws, we have the following theorem concerning the solution of Problem 4.3.1.

Theorem 4.6.2. *Consider system (4.3.1). Let Assumptions 4.3.1 and 4.4.1 hold. Given the set of data $(\Delta, b, \delta_0, g_0, D_0, \mathcal{W}, \mathcal{W}_0)$, admissible for output feedback, there exists an $\varepsilon^* \in (0, 1]$, for each $\varepsilon \in (0, \varepsilon^*]$ there exists a $\rho^*(\varepsilon) > 0$, for each $\rho \ge \rho^*(\varepsilon)$, $\varepsilon \in (0, \varepsilon^*]$, there exists a $\mu^*(\varepsilon, \rho) > 0$, and for each $\mu \in (0, \mu^*(\varepsilon, \rho)]$, $\rho \in (0, \rho^*(\varepsilon)]$, $\varepsilon \in (0, \varepsilon^*]$, there exists an $\ell^*(\varepsilon, \rho, \mu)$, such that, for all $\ell \in (0, \ell^*(\varepsilon, \rho, \mu)]$, $\mu \in (0, \mu^*(\varepsilon, \rho)]$, $\rho \ge \rho_0^*(\varepsilon)$, $\varepsilon \in (0, \varepsilon^*]$, the closed-loop system consisting of (4.3.1) and (4.6.9) has the following property: for any $\sigma \in \mathcal{S}_5(\Delta, b, \delta)$ and for all $g(x, t)$ satisfying Assumption 4.3.2, every trajectory that starts from $\mathcal{W}$ enters and remains in $\mathcal{W}_0$ after some finite time.*

⊤

Remark 4.6.2. *The comments of Remark 4.6.1 apply again, this time for output feedback.*

ℝ

Proof of Theorem 4.6.2. The proof involves the application of Lemma A.2.1. The procedure is similar to the proof of Theorems 4.4.2 and 4.5.2 and is omitted here.   ⊠

4.7. Concluding Remarks

Three design techniques are presented to solve some fundamental control problems for linear systems subject to actuator saturation, which are formulated as Problems 4.3.1 and 4.3.2. These three design techniques are ARE based low-and-high gain feedback design, eigenstructure assignment based low-and-high gain feedback design, and low gain based variable structure control design. The trade-offs between an ARE based low-and-high gain design and an eigenstructure assignment based low-and-high gain design are those between ARE based low gain design and eigenstructure assignment based low gain design, as discussed in Sections 2.4 and 3.5. The trade-offs between the two low-and-high gain designs and the low gain based variable structure control design include the well-known trade-offs between high gain feedback and variable structure control.

Chapter 5

Semi-Global Output Regulation for Linear Systems with Saturating Actuators

5.1. Introduction

In Chapter 3 we addressed the problem of semi-global asymptotic stabilization of linear systems with saturating actuators and established that, if a linear system is asymptotically null controllable with bounded controls, then, when subject to actuator magnitude saturation, it is semi-globally asymptotically stabilizable by linear feedback. If, in addition, the system is also detectable, then it is semi-globally asymptotically stabilizable via linear output feedback. These results were established by explicit construction of low gain feedback laws.

A natural problem following the stabilization problem is the problem of output regulation (or rejection) of references (or disturbances) generated by some external system, usually called the exosystem. In the linear literature, this is the classical output regulation problem (see, for example, [16]). In the context of linear systems subject to actuator saturation, the only recent work on this subject matter is [111] which deals with the global output regulation problem. In [111], a set of solvability conditions for the global output regulation problem was given. However, as shown in [70], it turns out that these solvability conditions are satisfied for only a few special cases and in general the global output regulation problem as formulated in [111] does not have a solution.

Moreover, in those special cases where the solutions do exist, one needs to use, in general, nonlinear feedback laws.

In this chapter, we show how low gain feedback design techniques can be utilized to solve semi-global output regulation problems for linear ANCBC systems subject to actuator saturation. We will study both continuous-time and discrete-time systems. The rationale behind the adoption of a semi-global framework for output regulation problem is two-fold. First, the semi-global framework allows us to use linear feedback laws, which is obviously very appealing; and second, the semi-global framework seems to be a natural choice when the global output regulation problem, in general, does not have a solution. We naturally extend the output regulator theory for linear systems in the absence of actuator saturation developed by several authors (e.g., [16] and [126]) to the class of linear ANCBC systems subject to actuator saturation. More specifically, we introduce the notion of semi-global output regulation problems. We provide a set of solvability conditions, which are also necessary for a fairly general class of systems [70].

This chapter is organized as follows. For the sake of completeness and to facilitate the comparison, we will in Section 5.2 briefly review the regulator theory for linear systems without actuator saturation and global output regulator theory for linear systems subject to actuator saturation [111]. We will then consider semi-global output regulation problems for linear systems subject to actuator saturation in Sections 5.3 and 5.4. Section 5.3 deals with continuous-time systems. Section 5.4 deals with discrete-time systems. In Section 5.5, we will formulate and solve the so-called generalized semi-global output regulation problems, for which an external driving signal to the exosystem is included. Finally, in Section 5.6, we make some brief remarks on the applicability of low-and-high gain feedback design techniques in the solution of semi-global output regulation problems.

5.2. Preliminaries

This section consists of two subsections. In the first subsection, we briefly review the linear regulator theory, while in the second subsection, we review the global output regulator theory for linear systems subject to actuator saturation as developed in [111].

5.2.1. Review of Linear Regulator Theory

Consider a linear system as given below,

$$\begin{cases} \dot{x} = Ax + Bu + Pw, \\ \dot{w} = Sw, \\ e = Cx + Qw, \end{cases} \tag{5.2.1}$$

where the first equation of this system describes a plant, with state $x \in \mathbb{R}^n$, and input $u \in \mathbb{R}^m$, subject to the effect of a *disturbance* represented by Pw. The third equation defines the error $e \in \mathbb{R}^p$ between the actual plant output Cx and a *reference* signal $-Qw$ which the plant output is required to track. The second equation describes an autonomous system, often called the *exosystem*, with state $w \in \mathbb{R}^s$. The exosystem models the class of disturbance or reference signals taken into consideration.

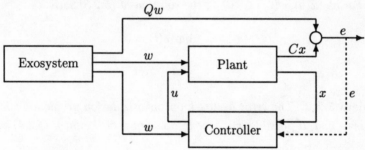

Figure 5.2.1: Configuration of a linear regulator

The control action to the plant, u, can be provided either by state feedback or by error feedback. A state feedback controller has the form

$$u = Fx + Gw. \tag{5.2.2}$$

Composing (5.2.1) and (5.2.2) yields a closed-loop system

$$\begin{cases} \dot{x} = (A + BF)x + (P + BG)w, \\ \dot{w} = Sw, \\ e = Cx + Qw. \end{cases} \tag{5.2.3}$$

An error feedback controller has the form

$$\begin{cases} \dot{z} = A_c z + B_c e, \quad z \in \mathbb{R}^l, \\ u = C_c z + D_c e. \end{cases} \tag{5.2.4}$$

The interconnection of (5.2.1) and (5.2.4) yields a closed-loop system

$$\begin{cases} \dot{x} = Ax + BC_c z + BD_c Cx + (P + BD_c Q)w, \\ \dot{z} = A_c z + B_c Cx + B_c Qw, \\ \dot{w} = Sw, \\ e = Cx + Qw. \end{cases} \tag{5.2.5}$$

The purpose of the control action is to achieve internal stability and output regulation. Internal stability means that, when the exosystem is disconnected (i.e., when w is set equal to 0), the closed-loop (5.2.3) [respectively, (5.2.5)] is asymptotically stable. Output regulation means that for the closed-loop system (5.2.3) [respectively, (5.2.5)] and for all initial conditions $(x(0), w(0))$ [respectively, $(x(0), z(0), w(0))$], we have $e(t) \to 0$ as $t \to \infty$. Formally, all of this can be summarized in the following two synthesis problems.

Problem 5.2,1. *The state feedback output regulation problem for linear systems is to find, if possible, a state feedback law of the form (5.2.2) such that*

1. *The system $\dot{x} = (A + BF)x$ is asymptotically stable;*

2. *For all $(x(0), w(0)) \in \mathbb{R}^{n+s}$, the solution of (5.2.3) satisfies*

$$\lim_{t \to \infty} e(t) = 0. \qquad (5.2.6)$$

P

Problem 5.2.2. *The error feedback output regulation problem for linear systems is to find, if possible, an error feedback law of the form (5.2.4) such that*

1. *The system*

$$\begin{cases} \dot{x} = Ax + BC_c z + BD_c Cx, \\ \dot{z} = A_c z + B_c Cx \end{cases} \qquad (5.2.7)$$

   is asymptotically stable;

2. *For all $(x(0), z(0), w(0)) \in \mathbb{R}^{n+l+s}$, the solution of (5.2.5) satisfies*

$$\lim_{t \to \infty} e(t) = 0. \qquad (5.2.8)$$

P

The solution of these two problems (see [16]) is based on the following three assumptions:

Assumption 5.2.1. *The eigenvalues of S have nonnegative real parts.*   H

Assumption 5.2.2. *The pair (A, B) is stabilizable.*   H

Assumption 5.2.3. *The pair $\left([C \quad Q], \begin{bmatrix} A & P \\ 0 & S \end{bmatrix} \right)$ is detectable.*   H

The first one of these assumptions does not involve a loss of generality because asymptotically stable modes in the exosystem do not affect the regulation of the output. The second one is indeed necessary for asymptotic stabilization of the closed loop via either state or error feedback. The third one is stronger than the assumption of detectability of the pair (C, A), that would be necessary for the asymptotic stabilization of the closed loop via error feedback, but again does not involve loss of generality, as discussed in detail by Francis in [16]. In fact, if the pair (C, A) is detectable and Assumption 5.2.3 does not hold, it is always possible to reduce the dimension of the exosystem which actually affects the error, and have, on the reduced system thus obtained, condition Assumption 5.2.3 satisfied.

The following results, due to Francis [16], describe necessary and sufficient conditions for the existence of solutions to the above two problems.

Theorem 5.2.1. *Suppose Assumptions 5.2.1 and 5.2.2 hold. Then the state feedback output regulation problem is solvable if and only if there exist matrices* Π *and* Γ *which solve the linear matrix equations*

$$\begin{cases} \Pi S = A\Pi + B\Gamma + P, \\ C\Pi + Q = 0. \end{cases} \qquad (5.2.9)$$

Moreover, a suitable state feedback is given by,

$$u = Fx + (-F\Pi + \Gamma)w, \qquad (5.2.10)$$

where F *is such that* $A + BF$ *is asymptotically stable.* ▣

Theorem 5.2.2. *Suppose Assumptions 5.2.1, 5.2.2 and 5.2.3 hold. Then the error feedback output regulation problem is solvable if and only if there exist matrices* Π *and* Γ *which solve the linear matrix equations (5.2.9).*

Moreover, a suitable error feedback is given by,

$$\begin{cases} \begin{bmatrix} \dot{\hat{x}} \\ \dot{\hat{w}} \end{bmatrix} = \begin{bmatrix} A & P \\ 0 & S \end{bmatrix} \begin{bmatrix} \hat{x} \\ \hat{w} \end{bmatrix} + \begin{bmatrix} B \\ 0 \end{bmatrix} u - \begin{bmatrix} L_A \\ L_S \end{bmatrix} \left(e - [C \quad Q] \begin{bmatrix} \hat{x} \\ \hat{w} \end{bmatrix} \right), \\ u = F\hat{x} + (-F\Pi + \Gamma)\hat{w}, \end{cases} \qquad (5.2.11)$$

where F *is such that* $A + BF$ *is asymptotically stable and* L_A *and* L_S *are such that the matrix*

$$\bar{A} = \begin{bmatrix} A + L_A C & P + L_A Q \\ L_S C & S + L_S Q \end{bmatrix}$$

is asymptotically stable. ▣

In summary, if Assumptions 5.2.1, 5.2.2 and 5.2.3 hold, then the error feedback output regulation problem is solvable if and only if the state feedback regulator problem is solvable, and the conditions for the existence of solutions can be expressed in terms of the solvability of certain linear matrix equations.

In [21], Hautus has proven that the possibility of solving these matrix equations can be characterized in terms of a comparison between the transmission polynomials of the system (5.2.1) (in which u is considered as the input and e as the output) and those of the system

$$\begin{cases} \dot{x} = Ax + Bu, \\ \dot{w} = Sw, \\ e = Cx. \end{cases} \quad (5.2.12)$$

The later can be interpreted as the system obtained from (5.2.1) by cutting the connections between the exosystem and the plant. More specifically, Hautus proved the following result.

Theorem 5.2.3. *The linear matrix equations (5.2.9) are solvable if and only if the system (5.2.1) and (5.2.12) have the same transmission polynomials.* ⊡

5.2.2. Review of Global Output Regulator Theory for Linear Systems Subject to Actuator Saturation

Consider a linear system with actuators that are subject to saturation together with an exosystem that generates disturbance and reference signals as described by the following system

$$\begin{cases} \dot{x} = Ax + B\sigma(u) + Pw, \\ \dot{w} = Sw, \\ e = Cx + Qw, \end{cases} \quad (5.2.13)$$

where $x \in \mathbb{R}^n$, $w \in \mathbb{R}^s$, $u \in \mathbb{R}^m$, $e \in \mathbb{R}^p$, and σ is a class S_1 saturation function as defined in Section 3.2.

Because of the presence of the saturation function σ, the system (5.2.13) is nonlinear. The global output regulation problems for such a nonlinear system were formulated in [111] as follows.

Problem 5.2.3. *The global state feedback output regulation problem for linear systems with saturating actuators is to find, if possible, a feedback $u = \alpha(x, w)$ such that*

1. The equilibrium $x = 0$ of

$$\dot{x} = Ax + B\sigma(\alpha(x, 0)) \quad (5.2.14)$$

is globally asymptotically stable and locally exponentially stable;

2. *For all* $(x(0), w(0)) \in \mathbb{R}^{n+s}$, *the solution of the closed-loop system satisfies*

$$\lim_{t \to \infty} e(t) = 0. \qquad (5.2.15)$$

P

Problem 5.2.4. *The global error feedback output regulation problem for linear systems with saturating actuator is to find, if possible, a dynamic error feedback* $u = \theta(z)$, $\dot{z} = \eta(z, e)$ *where* $z \in \mathbb{R}^l$ *such that*

1. *The equilibrium* $(x, z) = (0, 0)$ *of*

$$\begin{cases} \dot{x} = Ax + B\sigma(\theta(z)), \\ \dot{z} = \eta(z, Cx) \end{cases} \qquad (5.2.16)$$

   is globally asymptotically stable and locally exponentially stable.

2. *For all* $(x(0), z(0), w(0)) \in \mathbb{R}^{n+l+s}$, *the solution of the closed-loop system satisfies*

$$\lim_{t \to \infty} e(t) = 0. \qquad (5.2.17)$$

P

A set of sufficient conditions for the above global output regulation problems to have a solution is given in [111].

The global output regulation as defined in the above is clearly a very desirable property. Unfortunately it turns out that only in very special circumstances can we achieve global output regulation. In fact, the global error feedback output regulation problem as formulated in Problem 5.2.4 basically has no solution. This is established in the following lemma.

Lemma 5.2.1. *Suppose Assumptions 5.2.1, 5.2.2 and 5.2.3 hold. Then there exist initial conditions* w_0 *for* w *such that there exists no input* u *or initial condition* $x(0)$ *for which the closed-loop system satisfies* $\lim_{t \to \infty} e(t) = 0$. □

Proof of Lemma 5.2.1. See [70]. ⊠

The above clearly yields a good argument to restrict our attention to initial conditions $w(0)$ inside a given compact set. Moreover, regarding the initial conditions of the plant, in the context of stabilization of linear systems with saturating actuators, the step from global initial conditions to initial conditions inside a compact set has already been made. This has been referred to as semi-global stabilization. This yields good motivation to direct our attention to a semi-global framework. As seen in Chapter 3, semi-global framework allows for the use of linear feedback laws, while global framework generally requires nonlinear feedback laws.

5.3. Continuous-Time Systems

We split this section into two parts. In the first part we study the semi-global linear state feedback output regulation problem where all signals are available for feedback and it suffices to look at static feedbacks. In the second part we study the semi-global linear error feedback output regulation problem where only the error signal is available for feedback and we have to resort to dynamic feedback.

5.3.1. State Feedback Results

Problem 5.3.1. *Consider the system (5.2.13) with $\sigma \in S_1(\Delta)$ being bounded by a known number and a compact set $W_0 \subset \mathbb{R}^s$. The semi-global linear state feedback output regulation problem is defined as follows.*

For any a priori given (arbitrarily large) bounded set $X_0 \subset \mathbb{R}^n$, find, if possible, a linear static feedback law $u = Fx + Gw$ such that

1. *The equilibrium $x = 0$ of*

$$\dot{x} = Ax + B\sigma(Fx) \tag{5.3.1}$$

   is locally exponentially stable with X_0 contained in its basin of attraction;

2. *For all $\sigma \in S_1(\Delta)$ and for all $x(0) \in X_0$ and $w(0) \in W_0$, the solution of the closed-loop system satisfies*

$$\lim_{t \to \infty} e(t) = 0. \tag{5.3.2}$$

P

Remark 5.3.1. *We would like to emphasize that our formulation of semi-global linear state feedback output regulation problem does not view the set of initial conditions of the plant as given data. The set of given data consists of the models of the plant and the exosystem and the set of initial conditions for the exosystem. Therefore, any solvability conditions we obtain must be independent of the set of initial conditions of the plant, X_0.* R

The solvability conditions for semi-global linear state feedback output regulation problem are given in the following theorem.

Theorem 5.3.1. *Consider the system (5.2.13) with $\sigma \in S_1(\Delta)$ being bounded by a known number and the given compact set $W_0 \subset \mathbb{R}^s$. The semi-global linear state feedback output regulation problem is solvable if*

1. (A, B) *is stabilizable and A has all eigenvalues in the closed left half plane;*

2. *There exist matrices* Π *and* Γ *such that,*

    (a) *They solve the following linear matrix equations,*

    $$\begin{cases} \Pi S = A\Pi + B\Gamma + P, \\ C\Pi + Q = 0; \end{cases} \tag{5.3.3}$$

    (b) *There exist a* $\delta > 0$ *and a* $T \geq 0$ *such that* $\|\Gamma w\|_{\infty,\mathrm{T}} \leq \Delta - \delta$ *for all* w *with* $w(0) \in \mathcal{W}_0$. ▣

Proof of Theorem 5.3.1. We prove this theorem by first explicitly constructing a family of linear static state feedback laws, parameterized in ε, and then showing that for each given set $\mathcal{X}_0$, there exists an $\varepsilon^* > 0$ such that for all $\varepsilon \in (0, \varepsilon^*]$, both Items 1 and 2 of Problem 5.3.1 hold. The family of linear static state feedback laws we construct takes the following form

$$u = F(\varepsilon)x + (-F(\varepsilon)\Pi + \Gamma)w, \tag{5.3.4}$$

where $F(\varepsilon)$ is the low gain feedback gain matrix, parameterized in ε. It is constructed by using the ARE based low gain design technique of Section 2.2.2 or the eigenstructure assignment based low gain feedback design technique of Section 2.2.1. The existence of such an F is guaranteed by Condition 1 of the theorem. In what follows, we assume that $F(\varepsilon)$ has been designed using the ARE based low gain feedback design technique. Also to avoid notational confusion between P in (5.2.13) and $P(\varepsilon)$ in the ARE (2.2.51), in this proof, we will replace $P(\varepsilon)$ with $X(\varepsilon)$.

With this family of feedback laws, the system (5.3.1) is written as

$$\dot{x} = Ax + B\sigma(F(\varepsilon)x). \tag{5.3.5}$$

It then follows from Theorem 3.3.2 that there exists an $\varepsilon_1^* > 0$ such that, for all $\varepsilon \in (0, \varepsilon_1^*]$ and for all $\sigma \in \mathcal{S}_1(\Delta)$, the origin $x = 0$ is locally exponentially stable with $\mathcal{X}_0$ contained in its basin of attraction.

To show the rest of the theorem, let us introduce an invertible, triangular coordinate change $\xi = x - \Pi w$. Using Condition 2(a), we have

$$\begin{aligned} \dot{\xi} &= \dot{x} - \Pi\dot{w} \\ &= Ax + B\sigma(u) + Pw - \Pi S w \\ &= A\xi + B(\sigma(u) - \Gamma w). \end{aligned} \tag{5.3.6}$$

With the family of state feedback laws given above, the closed-loop system can be written as,

$$\dot{\xi} = A\xi + B\left[\sigma(\Gamma w + F(\varepsilon)\xi) - \Gamma w\right]. \tag{5.3.7}$$

By Condition 2(b), $\|\Gamma w\|_{\infty,\mathrm{T}} < 1 - \delta$. Moreover, $\xi(T)$ belongs to a bounded set, say $\mathcal{U}_{\mathrm{T}}$, independent of ε, since $\xi(0)$ is bounded and $\xi(T)$ is determined by a linear differential equation with bounded inputs $\sigma(u)$ and Γw.

We then pick a Lyapunov function $V(\xi) = \xi' X(\varepsilon)\xi$ and let $c > 0$ be such that

$$c \geq \sup_{\xi \in \mathcal{U}_{\mathrm{T}}, \varepsilon \in (0,1]} \xi' X(\varepsilon)\xi. \tag{5.3.8}$$

Such a c exists since $X(\varepsilon)$ is bounded due to Lemma 2.2.6 and $\mathcal{U}_{\mathrm{T}}$ is bounded and independent of ε. Let $\varepsilon_2^* \in (0,1]$ be such that $\xi \in L_V(c)$ implies that $|F(\varepsilon)\xi|_\infty \leq \delta$. The existence of such an ε_2^* is again due to Lemma 2.2.6. Hence, for $t \geq T$, and for $\xi \in L_V(c)$, the closed-loop system (5.3.7) reduces to

$$\dot{\xi} = (A + BF(\varepsilon))\xi. \tag{5.3.9}$$

Recall that $F(\varepsilon) = -B'X(\varepsilon)x$ with $X(\varepsilon)$ being the solution to the ARE

$$A'X + XA - X'BB'X + \varepsilon I = 0. \tag{5.3.10}$$

The evaluation of $\dot{V}$ along the trajectories of (5.3.9) then shows that

$$\dot{V} \leq -\varepsilon \xi'\xi. \tag{5.3.11}$$

This shows that any trajectory starting at $t = 0$ from $\{\xi = x - \Pi w : x \in \mathcal{X}_0, w \in \mathcal{W}_0\}$ remains inside the set $L_V(c)$ and approaches the equilibrium $\xi = 0$ as t goes to infinity, which implies that

$$\lim_{t \to \infty} e(t) = \lim_{t \to \infty} C\xi(t) = 0. \tag{5.3.12}$$

Finally, taking $\varepsilon^* = \min\{\varepsilon_1^*, \varepsilon_2^*\}$, we complete our proof.   ⊠

In view of Section 5.2, Condition 1 of Theorem 5.3.1 is necessary to guarantee solvability of the output regulation problem for the system in the absence of actuator saturation. Under Assumptions 5.2.1 and 5.2.3, Condition 2(a) is necessary for the existence of a linear stabilizing feedback for the system (5.2.13). The crucial condition for the solvability of the semi-global linear state feedback output regulation problem is Condition 2(b), which is also a sufficient condition. The necessity of Condition 2(b) is discussed in detail in [70].

It is interesting to observe that if $\Gamma w = 0$, then Condition 2(b) of Theorem 5.3.1 is automatically satisfied. The following remark examines the cases for which $\Gamma w = 0$ holds.

Remark 5.3.2. *Consider Condition 2(b) of Theorem 5.3.1. $\Gamma w = 0$ for all $w(0) \in \mathcal{W}_0$ if and only if $\mathcal{W}_0 \subset < \ker \Gamma \mid S >$, where $< \ker \Gamma \mid S >$ is the unobservable subspace of the pair (S, Γ).* ®

Note that according to the sufficient conditions in the above theorem regulation is possible for arbitrary compact sets $\mathcal{W}_0$ if $\Gamma = 0$. The following remark specifies when this can happen.

Remark 5.3.3. *Consider Condition 2(b) of Theorem 5.3.1. $\Gamma = 0$ if there exists a matrix Π which solves the following linear matrix equations*

$$\begin{cases} \Pi S = A\Pi + P, \\ C\Pi + Q = 0. \end{cases} \tag{5.3.13}$$

®

5.3.2. Error Feedback Results

Problem 5.3.2. *Consider the system (5.2.13) with $\sigma \in \mathcal{S}_1(\Delta)$ being bounded by a known number and a compact set $\mathcal{W}_0 \subset \mathbb{R}^s$. The semi-global linear observer based error feedback output regulation problem is defined as follows.*

For any a priori given (arbitrarily large) bounded sets $\mathcal{X}_0 \subset \mathbb{R}^n$ and $\mathcal{Z}_0 \subset \mathbb{R}^{n+s}$, find, if possible, a linear observer based error feedback law of the form,

$$\begin{cases} \begin{bmatrix} \dot{\hat{x}} \\ \dot{\hat{w}} \end{bmatrix} = \begin{bmatrix} A & P \\ 0 & S \end{bmatrix} \begin{bmatrix} \hat{x} \\ \hat{w} \end{bmatrix} + \begin{bmatrix} B \\ 0 \end{bmatrix} \sigma(u) - \begin{bmatrix} L_A \\ L_S \end{bmatrix} \left(e - \begin{bmatrix} C & Q \end{bmatrix} \begin{bmatrix} \hat{x} \\ \hat{w} \end{bmatrix} \right), \\ u = F\hat{x} + G\hat{w}, \end{cases} \tag{5.3.14}$$

such that

1. *The equilibrium $(x, \hat{x}, \hat{w}) = (0, 0, 0)$ of*

$$\begin{cases} \dot{x} = Ax + B\sigma(F\hat{x} + G\hat{w}), \\ \begin{bmatrix} \dot{\hat{x}} \\ \dot{\hat{w}} \end{bmatrix} = \begin{bmatrix} A & P \\ 0 & S \end{bmatrix} \begin{bmatrix} \hat{x} \\ \hat{w} \end{bmatrix} + \begin{bmatrix} B \\ 0 \end{bmatrix} (F\hat{x} + G\hat{w}) - \begin{bmatrix} L_A \\ L_S \end{bmatrix} \begin{bmatrix} C & Q \end{bmatrix} \begin{bmatrix} x - \hat{x} \\ -\hat{w} \end{bmatrix} \end{cases} \tag{5.3.15}$$

is locally exponential stable with $\mathcal{X}_0 \times \mathcal{Z}_0$ contained in its basin of attraction;

2. *For all $(x(0), \hat{x}(0), \hat{w}(0)) \in \mathcal{X}_0 \times \mathcal{Z}_0$ and $w(0) \in \mathcal{W}_0$, the solution of the closed-loop system satisfies*

$$\lim_{t \to \infty} e(t) = 0. \tag{5.3.16}$$

Ⓟ

Remark 5.3.4. *We would like to emphasize that our definition of the semi-global linear observer based error feedback output regulation problem does not view the set of initial conditions of the plant and the initial conditions of the controller dynamics as given data. The set of given data consists of the models of the plant and the exosystem and the set of initial conditions for the exosystem. Therefore, the solvability conditions must be independent of the set of initial conditions of the plant, $\mathcal{X}_0$, and the set of initial conditions for the controller dynamics, $\mathcal{Z}_0$.*   ℝ

The solvability conditions for semi-global linear observer based error feedback output regulation problem are given in the following theorem.

Theorem 5.3.2. *Consider the system (5.2.13) with $\sigma \in \mathcal{S}_1(\Delta)$ being bounded by a known number and the given compact set $\mathcal{W}_0 \subset \mathbb{R}^s$. The semi-global linear observer based error feedback output regulation problem is solvable if*

1. *(A, B) is stabilizable and A has all eigenvalues in the closed left-half plane. Moreover, the pair*

$$\left(\begin{bmatrix} C & Q \end{bmatrix}, \begin{bmatrix} A & P \\ 0 & S \end{bmatrix} \right)$$

   is detectable;

2. *There exist matrices Π and Γ such that,*

   (a) *They solve the following linear matrix equations,*

$$\begin{cases} \Pi S = A\Pi + B\Gamma + P, \\ C\Pi + Q = 0; \end{cases} \qquad (5.3.17)$$

   (b) *There exist a $\delta > 0$ and a $T \geq 0$ such that $\|\Gamma w\|_{\infty,\tau} \leq 1 - \delta$ for all w with $w(0) \in \mathcal{W}_0$.*   T

Proof of Theorem 5.3.2. We prove this theorem by first explicitly constructing a family of linear observer based error feedback laws of the form (5.3.14), parameterized in ε, and then showing that for each pair of sets $\mathcal{X}_0 \subset \mathbb{R}^n$ and $\mathcal{Z}_0 \subset \mathbb{R}^{n+s}$, there exists an $\varepsilon^* > 0$ such that for all $\varepsilon \in (0, \varepsilon^*]$, both Items 1 and 2 in Problem 5.3.2 are indeed satisfied. The family of linear error feedback laws we construct take the following form

$$\begin{cases} \dot{\hat{x}} = A\hat{x} + B\sigma(u) + P\hat{w} - L_A e + L_A(C\hat{x} + Q\hat{w}), \\ \dot{\hat{w}} = S\hat{w} - L_S e + L_S(C\hat{x} + Q\hat{w}), \\ u = F(\varepsilon)\hat{x} + (F(\varepsilon)\Pi + \Gamma)\hat{w}, \end{cases} \qquad (5.3.18)$$

where $F(\varepsilon)$ is the low gain feedback gain matrix, parameterized in ε. It is constructed by using either the ARE based low gain design technique of Section 2.2.2 or the eigenstructure assignment based low gain feedback design technique of Section 2.2.1. The existence of such an F is guaranteed by Condition 1 of the theorem. In what follows, we assume that $F(\varepsilon)$ has been designed using the ARE based low gain feedback design technique of Section 2.2.2. Also to avoid notational confusion between P in (5.2.13) and $P(\varepsilon)$ in the ARE (2.2.51), in this proof, we will replace $P(\varepsilon)$ with $X(\varepsilon)$. The matrices L_A and L_S are chosen such that the following matrix is asymptotically stable,

$$\bar{A} := \begin{bmatrix} A + L_A C & P + L_A Q \\ L_S C & S + L_S Q \end{bmatrix}. \tag{5.3.19}$$

The existence of such L_A and L_S is also guaranteed by Condition 1 of the theorem.

With this family of feedback laws, the closed-loop system consisting of the system (5.2.13) and the linear observer based error feedback laws (5.3.18) can be written as,

$$\begin{cases} \dot{x} = Ax + B\sigma(\Gamma\hat{w} + F(\varepsilon)(\hat{x} - \Pi\hat{w})) + Pw, \\ \dot{\hat{x}} = A\hat{x} + B[\Gamma\hat{w} + F(\varepsilon)(\hat{x} - \Pi\hat{w}] + P\hat{w} - L_A C(x - \hat{x}) - L_A Q(w - \hat{w}), \\ \dot{\hat{w}} = S\hat{w} - L_S C(x - \hat{x}) - L_S Q(w - \hat{w}). \end{cases} \tag{5.3.20}$$

We then adopt the invertible change of state variables,

$$\begin{cases} \xi = x - \Pi w, \\ \tilde{x} = x - \hat{x}, \\ \tilde{w} = w - \hat{w}, \end{cases} \tag{5.3.21}$$

and rewrite the closed loop system (5.3.20) as,

$$\begin{cases} \dot{\xi} = A\xi + B\sigma(F(\varepsilon)\xi + \Gamma w - \Gamma\tilde{w} - F(\varepsilon)\tilde{x} + F(\varepsilon)\Pi\tilde{w}) + (A\Pi - \Pi S + P)w, \\ \dot{\tilde{x}} = (A + L_A C)\tilde{x} + (P + L_A Q)\tilde{w}, \\ \dot{\tilde{w}} = L_S C\tilde{x} + (S + L_S Q)\tilde{w}. \end{cases} \tag{5.3.22}$$

To show that Item 1 of Problem 5.3.2 holds, we note that (5.3.15) is equal to (5.3.20) with $w = 0$. For $w = 0$, (5.3.22) reduces to

$$\begin{cases} \dot{\xi} = A\xi + B\sigma(F(\varepsilon)\xi - \Gamma\tilde{w} - F(\varepsilon)\tilde{x} + F(\varepsilon)\Pi\tilde{w}), \\ \dot{\tilde{x}} = (A + L_A C)\tilde{x} + (P + L_A Q)\tilde{w}, \\ \dot{\tilde{w}} = L_S C\tilde{x} + (S + L_S Q)\tilde{w}. \end{cases} \tag{5.3.23}$$

Denoting $\tilde{m} = [\tilde{x}', \tilde{w}']'$, we write (5.3.23) in the following compact form,

$$\begin{cases} \dot{\xi} = A\xi + B\sigma(F(\varepsilon)\xi + M\tilde{m}), \\ \dot{\tilde{m}} = \bar{A}\tilde{m}, \end{cases} \tag{5.3.24}$$

and (5.3.22) in the following form,

$$\begin{cases} \dot{\xi} = A\xi + B[\sigma(F(\varepsilon)\xi + M\tilde{m} + \Gamma w) - \Gamma w], \\ \dot{\tilde{m}} = \tilde{A}\tilde{m}, \end{cases} \quad (5.3.25)$$

where we have used Condition 2(a) of the theorem and defined

$$M = [\, -F(\varepsilon) \quad -\Gamma + F(\varepsilon)\Pi \,].$$

Recalling that the matrix $\tilde{A}$, defined in (5.3.19), is asymptotically stable, we readily see from the second equation of (5.3.24) that there exists a $T_1 \geq 0$ such that, for all possible initial conditions $(\tilde{x}(0), \tilde{w}(0))$,

$$\|M\tilde{m}\|_{\infty,T_1} \leq \frac{\Delta}{2}, \quad \|\tilde{m}\|_{\infty,T_1} \leq \frac{\Delta}{2}, \quad \varepsilon \in (0,1]. \quad (5.3.26)$$

Also let $\varepsilon_1^* \in (0,1]$ be such that, for all $\varepsilon \in (0, \varepsilon_1^*]$,

$$|F(\varepsilon)| \leq \frac{1}{2}, \quad |F(\varepsilon)\Pi| \leq \frac{1}{2}, \quad (5.3.27)$$

which ensures that $|M|^2 \leq \beta := (1 + |\Gamma|)^2$. The existence of such an ε_1^* is due to Lemma 2.2.6.

From the first equation of (5.3.24), $\xi(T_1)$ belongs to a bounded set, say $\mathcal{U}_{T_1}$, independent of ε, since $\xi(0)$ is bounded and since ξ is determined via a linear differential equation with bounded input $\sigma(u)$.

We next consider (5.3.24) and pick the Lyapunov function

$$V(\xi, \tilde{m}) = \xi' X(\varepsilon)\xi + (\beta + 1)\tilde{m}' \tilde{X}_m \tilde{m}, \quad (5.3.28)$$

where $\tilde{X}_m$ is such that

$$\tilde{A}' \tilde{X}_m + \tilde{X}_m \tilde{A} = -I. \quad (5.3.29)$$

We note that such an $\tilde{X}_m$ exists since $\tilde{A}$ is asymptotically stable.

Let $c_1 > 0$ be such that

$$c_1 \geq \sup_{\xi \in \mathcal{U}_{T_1}, |\tilde{m}| \leq \Delta/2, \varepsilon \in (0,1]} V(\xi, \tilde{m}). \quad (5.3.30)$$

Such a c_1 exists since $\lim_{\varepsilon \to 0} X(\varepsilon) = 0$ by Lemma 2.2.6, and $\mathcal{U}_{T_1}$ is bounded. Let $\varepsilon_2^* \in (0, \varepsilon_1^*]$ be such that $(\xi, \tilde{m}) \in L_V(c_1)$ implies that $|F(\varepsilon)\xi|_\infty \leq \Delta/2$. The existence of such an ε_2^* is again due to Lemma 2.2.6. Hence, for $(\xi, \tilde{m}) \in L_V(c_1)$, (5.3.24) takes the form,

$$\begin{cases} \dot{\xi} = (A + BF(\varepsilon))\xi + BM\tilde{m}, \\ \dot{\tilde{m}} = \tilde{A}\tilde{m}, \end{cases} \quad (5.3.31)$$

and the evaluation of $\dot{V}$ along its trajectories, using (2.2.51), shows that

$$
\begin{aligned}
\dot{V} &= -\varepsilon\xi'\xi - \xi'X(\varepsilon)BB'X(\varepsilon)\xi + 2\xi'X(\varepsilon)BM\tilde{m} - (\beta+1)\tilde{m}'\tilde{m} \\
&= -\varepsilon\xi'\xi - \xi F'(\varepsilon)F(\varepsilon)\xi + 2\xi'F'(\varepsilon)M\tilde{m} - (\beta+1)\tilde{m}'\tilde{m} \\
&\leq -\varepsilon\xi'\xi + |M|^2\tilde{m}'\tilde{m} - (\beta+1)\tilde{m}'\tilde{m} \\
&\leq -\varepsilon\xi'\xi - \tilde{m}'\tilde{m}.
\end{aligned}
\tag{5.3.32}
$$

Noting that for any $\varepsilon \in (0,\varepsilon_2^*]$ and for any $(x(0), \hat{x}(0), \hat{w}(0)) \in \mathcal{X}_0 \times \mathcal{Z}_0$, $(\xi(T_1), \tilde{m}(T_1)) \in L_V(c_1)$, we conclude that the equilibrium $(0,0,0)$ of (5.3.23) is locally exponentially stable with $\mathcal{X}_0 \times \mathcal{Z}_0$ contained in its basin of attraction.

We now proceed to show that Item 2 of Problem 5.3.2 also holds. To this end, we consider the closed-loop system (5.3.25). Recalling that the matrix $\bar{A}$ is asymptotically stable, it readily follows from the second equation of (5.3.25) that there exists a $T_2 \geq T$ such that, for all possible initial conditions $(\tilde{x}(0), \tilde{w}(0))$,

$$
\|M\tilde{m}\|_{\infty,\mathrm{T}_2} \leq \frac{\delta}{2}, \|\tilde{m}\|_{\infty,\mathrm{T}_2} \leq \frac{\delta}{2}, \ \forall\varepsilon \in (0,1].
\tag{5.3.33}
$$

We next consider the first equation of (5.3.25). For amy $x(0) \in \mathcal{X}_0$ and any $w(0) \in \mathcal{W}_0$, $\xi(T_2)$ belongs to a bounded set, say $\mathcal{U}_{\mathrm{T}_2}$, independent of ε, since $\xi(0)$ is bounded and since ξ is determined via a linear differential equation with bounded inputs $\sigma(u)$ and w.

Then using the Lyapunov function (5.3.28), let $c_2 > 0$ be such that

$$
c_2 \geq \sup_{\xi \in \mathcal{U}_{\mathrm{T}_2}, |\tilde{m}| \leq \delta/2, \varepsilon \in (0,1]} V(\xi, \tilde{m}).
\tag{5.3.34}
$$

Such a c_2 exists again since both $\mathcal{U}_{\mathrm{T}_2}$ and $X(\varepsilon)$ are bounded. Let $\varepsilon_3^* \in (0,\varepsilon_1^*]$ be such that $(\xi, \tilde{m}) \in L_V(c_2)$ implies that $|F(\varepsilon)\xi|_\infty \leq \delta/2$. The existence of such an ε_3^* is again due to Lemma 2.2.6. Recalling that $\|\Gamma w\|_{\infty,\mathrm{T}} \leq 1 - \delta$, we see that for all $(\xi, \tilde{m}) \in L_V(c_2)$ and all $t \geq T_2$, saturation does not occur in (5.3.25) and (5.3.25) is in the same form as (5.3.24). Using the same argument as used in establishing Item 1, we can show that the equilibrium $(0,0)$ of (5.3.20) is locally exponentially stable with $\mathcal{X}_0 \times \mathcal{Z}_0$ contained in its basin of attraction. Hence, for initial conditions in $\mathcal{X}_0 \times \mathcal{Z}_0$ and $\mathcal{W}_0$,

$$
\lim_{t\to\infty} e(t) = \lim_{t\to\infty} C\xi(t) = 0.
\tag{5.3.35}
$$

Finally, taking $\varepsilon^* = \min\{\varepsilon_2^*, \varepsilon_3^*\}$, we complete our proof. $\boxtimes$

As in the state feedback case, Condition 1 is necessary to guarantee solvability of the error feedback output regulation problem for the system in the absence of actuator saturation. Condition 2(a) is necessary for the existence of linear feedbacks that semi-globally stabilize the system (5.2.13), which is subject to actuator saturation. Clearly this time we also needed a detectability assumption. Finally, the crucial condition for the solvability of the semi-global linear observer based error feedback output regulation problem is Condition 2(b). This condition is sufficient.

Note also that, except for the detectability assumption, these conditions are the same for the cases of state feedback and error feedback.

5.4. Discrete-Time Systems

In this section, we consider the discrete-time counterpart of the results in Section 5.3. More specifically, we will consider a discrete-time linear system with actuators that are subject to saturation together with an exosystem that generates disturbance and reference signals as described by the following system

$$\begin{cases} x^+ = Ax + B\sigma(u) + Pw, \\ w^+ = Sw, \\ e = Cx + Qw, \end{cases} \tag{5.4.1}$$

where $x \in \mathbb{R}^n$, $w \in \mathbb{R}^s$, $u \in \mathbb{R}^m$, $e \in \mathbb{R}^p$, and σ is a class $\mathcal{S}_1$ saturation function as defined in Section 3.2.

As in the continuous-time case, our goal is to formulate and solve the semi-global output regulation problems for the above system. We split this section into two subsections. In the first subsection we solve the semi-global linear state feedback output regulation problem where all signals are available for feedback and it suffices to look at static feedbacks. In the second subsection we solve the semi-global error feedback output regulation problem where only the error signal is available for feedback, by designing a linear observer based controller.

5.4.1. State Feedback Results

Problem 5.4.1. *Consider the system (5.4.1) with $\sigma \in \mathcal{S}_1(\Delta)$ being bounded and a compact set $\mathcal{W}_0 \subset \mathbb{R}^s$. The semi-global linear state feedback output regulation problem is defined as follows.*

For any a priori given (arbitrarily large) bounded set $\mathcal{X}_0 \subset \mathbb{R}^n$, find, if possible, a linear static feedback law $u = Fx + Gw$ such that

1. The equilibrium $x = 0$ of

$$x^+ = Ax + B\sigma(Fx) \tag{5.4.2}$$

is locally exponentially stable with $\mathcal{X}_0$ contained in its basin of attraction;

2. For all $x(0) \in \mathcal{X}_0$ and $w(0) \in \mathcal{W}_0$, the solution of the closed-loop system satisfies

$$\lim_{k \to \infty} e(k) = 0. \tag{5.4.3}$$

P

Remark 5.4.1. We would like to emphasize that our definition of semi-global linear state feedback output regulation problem does not view the set of initial conditions of the plant as given data. The set of given data consists of the models of the plant and the exosystem and the set of initial conditions for the exosystem. Therefore, any solvability conditions we obtain must be independent of the set of initial conditions of the plant, $\mathcal{X}_0$.

R

The solvability conditions for semi-global linear state feedback output regulation problem are given in the following theorem.

Theorem 5.4.1. *Consider the system (5.4.1) with $\sigma \in \mathcal{S}_1(\Delta)$ being bounded and the given compact set $\mathcal{W}_0 \subset \mathbb{R}^s$. The semi-global linear state feedback output regulation problem is solvable if*

1. *(A, B) is stabilizable and A has all eigenvalues inside or on the unit circle;*

2. *There exist matrices Π and Γ such that,*

   (a) *They solve the following linear matrix equations,*

   $$\begin{cases} \Pi S = A\Pi + B\Gamma + P, \\ C\Pi + Q = 0; \end{cases} \tag{5.4.4}$$

   (b) *There exist a $\delta > 0$ and a $K \geq 0$ such that $\|\Gamma w\|_{\infty,\kappa} \leq 1 - \delta$ for all w with $w(0) \in \mathcal{W}_0$.*

T

Proof of Theorem 5.4.1. We prove this theorem by first explicitly constructing a family of linear static state feedback laws, parameterized in ε, and then showing that for each given set $\mathcal{X}_0$, there exists an $\varepsilon^* > 0$ such that for all $\varepsilon \in (0, \varepsilon^*]$, both Items 1 and 2 of Problem 5.4.1 hold. The family of linear state feedback laws we construct takes the following form

$$u = F(\varepsilon)x + (-F(\varepsilon)\Pi + \Gamma)w, \tag{5.4.5}$$

where $F(\varepsilon)$ is the low gain feedback gain matrix, parameterized in ε. It is constructed by using either the ARE based low gain design technique of Section 2.3.2 or the eigenstructure assignment based low gain feedback design technique of Section 2.3.1. The existence of such an F is guaranteed by Condition 1 of the theorem. In what follows, we assume that $F(\varepsilon)$ has been designed using the ARE based low gain feedback design technique of Section 2.3.2. Also to avoid notational confusion between P in (5.4.1) and $P(\varepsilon)$ in the ARE (2.3.69), in this proof, we will replace $P(\varepsilon)$ with $X(\varepsilon)$.

With this family of feedback laws, the system (5.4.2) is written as

$$x^+ = Ax + B\sigma(F(\varepsilon)x). \tag{5.4.6}$$

The fact that there exists an $\varepsilon_1^* > 0$ such that for all $\varepsilon \in (0, \varepsilon_1^*]$, the equilibrium $x = 0$ of (5.4.6) is locally exponentially stable with $\mathcal{X}_0$ contained in its basin of attraction has been established in Section 3.4.

Next, we show that there exists an $\varepsilon_2^* \in (0, 1]$ such that for each $\varepsilon \in (0, \varepsilon_2^*]$, Item 2 of Problem 5.4.1 holds. To this end, let us introduce an invertible, triangular coordinate change $\xi = x - \Pi w$. Using Condition 2(a) of the theorem, we have

$$\xi^+ = A\xi + B[\sigma(u) - \Gamma w]. \tag{5.4.7}$$

With the family of state feedback laws given above, the closed-loop system can be written as

$$\xi^+ = A\xi + B[\sigma(F(\varepsilon)\xi + \Gamma w) - \Gamma w]. \tag{5.4.8}$$

By Condition 2(b), $\|\Gamma w\|_{\infty,\kappa} \leq 1 - \delta$. Moreover, for any $x(0) \in \mathcal{X}_0$ and any $w(0) \in \mathcal{W}_0$, $\xi(K)$ belongs to a bounded set, say $\mathcal{U}_K$, independent of ε, since $\mathcal{X}_0$ and $\mathcal{W}_0$ are both bounded and $\xi(K)$ is determined by a linear difference equation with bounded inputs $\sigma(\cdot)$ and Γw.

It follows from (2.3.69) that

$$(A + BF(\varepsilon)'X(\varepsilon)(A + BF(\varepsilon)) - X(\varepsilon) = -\varepsilon I - F(\varepsilon)'F(\varepsilon). \tag{5.4.9}$$

We then pick the Lyapunov function

$$V(\xi) = \xi'X(\varepsilon)\xi \tag{5.4.10}$$

and let $c > 0$ be such that

$$c \geq \sup_{\xi \in \mathcal{U}_K, \varepsilon \in (0,1]} \xi'X(\varepsilon)\xi. \tag{5.4.11}$$

Such a c exists since $\lim_{\varepsilon \to 0} X(\varepsilon) = 0$ and $\mathcal{U}_K$ is bounded. Let $\varepsilon_2^* \in (0, 1]$ be such that $\xi \in L_V(c)$ implies that $|F(\varepsilon)\xi|_\infty \leq \delta$. The existence of such an ε_2^* is again due to Lemma 2.3.6. Hence, for $k \geq K$, and for all $\xi \in L_V(c)$, (5.4.8) takes the form,

$$\xi^+ = (A + BF(\varepsilon))\xi. \tag{5.4.12}$$

The evaluation of the deference of V, inside the set $L_V(c)$ and for $k \geq K$, using (5.4.9), now shows that for all $\xi \in L_V(c)$,

$$\Delta V = -\xi' \left(\varepsilon I + F(\varepsilon)' F(\varepsilon) \right) \xi. \tag{5.4.13}$$

This shows that any trajectory of (5.4.8) starting at $k = 0$ from $\{\xi = x - \Pi w : x \in \mathcal{X}_0, w \in \mathcal{W}_0\}$ remains inside the set $L_V(c)$ and approaches the equilibrium $\xi = 0$ as $k \to \infty$, which implies that

$$\lim_{k \to \infty} e(k) = \lim_{k \to \infty} C\xi(k) = 0. \tag{5.4.14}$$

Finally, setting $\varepsilon^* = \min\{\varepsilon_1^*, \varepsilon_2^*\}$, we conclude our proof of Theorem 5.4.1.⊠

Remark 5.4.2. *In view of Yang's results ([128]) and the solvability conditions for the state feedback output regulation problem for linear systems in the absence of actuator saturation as given by Theorem 5.2.1, it is obvious to observe that Conditions 1 and 2(a) of Theorem 5.4.1 are necessary. The crucial condition for the solvability of this semi-global linear state feedback output regulation problem is Condition 2(b), which is a sufficient condition. The necessity of Condition 2(b) was discussed in detail in [72].* ℝ

It is interesting to observe that if $\Gamma w = 0$, then Condition 2(b) of Theorem 5.4.1 is automatically satisfied. The following remark examines the cases for which $\Gamma w = 0$ holds.

Remark 5.4.3. *Consider Condition 2(b) of Theorem 5.4.1. $\Gamma w = 0$ for all $w(0) \in \mathcal{W}_0$ if and only if $\mathcal{W}_0 \subset\subset \ker \Gamma \mid S >$, where $< \ker \Gamma \mid S >$ is the unobservable subspace of the pair (S, Γ).* ℝ

Note that according to the sufficient conditions in the above theorem regulation is possible for arbitrary compact sets $\mathcal{W}_0$ if $\Gamma = 0$. The following remark specifies when this can happen.

Remark 5.4.4. *Consider Condition 2(b) of Theorem 5.4.1. $\Gamma = 0$ if there exists a matrix Π which solves the following matrix equations,*

$$\begin{cases} \Pi S = A\Pi + P, \\ C\Pi + Q = 0. \end{cases} \tag{5.4.15}$$

ℝ

5.4.2. Error Feedback Results

Problem 5.4.2. *Consider the system (5.4.1) with $\sigma \in S_1(\Delta)$ being bounded by a known number and a compact set $\mathcal{W}_0 \subset \mathbb{R}^s$. The semi-global linear observer based error feedback output regulation problem is defined as follows.*

For any a priori given (arbitrarily large) bounded sets $\mathcal{X}_0 \subset \mathbb{R}^n$ and $\mathcal{Z}_0 \subset \mathbb{R}^{n+s}$, find, if possible, a linear observer based error feedback law of the form,

$$\begin{cases} \begin{bmatrix} \hat{x}^+ \\ \hat{w}^+ \end{bmatrix} = \begin{bmatrix} A & P \\ 0 & S \end{bmatrix} \begin{bmatrix} \hat{x} \\ \hat{w} \end{bmatrix} + \begin{bmatrix} B \\ 0 \end{bmatrix} \sigma(u) - \begin{bmatrix} L_A \\ L_S \end{bmatrix} \left(e - [C \quad Q] \begin{bmatrix} \hat{x} \\ \hat{w} \end{bmatrix} \right), \\ u = F\hat{x} + G\hat{w}, \end{cases}$$

(5.4.16)

such that

1. *The equilibrium $(x, \hat{x}, \hat{w}) = (0, 0, 0)$ of*

$$\begin{cases} x^+ = Ax + B\sigma(F\hat{x} + G\hat{w}), \\ \begin{bmatrix} \hat{x}^+ \\ \hat{w}^+ \end{bmatrix} = \begin{bmatrix} A & P \\ 0 & S \end{bmatrix} \begin{bmatrix} \hat{x} \\ \hat{w} \end{bmatrix} + \begin{bmatrix} B \\ 0 \end{bmatrix} \sigma(F\hat{x} + G\hat{w}) - \begin{bmatrix} L_A \\ L_S \end{bmatrix} [C \quad Q] \begin{bmatrix} x - \hat{x} \\ -\hat{w} \end{bmatrix} \end{cases}$$

(5.4.17)

is locally exponentially stable with $\mathcal{X}_0 \times \mathcal{Z}_0$ contained in its basin of attraction;

2. *For all $(x(0), p(0)) \in \mathcal{X}_0 \times \mathcal{Z}_0$ and $w(0) \in \mathcal{W}_0$, the solution of the closed-loop system satisfies*

$$\lim_{k \to \infty} e(k) = 0. \qquad (5.4.18)$$

                                                                  ⊤

Remark 5.4.5. We would like to emphasize that our definition of the semi-global linear observer based error feedback output regulation problem does not view the set of initial conditions of the plant and the initial conditions of the controller dynamics as given data. The set of given data consists of the models of the plant and the exosystem and the set of initial conditions for the exosystem. Therefore, the solvability conditions must be independent of the set of initial conditions of the plant, $\mathcal{X}_0$, and the set of initial conditions for the controller dynamics, $\mathcal{Z}_0$.   ℝ

The solvability conditions for semi-global linear observer based error feedback output regulation problem are given in the following theorem.

Theorem 5.4.2. *Consider the system (5.4.1) with $\sigma \in S_1(\Delta)$ being bounded by a known number and the given compact set $\mathcal{W}_0 \subset \mathbb{R}^s$. The semi-global linear observer based error feedback output regulation problem is solvable if*

1. (A, B) is stabilizable and A has all eigenvalues inside or on the unit circle; Moreover, the pair

$$\left([\, C \quad Q \,], \begin{bmatrix} A & P \\ 0 & S \end{bmatrix} \right)$$

   is detectable;

2. There exist matrices Π and Γ such that

   (a) They solve the following linear matrix equations

   $$\begin{cases} \Pi S = A\Pi + B\Gamma + P, \\ C\Pi + Q = 0; \end{cases} \tag{5.4.19}$$

   (b) There exists a $\delta > 0$ and a $K \geq 0$ such that $\|\Gamma w\|_{\infty, \kappa} \leq 1 - \delta$ for all w with $w(0) \in \mathcal{W}_0$. $\quad\boxed{\text{T}}$

Proof of Theorem 5.4.2. We again prove this theorem by first explicitly constructing a family of linear observer based error feedback laws of the form (5.4.16), parameterized in ε, and then showing that for each pair of sets $\mathcal{X}_0 \subset \mathbb{R}^n$ and $\mathcal{Z}_0 \subset \mathbb{R}^{n+s}$, there exists an $\varepsilon^* > 0$ such that for all $\varepsilon \in (0, \varepsilon^*]$, both Items 1 and 2 in Problem 5.4.2 are indeed satisfied. The family of linear observer based error feedback laws we construct take the following form,

$$\begin{cases} \hat{x}^+ = A\hat{x} + B\sigma(u) + P\hat{w} - L_A e + L_A(C\hat{x} + Q\hat{w}), \\ \hat{w}^+ = S\hat{w} - L_S e + L_S(C\hat{x} + Q\hat{w}), \\ u = F(\varepsilon)\hat{x} + (-F(\varepsilon)\Pi + \Gamma)\hat{w}, \end{cases} \tag{5.4.20}$$

where $F(\varepsilon)$ is the low gain feedback gain matrix, parameterized in ε. It is constructed by using either the ARE based low gain design technique of Section 2.3.2 or the eigenstructure assignment based low gain feedback design technique of Section 2.3.1. The existence of such an F is guaranteed by Condition 1 of the theorem. In what follows, we assume that $F(\varepsilon)$ has been designed using the ARE based low gain feedback design technique of Section 2.3.2. Also to avoid notational confusion between P in (5.4.1) and $P(\varepsilon)$ in the ARE (2.3.69), in this proof, we will replace $P(\varepsilon)$ with $X(\varepsilon)$. The matrices L_A and L_S are chosen such that the following matrix is asymptotically stable,

$$\bar{A} := \begin{bmatrix} A + L_A C & P + L_A Q \\ L_S C & S + L_S Q \end{bmatrix}. \tag{5.4.21}$$

With this family of feedback laws, the closed-loop system consisting of the system (5.4.1) and the linear observer based error feedback laws (5.4.20) can be written as,

$$\begin{cases} x^+ = Ax + B\sigma(\Gamma\hat{w} + F(\varepsilon)(\hat{x} - \Pi\hat{w})) + Pw, \\ \hat{x}^+ = A\hat{x} + B\sigma(\Gamma\hat{w} + F(\varepsilon)(\hat{x} - \Pi\hat{w})) + P\hat{w} - L_AC(x - \hat{x}) - L_AQ(w - \hat{w}), \\ \hat{w}^+ = S\hat{w} - L_SC(x - \hat{x}) - L_SQ(w - \hat{w}). \end{cases}$$

$$(5.4.22)$$

We then adopt the invertible change of state variables,

$$\xi = x - \Pi w, \quad \tilde{x} = x - \hat{x}, \quad \tilde{w} = w - \hat{w}, \tag{5.4.23}$$

and rewrite the closed loop system (5.4.22) as,

$$\begin{cases} \xi^+ = A\xi + B\sigma(F(\varepsilon)\xi + \Gamma w - \Gamma\tilde{w} - F(\varepsilon)\tilde{\xi}) + (A\Pi - \Pi S + P)w, \\ \tilde{x}^+ = (A + L_AC)\tilde{x} + (P + L_AQ)\tilde{w}, \\ \tilde{w}^+ = L_SC\tilde{x} + (S + L_SQ)\tilde{w}, \end{cases} \tag{5.4.24}$$

where we have denoted $\tilde{\xi} = \tilde{x} - \Pi\tilde{w}$.

To show that item 1 of Problem 5.4.2 holds, we note that (5.4.17) is the same as (5.4.22) with $w = 0$. We know (5.4.22) is equivalent to (5.4.24) which for $w = 0$ reduces to

$$\begin{cases} \xi^+ = A\xi + B\sigma(F(\varepsilon)\xi - \Gamma\tilde{w} - F(\varepsilon)\tilde{\xi}), \\ \tilde{x}^+ = (A + L_AC)\tilde{x} + (P + L_AQ)\tilde{w}, \\ \tilde{w}^+ = +L_SC\tilde{x} + (S + L_SQ)\tilde{w}. \end{cases} \tag{5.4.25}$$

Denoting $\tilde{m} = [\tilde{x}', \tilde{w}']'$, we write (5.4.25) in the following compact form,

$$\begin{cases} \xi^+ = A\xi + B[\sigma(F(\varepsilon)\xi + M\tilde{m})], \\ \tilde{m}^+ = \tilde{A}\tilde{m}, \end{cases} \tag{5.4.26}$$

and (5.4.24) in the following form,

$$\begin{cases} \xi^+ = A\xi + B[\sigma(F(\varepsilon)\xi + M\tilde{m} + \Gamma w) - \Gamma w], \\ \tilde{m}^+ = \tilde{A}\tilde{m}, \end{cases} \tag{5.4.27}$$

where

$$M = [-F(\varepsilon) \quad -\Gamma + F(\varepsilon)\Pi].$$

Recalling that the matrix $\tilde{A}$, as defined in (5.4.21), is asymptotically stable, we readily see from the second equation of (5.4.26) that there exists a $K_1 \geq 0$ such that, for all possible initial conditions $(\tilde{x}(0), \tilde{w}(0))$,

$$\|M\tilde{m}\|_{\infty,K_1} \leq \frac{\Delta}{2}, \quad \|\tilde{m}\|_{\infty,K_1} \leq \frac{\Delta}{2}, \quad \forall\varepsilon \in (0,1]. \tag{5.4.28}$$

For any $x(0) \in \mathcal{X}_0$ and $w(0) \in \mathcal{W}_0$, $\xi(K_1)$ belongs to a bounded set, say $\mathcal{U}_{K_1}$, independent of ε, since $\mathcal{X}_0$ and $\mathcal{W}_0$ are both bounded and $\xi(K_1)$ is determined by a linear difference equation (5.4.26) with bounded input $\sigma(\cdot)$. Let $\varepsilon_1^* \in (0,1]$ be chosen such that for all $\varepsilon \in (0, \varepsilon_1^*]$,

$$|F(\varepsilon)| \leq \frac{1}{2}, \quad |F(\varepsilon)\Pi| \leq \frac{1}{2}. \tag{5.4.29}$$

This ensures that $|M|^2 \leq (1 + |\Gamma|)^2$. Let's define $\beta := (1 + |\Gamma|)^2(1 + |B|^2)$ for later use. Let $\tilde{X}$ be the unique positive definite solution to the Lyapunov equation

$$\tilde{X} = \bar{A}'\tilde{X}\bar{A} + I. \tag{5.4.30}$$

Such a $\tilde{X}$ exists since $\bar{A}$ is asymptotically stable.

We next define the Lyapunov function

$$V(\xi, \tilde{m}) = \xi'X(\varepsilon)\xi + (\beta + 1)\tilde{m}'\tilde{X}\tilde{m}, \tag{5.4.31}$$

and let $c_1 > 0$ be such that

$$c_1 \geq \sup_{\xi \in \mathcal{U}_{K_1}, |\tilde{m}| \leq \Delta/2, \varepsilon \in (0,1]} V(\xi, \tilde{m}). \tag{5.4.32}$$

Such a c_1 exists since $\lim_{\varepsilon \to 0} X(\varepsilon) = 0$ by Lemma 2.3.6 and $\mathcal{U}_{K_1}$ is bounded. Let $\varepsilon_2^* \in (0, \varepsilon_1^*]$ be such that $\xi \in L_V(c_1)$ implies that $|F(\varepsilon)\xi|_\infty \leq \Delta/2$. The existence of such an ε_2^* is again due to Lemma 2.3.6. Hence, for $k \geq K_1$, and for $(\xi, \tilde{m}) \in L_V(c_1)$, (5.4.26) takes the form,

$$\begin{cases} \xi^+ = (A + BF(\varepsilon))\xi + BM\tilde{m}, \\ \tilde{m}^+ = \bar{A}\tilde{m}, \end{cases} \tag{5.4.33}$$

and the evaluation of ΔV along its trajectories, using (2.3.69), shows that for all $\xi \in L_V(c_1)$,

$$\begin{aligned} \Delta V &= [(A + BF(\varepsilon))\xi + BM\tilde{m}]'X(\varepsilon)[(A + BF(\varepsilon))\xi + BM\tilde{m}] - \xi'X(\varepsilon)\xi \\ &\quad - (\beta + 1)\tilde{m}'\tilde{m} \\ &= -\varepsilon\xi'\xi - \xi'F'(\varepsilon)F(\varepsilon)\xi + 2\tilde{m}'M'F(\varepsilon)\xi + \tilde{m}'M'B'BM\tilde{m} - (\beta + 1)\tilde{m}'\tilde{m} \\ &\leq -\varepsilon\xi'\xi - |F(\varepsilon)\xi|^2 + |F(\varepsilon)\xi|^2 + |M|^2\tilde{m}'\tilde{m} + |B|^2|M|^2\tilde{m}'\tilde{m} - (\beta + 1)\tilde{m}'\tilde{m} \\ &\leq -\varepsilon\xi'\xi - \tilde{m}'\tilde{m}. \end{aligned} \tag{5.4.34}$$

Noting that for any $\varepsilon \in (0, \varepsilon_2^*]$, any $(x(0), \hat{x}(0), \hat{w}(0)) \in \mathcal{X}_0 \times \mathcal{Z}_0$, and any $w(0) \in \mathcal{W}_0$, $(\xi(K_1), \tilde{m}(K_1)) \in L_V(c_1)$, we conclude that, the equilibrium $(0,0,0)$ of (5.4.17) is locally exponentially stable with $\mathcal{X}_0 \times \mathcal{Z}_0$ contained in its basin of attraction.

We now proceed to show that Item 2 of Problem 5.4.2 also holds. To this end, we consider the closed-loop system (5.4.24). Recalling that the matrix $\bar{A}$ is asymptotically stable, and using (5.4.24), which is equivalent to the system (5.4.27), it readily follows from the second equation of (5.4.27) that there exists a $K_2 \geq K$ such that, for all possible initial conditions $(\tilde{x}(0), \tilde{w}(0))$,

$$\|M\tilde{m}\|_{\infty,K_2} \leq \frac{\delta}{2}, \quad \|\tilde{m}\|_{\infty,K_2} \leq \frac{\delta}{2}, \quad \forall \varepsilon \in (0,1]. \tag{5.4.35}$$

We then consider the first equation of (5.4.27). By Condition 2(b), $\|\Gamma w\|_{\infty,K} \leq 1 - \delta$. Moreover, for any $x(0) \in \mathcal{X}_0$ and any $w(0) \in \mathcal{W}_0$, $\xi(K_2)$ belongs to a bounded set, say $\mathcal{U}_{K_2}$, independent of ε, since $\mathcal{X}_0$ and $\mathcal{W}_0$ are both bounded and $\xi(K_2)$ is determined by a linear difference equation with bounded inputs $\sigma(u)$ and Γw. Then, using the Lyapunov function (5.4.31), let $c_2 > 0$ be such that

$$c_2 \geq \sup_{\xi \in \mathcal{U}_{K_2}, |\tilde{m}| \leq \delta/2, \varepsilon \in (0,1]} V(\xi, \tilde{m}). \tag{5.4.36}$$

Such a c_2 exists since $\lim_{\varepsilon \to 0} X(\varepsilon) = 0$ by Lemma 2.3.6 and $\mathcal{U}_{K_2}$ is bounded. Let $\varepsilon_3^* \in (0, \varepsilon_1^*]$ be such that $\xi \in L_V(c_2)$ implies that $|F(\varepsilon)\xi|_{\infty} \leq \delta/2$. The existence of such an ε_3^* is again due to Lemma 2.3.6.

Recalling that $\|\Gamma w\|_{\infty,K} \leq 1 - \delta$, we see that for all $(\xi, \tilde{m}) \in L_V(c_2)$ and $k \geq K_2$, saturation does not occur in (5.4.27) and (5.4.27) is in the same form as (5.4.26). Using the same argument as used in establishing Item 1, we can show that the equilibrium $(0,0)$ of (5.4.27) is locally exponentially stable with $\mathcal{X}_0 \times \mathcal{Z}_0$ contained in its basin of attraction. Hence, for initial conditions in $\mathcal{X}_0 \times \mathcal{Z}_0$,

$$\lim_{k \to \infty} e(k) = \lim_{k \to \infty} C\xi(k) = 0. \tag{5.4.37}$$

Finally, taking $\varepsilon^* = \min\{\varepsilon_2^*, \varepsilon_3^*\}$, we complete our proof.   ⊠

Remark 5.4.6. *As in the state feedback case, in view of Yang's results ([128]) and the solvability conditions for the error feedback output regulation problem for linear systems in the absence of input saturation as given by Theorem 5.2.2, it is obvious to observe that Conditions 1 and 2(a) of Theorem 5.4.2 are necessary. The crucial condition for the solvability of this semi-global linear observer based error feedback regulator problem is Condition 2(b), which is a sufficient condition. The necessity of Condition 2(b) was discussed in detail in [72].*   ▣

5.5. Generalized Semi-Global Output Regulation Problems

5.5.1. Introduction and Problem Statement

In Sections 5.3 and 5.4, we have formulated the semi-global linear feedback output regulation problems for linear systems subject to actuator saturation following the traditional formulation of linear output regulation problems where the exosystem is autonomous. As a result, the disturbances and the references generated by the exosystem contain only the frequency components of the exosystem. In an effort to broaden the class of disturbance and reference signals, we formulate in this section the generalized semi-global linear feedback output regulation problems, for which an external driving signal to the exosystem is included (Fig. 5.5.1). While we only focus on continuous-time linear systems subject to actuator saturation, the discrete-time counterpart can be treated in the same manner.

More specifically, we consider a linear system with actuators that are subject to saturation together with an exosystem that generates disturbance and reference signals as described by the following system

$$\begin{cases} \dot{x} = Ax + B\sigma(u) + Pw, \\ \dot{w} = Sw + r, \\ e = Cx + Qw, \end{cases} \tag{5.5.1}$$

where $x \in \mathbb{R}^n$, $w \in \mathbb{R}^s$, $u \in \mathbb{R}^m$, $e \in \mathbb{R}^p$, $r \in \mathcal{C}^0$ is an external signal to the exosystem, and σ is a class $\mathcal{S}_1$ saturation function as defined in Section 3.2.

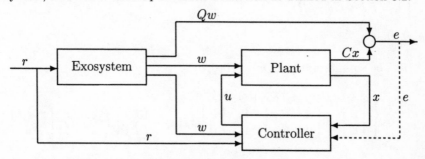

Figure 5.5.1: Configuration of a generalized regulator

The generalized semi-global linear state feedback output regulation problem and the generalized error feedback output regulation problem are formulated as follows.

Problem 5.5.1. *Consider the system (5.5.1) with $\sigma \in \mathcal{S}_1(\Delta)$ being bounded by a known number, a compact set $\mathcal{W}_0 \subset \mathbb{R}^s$ and a compact set $\mathcal{R} \subset \mathcal{C}^0$.*

The problem of generalized semi-global linear state feedback output regulation problem is defined as follows.

For any a priori given (arbitrarily large) bounded set $\mathcal{X}_0 \subset \mathbb{R}^n$, find, if possible, a linear feedback law $u = Fx + Gw + Hr$, such that

1. The equilibrium $x = 0$ of

$$\dot{x} = Ax + B\sigma(Fx) \tag{5.5.2}$$

   is locally exponentially stable with $\mathcal{X}_0$ contained in its basin of attraction;

2. For all $x(0) \in \mathcal{X}_0$, $w(0) \in \mathcal{W}_0$ and $r \in \mathcal{R}$, the solution of the closed-loop system satisfies

$$\lim_{t \to \infty} e(t) = 0. \tag{5.5.3}$$

   ☐

Problem 5.5.2. *Consider the system (5.5.1) with $\sigma \in \mathcal{S}_1(\Delta)$ being bounded and two compact sets $\mathcal{W}_0 \subset \mathbb{R}^s$ and $\mathcal{R} \subset \mathcal{C}^0$. The semi-global linear observer based error feedback output regulation problem is defined as follows.*

For any a priori given (arbitrarily large) bounded sets $\mathcal{X}_0 \subset \mathbb{R}^n$ and $\mathcal{Z}_0 \subset \mathbb{R}^{n+s}$, find, if possible, a linear observer based error feedback law of the form,

$$\begin{cases} \begin{bmatrix} \dot{\hat{x}} \\ \dot{\hat{w}} \end{bmatrix} = \begin{bmatrix} A & P \\ 0 & S \end{bmatrix} \begin{bmatrix} \hat{x} \\ \hat{w} \end{bmatrix} + \begin{bmatrix} B \\ 0 \end{bmatrix} \sigma(u) - \begin{bmatrix} L_A \\ L_S \end{bmatrix} \left(e - [C \quad Q] \begin{bmatrix} \hat{x} \\ \hat{w} \end{bmatrix} \right), \\ u = F\hat{x} + G\hat{w} + Hr, \end{cases} \tag{5.5.4}$$

such that

1. The equilibrium $(x, \hat{x}, \hat{w}) = (0, 0, 0)$ of

$$\begin{cases} \dot{x} = Ax + B\sigma_h(F\hat{x} + G\hat{w})), \\ \begin{bmatrix} \dot{\hat{x}} \\ \dot{\hat{w}} \end{bmatrix} = \begin{bmatrix} A & P \\ 0 & S \end{bmatrix} \begin{bmatrix} \hat{x} \\ \hat{w} \end{bmatrix} + \begin{bmatrix} B \\ 0 \end{bmatrix} \sigma_h(u) - \begin{bmatrix} L_A \\ L_S \end{bmatrix} \left([C \quad Q] \begin{bmatrix} x - \hat{x} \\ -\hat{w} \end{bmatrix} \right) \end{cases} \tag{5.5.5}$$

   is locally exponentially stable with $\mathcal{X}_0 \times \mathcal{Z}_0$ contained in its basin of attraction;

2. For all $(x(0), \hat{x}(0), \hat{w}(0)) \in \mathcal{X}_0 \times \mathcal{Z}_0$, $w(0) \in \mathcal{W}_0$, and all $r \in \mathcal{R}$, the solution of the closed-loop system satisfies

$$\lim_{t \to \infty} e(t) = 0. \tag{5.5.6}$$

   ☐

Remark 5.5.1. *We would like to emphasize here that our definition of the generalized semi-global linear state feedback [respectively, linear observer based error feedback] output regulation problem does not view the set of initial conditions of the plant as given data. The set of given data consists of the models of the plant and the exosystem, the set of initial conditions for the exosystem and the set of external input to the exosystem. Moreover, the generalized semi-global linear state feedback [respectively, error feedback] output regulation problem reduces to the semi-global linear state feedback [respectively, error feedback] output regulation problem as formulated in Problem 5.3.1 [respectively, Problem 5.3.2] when the external input r to the exosystem is nonexistent.* Ⓡ

We will give solvability conditions for the above two problems. For clarity, we present these solvability conditions in two separate subsections, one for the state feedback case and the other for the error feedback case. As a special case of the generalized semi-global linear feedback output regulation problems, the solvability conditions for semi-global linear feedback restricted tracking problems for a chain of integrators are thus obtained readily. The same set of solvability conditions for the global state feedback restricted tracking problem for a chain of integrators were given earlier in [111], where nonlinear feedbacks were resorted to.

5.5.2. State Feedback Results

The solvability conditions for the generalized semi-global linear state feedback output regulation problem is given in the following theorem.

Theorem 5.5.1. *Consider the system (5.5.1) with $\sigma \in \mathcal{S}_1(\Delta)$ being bounded by a known number and given compact sets $\mathcal{W}_0 \subset \mathbb{R}^s$ and $\mathcal{R} \subset \mathcal{C}^0$. The generalized semi-global linear state feedback output regulation problem is solvable if*

1. *(A, B) is stabilizable and A has all its eigenvalues in the closed left half plane;*

2. *There exist matrices Π and Γ such that*

   (a) *They solve the following linear matrix equations*

   $$\begin{cases} \Pi S = A\Pi + B\Gamma + P, \\ C\Pi + Q = 0; \end{cases} \qquad (5.5.7)$$

   (b) *For each $r \in \mathcal{R}$, there exists a function $\tilde{r} \in \mathcal{C}^0$ such that $\Pi r = B\tilde{r}$;*

(c) *There exists a $\delta > 0$ and a $T \geq 0$ such that $\|\Gamma w + \tilde{r}\|_{\infty,T} \leq 1 - \delta$ for all w with $w(0) \in \mathcal{W}_0$ and all $r \in \mathcal{R}$.*   ⊤

Remark 5.5.2. *We would like to make the following observations on the solvability conditions as given in the above theorem,*

1. *As expected, the solvability conditions for the generalized semi-global linear state feedback output regulation problem as given in the above theorem reduces to those for the semi-global linear state feedback regulator problem as formulated in Problem 5.3.1 where the external input to the exosystem is nonexistent.*

2. *If $\mathrm{Im}\,\Pi \subset \mathrm{Im}\,B$, then Condition 2(b) is automatically satisfied for any given set $\mathcal{R}$.*

3. *If $\mathrm{Im}\,\Pi \cap \mathrm{Im}\,B = \{0\}$, then Condition 2(b) can never be satisfied for any given $\mathcal{R}$ except for $\mathcal{R} = \{0\}$.*   ℝ

Proof of Theorem 5.5.1. The proof of this theorem is similar, *mutatis mutandis*, to that of Theorem 5.3.1. As in the proof of Theorem 5.3.1, we prove this theorem by first constructing a family of linear state feedback laws, parameterized in ε, and then showing that for each given set $\mathcal{X}_0$, there exists an $\varepsilon^* > 0$ such that for all $\varepsilon \in (0, \varepsilon^*]$, both Items 1 and 2 of Problem 5.5.1 hold. The family of linear state feedback laws we construct takes the following form

$$u = F(\varepsilon)x + (-F(\varepsilon)\Pi + \Gamma)w + \tilde{r}, \qquad (5.5.8)$$

where $F(\varepsilon)$ is designed the same way as in the proof of Theorem 5.3.1. The rest of the proof is the same as that of Theorem 5.3.1 except that (5.3.6) takes the following slightly different form

$$\dot{\xi} = A\xi + B(\sigma(u) - \Gamma w - \tilde{r}). \qquad (5.5.9)$$

⊠

As a corollary to Theorem 5.5.1, we give the solvability conditions for the semi-global linear state feedback restricted tracking problem for a chain of integrators subject to actuator saturation. It is interesting to note that the same solvability condition was given in [111] for the global nonlinear state feedback restricted tracking problem for a chain of integrators.

Corollary 5.5.1. *Consider a system consisting of a chain of integrators*

$$\begin{cases} \dot{x}_i = x_{i+1}, \quad i = 1, 2, \cdots, n-1, \\ \dot{x}_n = \sigma(u), \\ y = x_1, \end{cases} \qquad (5.5.10)$$

where σ is the same as in (5.5.1). Let a desired reference trajectory be given by $y_d \in C^n$. Assume that there exists a $T \geq 0$ and a $\delta > 0$ such that $|y_d^{(n)}| \leq 1 - \delta$ for all $t \geq T$. Then the reference signal y_d can be semi-globally tracked by y via linear static state feedback. More specifically, for any given (arbitrarily large) bounded set $\mathcal{X}_0 \subset \mathbb{R}^n$, there exists a linear state feedback law $u = \sum_{i=1}^n F_i x_i + y_d^{(n)}$ such that

1. The equilibrium $(x_1, x_2, \cdots, x_n) = (0, 0, \cdots, 0)$ of

$$\begin{cases} \dot{x}_i = x_{i+1}, \ i = 1, 2, \cdots, n-1, \\ \dot{x}_n = \sigma \left(\sum_{i=1}^n F_i x_i \right) \end{cases} \tag{5.5.11}$$

is locally exponentially stable with $\mathcal{X}_0$ contained in its basin of attraction;

2. For any $x(0) \in \mathcal{X}_0$, the solution of the closed-loop system satisfies

$$\lim_{t \to \infty} y(t) = y_d(t). \tag{5.5.12}$$

□

Proof Corollary 5.5.1. The desired reference trajectory can be modeled as that of the following exogenous system

$$\begin{cases} \dot{w}_i = w_{i+1}, \ w_i(0) = y_d^{(i-1)}(0), \ i = 1, 2, \cdots, n-1, \\ \dot{w}_n = y_d^{(n)}, \ w_n(0) = y_d^{(n-1)}(0), \end{cases} \tag{5.5.13}$$

and hence the tracking problem can be cast into a generalized linear state feedback output regulation problem as defined in Problem 5.5.1, with $P = 0$, $\mathcal{R} = \{[0\,0\cdots 0\,r_n(t)]' \in C^0 : |r_n^{(n)}| \leq 1 - \delta\}$, $\mathcal{W}_0 := \{(y_d(0), y_d'(0), \cdots, y_d^{(n-1)}(0))\}$ and

$$A = S = \begin{bmatrix} 0 & 1 & 0 & \cdots & 0 \\ 0 & 0 & 1 & \cdots & 0 \\ \vdots & \vdots & \vdots & \ddots & \vdots \\ 0 & 0 & 0 & \cdots & 1 \\ 0 & 0 & 0 & \cdots & 0 \end{bmatrix}, \quad C = Q = \begin{bmatrix} 1 \\ 0 \\ \vdots \\ 0 \\ 0 \end{bmatrix}'. \tag{5.5.14}$$

Clearly, $\Pi = I$ and $\Gamma = 0$ solve the matrix equations (5.5.7). Condition 1 of Theorem 5.5.1 is clearly satisfied, while Condition 2 is also satisfied trivially with $\tilde{r} = y_d^{(n)}$. The results of this corollary thus follow from that of Theorem 5.5.1. ⊠

5.5.3. Error Feedback Results

The solvability conditions for generalized semi-global linear observer based error feedback output regulation problem is given in the following theorem.

Theorem 5.5.2. *Consider the system (5.5.1) with $\sigma \in \mathcal{S}_1(\Delta)$ being bounded by a known number and the given compact sets $\mathcal{W}_0 \subset \mathbb{R}^s$ and $\mathcal{R} \subset \mathcal{C}^0$. The generalized semi-global linear observer based error feedback output regulation problem is solvable if*

1. *(A, B) is stabilizable and A has all its eigenvalues in the closed left-half plane. Moreover, the pair*

$$\left(\begin{bmatrix} C & Q \end{bmatrix}, \begin{bmatrix} A & P \\ 0 & S \end{bmatrix} \right)$$

   is detectable;

2. *There exist matrices Π and Γ such that*

   (a) *They solve the following linear matrix equations*

$$\begin{cases} \Pi S = A\Pi + B\Gamma + P, \\ C\Pi + Q = 0; \end{cases} \qquad (5.5.15)$$

   (b) *For each $r \in \mathcal{R}$, there exists a function $\tilde{r}(t) \in \mathcal{C}^0$ such that $\Pi r(t) = B\tilde{r}(t)$ for all $t \geq 0$;*

   (c) *There exist a $\delta > 0$ and a $T \geq 0$ such that $\|\Gamma w + \tilde{r}\|_{\infty, \text{T}} \leq 1 - \delta$ for all w with $w(0) \in \mathcal{W}_0$ and all $r \in \mathcal{R}$.* $\quad$ ⊤

Remark 5.5.3. *As expected, the solvability conditions for the generalized semi-global linear observer based error feedback output regulation problem as given in the above theorem reduces to those for the semi-global linear observer based error feedback output regulation problem as formulated in Problem 5.3.2 where the external input to the exosystem is nonexistent.* $\quad$ ⓡ

Proof of Theorem 5.5.2. The proof of this theorem is similar, *mutatis mutandis* to that of Theorem 5.3.2. As in the proof of Theorem 5.3.2, we prove this theorem by first constructing a family of linear observer based error feedback laws, parameterized in ε, and then showing that both Items 1 and 2 of Problem 5.5.2 indeed hold. The family of linear observer based error feedback laws we construct takes the following form

$$\begin{cases} \dot{\hat{x}} = A\hat{x} + B\sigma(u) + P\hat{w} - L_A C(x - \hat{x}) - L_A Q(w - \hat{w}), \\ \dot{\hat{w}} = S\hat{w} - L_S C(x - \hat{x}) - L_S Q(w - \hat{w}), \\ u = F(\varepsilon)\hat{x} + (-F(\varepsilon)\Pi + \Gamma)\hat{w} + \tilde{r}, \end{cases} \qquad (5.5.16)$$

where L_A and L_S are such that the following matrix is asymptotically stable,

$$\bar{A} := \begin{bmatrix} A + L_A C & P + L_A Q \\ L_S C & S + L_S Q \end{bmatrix}.$$

The rest of the proof is the same as that of Theorem 5.3.2 except that (5.3.22) takes the following form instead

$$\begin{cases} \dot{\xi} = A\xi + B\sigma(F(\varepsilon)\xi + \Gamma w - \Gamma \tilde{w} - F(\varepsilon)\tilde{x} + F(\varepsilon)\Pi \tilde{w} + \tilde{r}) \\ \quad + (A\Pi - \Pi S + P)w - \Pi r, \\ \dot{\tilde{x}} = (A + L_A C)\tilde{x} + (P + L_A Q)\tilde{w}, \\ \dot{\tilde{w}} = L_S C\tilde{x} + (S + L_S Q)\tilde{w}. \end{cases} \qquad (5.5.17)$$

⊠

Remark 5.5.4. *From the above proof of Theorem 5.5.2, we note that the linear state feedback law (5.5.8) interconnected with any exponentially stable observer (where x and w are replaced by their respective estimates which converge exponentially to the real x and w as $t \to \infty$) will solve the semi-global error feedback regulator problem.* ℝ

As a corollary to Theorem 5.5.2, we obtain the solvability conditions for the semi-global linear observer based output feedback restricted tracking problem.

Corollary 5.5.2. *Consider the system of a chain of integrators*

$$\begin{cases} \dot{x}_i = x_{i+1}, \quad i = 1, 2, \cdots, n-1, \\ \dot{x}_n = \sigma(u), \\ y = x_1, \end{cases} \qquad (5.5.18)$$

where σ is the same as in 5.5.1. Let a desired reference trajectory be given by $y_d \in \mathcal{C}^n$. If there exist a $T \geq 0$ and a $\delta > 0$ such that $\|y_d^{(n)}\|_{\infty,\mathrm{T}} \leq 1 - \delta$. The reference signal $y_d(t)$ can then be semi-globally tracked by y via linear observer based output feedback.

More specifically, for any given (arbitrarily large) bounded sets $\mathcal{X}_0 \subset \mathbb{R}^n$ and $\mathcal{Z}_0 \subset \mathbb{R}^n$, there exists a linear observer based output feedback law

$$\begin{cases} \dot{\hat{x}} = A\hat{x} + B\sigma(u) - L(y - C\hat{x}), \\ u = F\hat{x} + y_d^{(n)}, \end{cases} \qquad (5.5.19)$$

where

$$A = \begin{bmatrix} 0 & 1 & 0 & \cdots & 0 \\ 0 & 0 & 1 & \cdots & 0 \\ \vdots & \vdots & \vdots & \ddots & \vdots \\ 0 & 0 & 0 & \cdots & 1 \\ 0 & 0 & 0 & \cdots & 0 \end{bmatrix}, \quad B = \begin{bmatrix} 0 \\ 0 \\ \vdots \\ 0 \\ 1 \end{bmatrix},$$

such that

1. The equilibrium $(x_1, x_2, \cdots, x_n, \hat{x}) = (0, 0, \cdots, 0, 0)$ of

$$
\begin{cases}
\dot{x}_i = x_{i+1}, \ i = 1, 2, \cdots, n-1, \\
\dot{x}_n = \sigma(F\hat{x} + y_d^{(n)}), \\
\dot{\hat{x}} = A\hat{x} + B\sigma(-F\hat{x} + y_d^{(n)}) - L(y - C\hat{x})
\end{cases}
\tag{5.5.20}
$$

is locally exponentially stable with $\mathcal{X}_0 \times \mathcal{Z}_0$ contained in its basin of attraction;

2. For any $(x(0), \hat{x}(0)) \in \mathcal{X}_0 \times \mathcal{Z}_0$, the solution of the closed-loop system satisfies

$$
\lim_{t \to \infty} y(t) = y_d(t).
\tag{5.5.21}
$$

□

Proof of Corollary 5.5.2. The desired reference trajectory can be modeled as that of the following exosystem

$$
\begin{cases}
\dot{w}_i = w_{i+1}, \ w_i(0) = y_d^{(i-1)}(0), \ i = 1, 2, \cdots, n-1, \\
\dot{w}_n = y_d^{(n)}, \ w_n(0) = y_d^{(n-1)}(0),
\end{cases}
\tag{5.5.22}
$$

and hence the output feedback tracking problem can be cast into a generalized semi-global linear observer-based error feedback output regulation problem as defined in Problem 5.5.2, with $P = 0$, $\mathcal{R} = \{[0\,0\cdots 0\,r_n(t)]' \in \mathcal{C}^0 : |r_n^{(n)}| \leq 1 - \delta\}$, $W_0 = \{(y_d(0), y'_d(0), \cdots, y_d^{(n-1)}(0))\}$ and

$$
A = S = \begin{bmatrix} 0 & 1 & 0 & \cdots & 0 \\ 0 & 0 & 1 & \cdots & 0 \\ \vdots & \vdots & \vdots & \ddots & \vdots \\ 0 & 0 & 0 & \cdots & 1 \\ 0 & 0 & 0 & \cdots & 0 \end{bmatrix}, \quad C = Q = \begin{bmatrix} 1 \\ 0 \\ \vdots \\ 0 \\ 0 \end{bmatrix}'.
$$

Clearly, $\Pi = I$ and $\Gamma = 0$ solve the matrix equations (5.5.15) and hence Condition 2 of Theorem 5.5.2 is satisfied trivially with $\tilde{r} = y_d^{(n)}$. We note that Condition 1 is not satisfied since the pair

$$
\left([C \ \ Q], \begin{bmatrix} A & P \\ 0 & S \end{bmatrix}\right)
$$

is not detectable. From the proof of Theorem 5.5.2, we note that the linear state feedback law (5.5.8) interconnected with any stable observer will solve the semi-global error feedback output regulation problem (see Remark 5.5.4). We know the disturbance $w = \begin{bmatrix} y_d & \dot{y}_d & \cdots & y_d^{(n-1)} \end{bmatrix}$ exactly and hence we only need an asymptotic estimate of the state x. To this end, we build a state estimator to estimate the state x using the measurement $y = e + y_d$. This estimator takes the form

$$\dot{\hat{x}} = A\hat{x} + B\sigma(u) - L(y - C\hat{x}), \qquad (5.5.23)$$

where L is such that the matrix $(A + LC)$ is asymptotically stable. We then implement the state feedback law as given in Corollary 5.5.1 with the estimated state $\hat{x}$ and obtain the following output feedback law that solves the output feedback restricted tracking problem

$$\begin{cases} \dot{\hat{x}} = A\hat{x} + B\sigma(u) - L(y - C\hat{x}), \\ u = F\hat{x} + y_d^{(n)}. \end{cases} \qquad (5.5.24)$$

This completes the proof of Corollary 5.5.2. ⊠

5.6. Concluding Remarks

In this chapter we have studied the semi-global output regulation problems for linear systems, both continuous-time and discrete-time, subject to actuator saturation and identified sets of solvability conditions for these problems. Under these solvability conditions, we show how low gain feedback design techniques as developed in Chapter 2 can be utilized to construct linear feedback laws that solve these problems. The low-and-high gain feedback design techniques of Chapter 4 can also be utilized to construct feedback laws that solve these problems. As shown in [51], the low-and-high gain feedback laws result in closed-loop systems that have much better transience performance.

Finally, in an effort to broaden the class of disturbance and reference signals, we formulated and solved the generalized semi-global linear feedback output regulation problems, for which an external driving signal to the exosystem is included. These generalized semi-global output regulation problems were shown to include the so-called semi-global restricted tracking problems as special cases.

Chapter 6

Semi-Global Almost Disturbance Decoupling with Stability for Linear Systems with Saturating Actuators

6.1. Introduction

The problem of disturbance decoupling or almost disturbance decoupling has a vast history behind it, occupying a central part of classical as well as modern control theory. Several important problems, such as robust control, decentralized control, non-interacting control, model reference or tracking control can all be recast as an almost disturbance decoupling problem. Regardless of from where the problem arises, the basic almost disturbance decoupling problem is to find an output feedback control law such that in the closed-loop system the disturbances are quenched, say in the L_p-gain sense, up to any pre-specified degree of accuracy while maintaining internal stability. In the linear system setting, the above problem is labeled by Willems [124,125] as ADDPMS (the almost disturbance decoupling problem with measurement feedback and internal stability). In the case that, instead of a measurement feedback, a state feedback is used, the above problem is termed as ADDPS (the almost disturbance decoupling problem with internal stability).

133

The ADDPMS for linear systems is now very well understood (see for example, [82,122] and Chapter 11). The notion of ADDPMS was also extended to some nonlinear systems having in global normal form (see Chapter 11 and the references therein). Equipped with the low-and-high gain feedback design techniques of Chapter 4, in this chapter, we make an attempt to extend this notion to linear systems subject to actuator saturation.

The recent renewed interest in the linear system subject to actuator saturation is mainly due to the wide recognition of the inherent constraints on the control input. This renewed interest has led to many interesting results on linear systems subject to actuator saturation. Most of these results, however, pertain only to the issues related to global and semi-global *internal* stabilization and output regulation of such a class of systems (see Chapters 3, 4 and 5 and the references therein). As a natural development over internal stabilization, Liu, Chitour and Sontag [71] recently studied the input/output stabilizability of linear systems subject to actuator saturation. Their results show that under the assumptions that all the poles of the open-loop system are in the closed left-half plane with those on the $j\omega$-axis having Jordan blocks of size one and the system in the absence of actuator saturation is stabilizable and detectable, the linear system subject to actuator saturation is simultaneously finite gain L_p-stabilizable and globally asymptotically stabilizable.

In this chapter, we exploit the low-and-high gain design technique of Chapter 4 to study the almost disturbance decoupling problem with internal stability for linear systems subject to actuator saturation and input-additive disturbance. Our main results show that the problem of semi-global almost disturbance decoupling with local asymptotic stability (semi-global ADDP/LAS) is always solvable as long as the system in the absence of actuator saturation is stabilizable, no matter where the poles of the open-loop system are, and the locations of these poles play a role only when we need to solve the problem of semi-global almost disturbance decoupling with semi-global asymptotic stability (semi-global ADDP/SGAS) or the problem of semi-global almost disturbance decoupling with global asymptotic stability (semi-global ADDP/GAS). Here by semi-global almost disturbance decoupling we mean that the disturbances are bounded either in magnitude or in energy by any *a priori* given (arbitrarily large) bounded number, and by semi-global asymptotic stability we actually mean semi-global asymptotic stabilization in the sense of Chapter 3.

As will become clear in the next section when we precisely define it, almost disturbance decoupling is a problem of finite gain L_p-stabilization with the extra requirement that the L_p-gain is smaller or equal to any pre-specified value. Some other related work can be found in [80,105] and the references therein.

The remainder of the chapter is organized as follows. In Section 6.2, we define the problems of semi-global ADDP/LAS, semi-global ADDP/SGAS and semi-global ADDP/GAS. Section 6.3 contains the solutions to these three problems. Some brief concluding remarks are made in Section 6.4.

6.2. Problem Statement

Consider a linear system subject to actuator saturation and input-additive disturbance d,

$$\dot{x} = Ax + B\sigma(u + d), \quad x \in \mathbf{R}^n, u \in \mathbf{R}^m, d \in \mathbf{R}^m, \qquad (6.2.1)$$

where (A, B) is stabilizable, and $\sigma : \mathbf{R}^m \to \mathbf{R}^m$ is a class S_5 saturation function as defined in Section 4.2, i.e., $\sigma \in S_5(\Delta, 0, \delta)$ for some known constants $\Delta > 0$ and $\delta > 0$. For a detailed description of class S_5 saturation function, refer to Section 4.2.

Before we state the problem to be solved in this chapter, we need the follow preliminary definition.

Definition 6.2.1. *Let $D > 0$. For any $p = [1, \infty]$, we define the set $L_p^n(D)$ as*

$$L_p^n(D) = \left\{ x \in L_p^n : \|x\|_{L_2} \le D \text{ or } \|x\|_{L_\infty} \le D \right\}. \qquad (6.2.2)$$

▣

Clearly, $L_p^n(D)$ is the set of all L_p^n signals that are bounded either in magnitude or in energy by the constant D.

Problem 6.2.1. *Given a $p \in [1, \infty]$, the semi-global almost disturbance decoupling problem with local asymptotic stability via state feedback (semi-global ADDP/LAS) is defined as follows. For any a priori given (arbitrarily large) $D > 0$ and (arbitrarily small) $\gamma > 0$, find, if possible, a state feedback law $u = F(x; D, \gamma)$ such that, for any $\sigma \in S_5(\Delta, 0, \delta)$,*

1. *In the absence of the disturbance d, the equilibrium $x = 0$ of the closed-loop system is locally asymptotically stable;*

2. *The closed-loop system is semi-global finite gain L_p-stable and its L_p-gain from d to x is less than or equal to γ, i.e., for $x(0) = 0$,*

$$\|x\|_{L_p} \le \gamma \|d\|_{L_p}, \quad \forall d \in L_p^m(D). \qquad (6.2.3)$$

▣

Problem 6.2.2. *Given a* $p \in [1, \infty]$, *the semi-global almost disturbance decoupling problem with semi-global asymptotic stability via state feedback (semi-global ADDP/SGAS) is defined as follows. For any a priori given (arbitrarily large)* $D > 0$, *(arbitrarily large) bounded set* $\mathcal{X} \subset \mathbb{R}^n$ *and (arbitrarily small)* $\gamma > 0$, *find, if possible, a state feedback law* $u = F(x; \mathcal{X}, D, \gamma)$ *such that, for any* $\sigma \in \mathcal{S}_5(\Delta, 0, \delta)$,

1. *In the absence of the disturbance* d, *the equilibrium* $x = 0$ *of the closed-loop system is locally asymptotically stable with* $\mathcal{X}$ *contained in its basin of attraction;*

2. *The closed-loop system is semi-global finite gain* L_p-*stable and its* L_p-*gain from* d *to* x *is less than or equal to* γ, *i.e., for* $x(0) = 0$,

$$\|x\|_{L_p} \leq \gamma \|d\|_{L_p}, \quad \forall d \in L_p^m(D). \tag{6.2.4}$$

P

Problem 6.2.3. *Given a* $p \in [1, \infty]$, *the semi-global almost disturbance decoupling problem with global asymptotic stability via state feedback (semi-global ADDP/GAS) is defined as follows. For any a priori given (arbitrarily large)* $D > 0$ *and (arbitrarily small)* $\gamma > 0$, *find, if possible, a state feedback law* $u = F(x; D, \gamma)$ *such that, for any* $\sigma \in \mathcal{S}_5(\Delta, 0, \delta)$,

1. *In the absence of the disturbance* d, *the equilibrium* $x = 0$ *of the closed-loop system is globally asymptotically stable;*

2. *The closed-loop system is global finite gain* L_p-*stable and its* L_p-*gain from* d *to* x *is less than or equal to* γ, *i.e., for* $x(0) = 0$,

$$\|x\|_{L_p} \leq \gamma \|d\|_{L_p}, \quad \forall d \in L_p^m. \tag{6.2.5}$$

P

We will also refer to the above three problems collectively as the problems of semi-global almost disturbance decoupling problems with stability (semi-global ADDPS). Also, when (6.2.3), (6.2.4) or (6.2.5) holds for all any $d \in L_p^m$ instead of $d \in L_p^m(D)$, the semi-global ADDPS become their respective global ADDPS. Furthermore if the L_p norm needs to be specify, the problems are referred to as semi-global or global L_p-ADDPS.

The objective of this chapter is to identify the conditions under which the above three problems are solvable and to utilize the low-and-high gain feedback design techniques of Chapter 4 to construct feedback laws that actually solve these problems.

In the next chapter when we introduce some techniques for scheduling the low gain and high gain parameters, we will show how the high gain parameter can be scheduled as a function of the state to solve the global ADDPS.

6.3. Solutions of Semi-Global Almost Disturbance Decoupling Problems with Stability

The main results of this chapter concern the solvability conditions of the semi-global ADDPS and are given in the following three theorems. The proofs of these theorems provide explicit construction of feedback laws that solve these problems.

Theorem 6.3.1. *Consider the system (6.2.1). The semi-global ADDP/LAS is solvable for any $p \in (1, \infty]$ by linear feedback.* ⊺

Theorem 6.3.2. *Consider the system (6.2.1) and assume that all the eigenvalues of A are in the closed left-half plane. Then, the semi-global ADDP/SGAS is solvable for any $p \in (1, \infty]$ by linear feedback.* ⊺

Theorem 6.3.3. *Consider the system (6.2.1) and assume that all the eigenvalues of A are in the open left-half plane. Then, the global ADDP/GAS is solvable for any $p \in (1, \infty]$ by linear feedback.* ⊺

Remark 6.3.1. *Although the semi-global ADDPS can be solved by linear feedback only for $p \in (1, \infty]$, the weaker problem of semi-global finite gain L_p-stabilization can be achieved for any $p \in [1, \infty]$ by linear feedback. Remark 6.3.2 shows how this can be done.* ⓡ

Proof of Theorem 6.3.1. We prove this theorem by first explicitly constructing a family of state feedback laws $u = F(\rho)x$, parameterized in ρ, and then showing that, for any given $\gamma > 0$ and any given $D > 0$, there exists a ρ^* such that, for each $\rho \geq \rho^*$, the closed-loop system has the two properties in the statement of Problem 6.2.1. This family of state feedback laws takes the form of

$$u = -(1 + \rho)B'Px, \quad \rho \geq 0, \tag{6.3.1}$$

where P is the unique positive definite solution to the following algebraic Riccati equation (ARE),

$$A'P + PA - 2PBB'P + Q = 0, \tag{6.3.2}$$

and where Q is any positive definite matrix.

With this family of state feedback laws, the closed-loop system takes the form of

$$\dot{x} = Ax + B\sigma(-(1+\rho)B'Px + d). \qquad (6.3.3)$$

We now pick the Lyapunov function

$$V = x'Px. \qquad (6.3.4)$$

Let c be such that $x \in L_V(c)$ implies that $|B'Px| \leq \Delta$.

We first consider the first property, local asymptotic stability when $d \equiv 0$. In this case, the evaluation of the derivative of V along the trajectories of the closed-loop system, using Remark 4.2.1, gives that for all $x \in L_V(c)$,

$$\begin{aligned}
\dot{V} &= -x'Qx + 2x'PB[\sigma(-(1+\rho)B'Px) + B'Px] \\
&= -x'Qx - 2\sum_{i=1}^{m} v_i[\sigma_i((1+\rho)v_i) - \mathrm{sat}_\Delta(v_i)] \\
&\leq -x'Qx, \qquad (6.3.5)
\end{aligned}$$

where we have defined $v \in \mathbb{R}^m$ by $v = -B'Px$. This shows that the equilibrium $x = 0$ of the closed-loop system (6.3.3) with $d \equiv 0$ is locally asymptotically stable for all $\rho \geq 0$.

It remains to show that there exists a $\rho^* > 0$ such that, for all $\rho \geq \rho^*$, property (6.2.3) also holds. To this end, we evaluate the derivative of this V along the trajectories of the system (6.3.3) yielding that for all $x \in L_V(c)$,

$$\begin{aligned}
\dot{V} &= -x'Qx + 2x'PB[\sigma(-(1+\rho)B'Px + d) + B'Px] \\
&= -x'Qx - 2\sum_{i=1}^{m} v_i[\sigma_i((1+\rho)v_i + d_i) - \mathrm{sat}_\Delta(v_i)], \qquad (6.3.6)
\end{aligned}$$

where d_i is the ith element of d.

Recalling Remark 4.2.1 on class S_5 saturation functions, we have that,

$$|\rho v_i| \geq |d_i| \implies -v_i[\sigma_i((1+\rho)v_i + d_i) - \mathrm{sat}_\Delta(v_i)] \leq 0,$$

and

$$\begin{aligned}
|\rho v_i| \leq |d_i| \implies &-v_i[\sigma_i((1+\rho)v_i + d_i) - \sigma_i(v_i) + \sigma_i(v_i) - \mathrm{sat}_\Delta(v_i)] \\
&\leq \frac{|d_i|}{\rho}|\sigma_i((1+\rho)v_i + d_i) - \sigma_i(v_i)| \\
&\leq \frac{2}{\rho}\delta d_i^2.
\end{aligned}$$

Hence, we conclude that

$$\dot{V} \leq -\alpha_0 V + \frac{\beta_0}{\rho}|d|^2, \tag{6.3.7}$$

for some positive constants α_0 and β_0 independent from ρ.

In the case that d is such that $\|d\|_{L_2} \leq D$, choosing

$$\rho_{1a}^* = \frac{\beta_0 D^2}{c} \tag{6.3.8}$$

and integrating both sides of (6.3.7) from 0 to t, we have that, for all $\rho \geq \rho_{1a}^*$,

$$V(x) \leq c, \quad x(0) = 0, \tag{6.3.9}$$

and hence trajectories starting from $x(0) = 0$ will remain in $L_V(c)$ forever.

In the case that d is such that $\|d\|_{L_\infty} \leq D$, choosing

$$\rho_{1b}^* = \frac{\beta_0 D^2}{\alpha_0 c}, \tag{6.3.10}$$

we have that, for all $\rho \geq \rho_{1b}^*$,

$$V(x) \leq 0, \quad \forall x \notin L_V(c), \tag{6.3.11}$$

which shows that $L_V(c)$ is an invariant set and trajectories starting from inside it will remain in it forever.

In conclusion, we have shown that for any $\rho \geq \rho_1^* = \max\{\rho_{1a}^*, \rho_{1b}^*\}$,

$$x \in L_V(c), \quad \forall d \in L_p^m(D) \text{ and } x(0) = 0. \tag{6.3.12}$$

Remark 6.3.2. *For $x \in L_V(c)$, we rewrite the closed-loop system equation (6.3.3) as,*

$$\dot{x} = Ax + B\sigma(-(1+\rho)B'Px) + B[\sigma(-(1+\rho)B'Px + d) - \sigma(-(1+\rho).B'Px)] \tag{6.3.13}$$

Using (6.3.5), we have

$$\dot{V} \leq -x'Qx + 2\delta|x|\|PB\||d|. \tag{6.3.14}$$

Letting $W = V^{\frac{1}{2}}$, we have

$$\dot{W} \leq -\pi W + \kappa|d|, \tag{6.3.15}$$

from which we conclude by standard comparison theorems [120] that, for any $p \in [1, \infty]$,

$$\|x\|_{L_p} \leq \gamma_p \|d\|_{L_p}, \quad \forall d \in L_p^m(D), \tag{6.3.16}$$

for some constant $\gamma_p > 0$. This shows that, unlike semi-global ADDP/LAS which is solvable by linear feedback only for $p \in (1, \infty]$, semi-global finite gain L_p-stabilization for is solvable for all $p \in [1, \infty]$. Ⓡ

To complete the proof of the property (6.2.3), we consider only $\rho \geq \rho_1^*$, in which case, all the trajectories starting from $x(0)$ will remain in $L_V(c)$ forever. Again using Remark 4.2.1 on the property of class $\mathcal{S}_5$ saturation functions, we have,

$$|\rho v_i| \geq |d_i| \implies -v_i[\sigma_i((1+\rho)v_i + d_i) - \text{sat}_\Delta(v_i)] \leq 0,$$

and

$$|\rho v_i| \leq |d_i| \implies -v_i[\sigma_i((1+\rho)v_i + d_i) - \sigma_i(v_i) + \sigma_i(v_i) - \text{sat}_\Delta(v_i)]$$

$$\leq \frac{|d_i|^\theta}{\rho^\theta}|v_i|^{1-\theta}|\sigma_i((1+\rho)v_i + d_i) - \sigma_i(v_i)|$$

$$\leq \frac{2}{\rho^\theta}|B'P|^{1-\theta}\delta|x|^{1-\theta}|d_i|^{1+\theta},$$

where $\theta \in (0,1)$ is such that $1 + \theta \leq p$.

It then follows from (6.3.6) that

$$\dot{V} \leq -\alpha_1 V + \frac{1}{\rho^\theta}\beta_1 V^{\frac{1-\theta}{2}}|d|^{1+\theta}, \qquad (6.3.17)$$

for some positive constants α_1 and β_1 independent of ρ.

Letting $W = V^{\frac{1+\theta}{2}}$, it follows from (6.3.17) that

$$\dot{V} = \frac{2}{1+\theta}W^{\frac{1-\theta}{1+\theta}}\dot{W} \leq -\alpha_1 W^{\frac{2}{1+\theta}} + \frac{1}{\rho^\theta}\beta_1 W^{\frac{1-\theta}{1+\theta}}|d|^{1+\theta}, \qquad (6.3.18)$$

which, in turn, shows that, for $x \neq 0$,

$$\dot{W} \leq -\alpha_2 W + \frac{1}{\rho^\theta}\beta_2|d|^{1+\theta}, \qquad (6.3.19)$$

for some positive constants α_2 and β_2 independent of ρ. As will be seen shortly, we work with W because its behavior can be compared to that of a stable linear system through standard comparison theorems [120]. A similar reasoning was also used in [119, pp. 286-289].

Recall that $d \in L_p^m$ for $p \in (1, \infty]$. This means that, if $p \in (1, \infty)$,

$$\|d\|_{L_p} = \left(\int_0^\infty |d(t)|^p dt\right)^{\frac{1}{p}} < \infty, \qquad (6.3.20)$$

or, if $p = \infty$,

$$\|d\|_{L_\infty} = \text{ess sup}_{0 \leq t < \infty}|d| < \infty. \qquad (6.3.21)$$

It is then clear that $|d|^{1+\theta} \in L_{\frac{p}{1+\theta}}$. Furthermore, if $p \in (1, \infty)$,

$$\||d|^{1+\theta}\|_{L_{\frac{p}{1+\theta}}} = \left(\int_0^\infty (|d|^{1+\theta})^{\frac{p}{1+\theta}} dt\right)^{\frac{1+\theta}{p}} = \|d\|_{L_p}^{1+\theta}, \qquad (6.3.22)$$

or, if $p = \infty$,

$$\left\| |d|^{1+\theta} \right\|_{L_\infty} = \text{ess sup}_{0 \leq t < \infty} |d|^{1+\theta} = \|d\|_{L_\infty}^{1+\theta}. \tag{6.3.23}$$

Using standard comparison theorems on (6.3.19), we easily get that $W \in L_{\frac{p}{1+\theta}}$, and furthermore that

$$\|W\|_{L_{\frac{p}{1+\theta}}} \leq \frac{\beta_2}{\rho^\delta \alpha_2} \left\| |d|^{1+\theta} \right\|_{L_{\frac{p}{1+\theta}}} = \frac{\beta_2}{\rho^\theta \alpha_2} \|d\|_{L_p}^{1+\theta}. \tag{6.3.24}$$

Finally, recalling that $W = V^{\frac{1+\delta}{2}}$, we have

$$|x| \leq \alpha_3 W^{\frac{1}{1+\theta}}, \tag{6.3.25}$$

for some positive constant α_3 independent of ρ. Hence, we have

$$\|x\|_{L_p} = \left(\int_0^\infty |x|^p dt \right)^{\frac{1}{p}} \leq \alpha_3 \left(\int_0^\infty W^{\frac{p}{1+\theta}} dt \right)^{\frac{1}{p}} = \alpha_3 \|W\|_{L_{\frac{p}{1+\theta}}}^{\frac{1}{1+\theta}}$$

$$\leq \alpha_3 \left(\frac{\beta_2}{\rho^\theta \alpha_2} \right)^{\frac{1}{1+\theta}} \|d\|_{L_p}, \tag{6.3.26}$$

which shows that

$$\|x\|_{L_p} \leq \frac{\alpha_4}{\rho^{\frac{\theta}{1+\theta}}} \|d\|_{L_p}, \tag{6.3.27}$$

for some positive constant α_4 independent from ρ. This shows the finite gain L_p-stability of the closed-loop system.

Choosing

$$\rho^* = \max \left\{ \rho_1^*, \left(\frac{\alpha_4}{\gamma} \right)^{1+\frac{1}{\theta}} \right\},$$

we have that

$$\|x\|_{L_p} \leq \gamma \|d\|_{L_p}, \quad \forall d \in L_p^m(D) \text{ and } \rho \geq \rho^*.$$

This completes our proof. ⊠

Proof of Theorem 6.3.2. We prove this theorem by first explicitly constructing a family of state feedback laws $u = F(\varepsilon, \rho)x$, parameterized in ε and ρ, and then showing that, for any *a priori* given (arbitrarily large) bounded set $\mathcal{X}$ and any *a priori* given (arbitrarily small) number $\gamma > 0$, there exists an $\varepsilon^* > 0$ and for each $\varepsilon \in (0, \varepsilon^*]$, there exists a $\rho^*(\varepsilon) > 0$ such that, for each $p \in (1, \infty]$, each $D > 0$ and each $\rho \geq \rho^*(\varepsilon)$, $\varepsilon \in (0, \varepsilon^*]$, the two properties in the statement of Problem 6.2.2 hold. This family of state feedback laws takes the form of

$$u = -(1 + \rho)B'P(\varepsilon)x, \quad \varepsilon > 0, \rho \geq 0, \tag{6.3.28}$$

where $P(\varepsilon)$ is the unique positive definite solution to the following algebraic Riccati equation (ARE),

$$A'P + PA - 2PBB'P + \varepsilon Q = 0, \qquad (6.3.29)$$

where Q is any positive definite matrix. We recall here that this family of feedback laws is the ARE based low-and-high gain feedback laws of Chapter 4.

With the family of state feedback laws (6.3.28), the closed-loop system takes the form of

$$\dot{x} = Ax + B\sigma(-(1 + \rho)B'P(\varepsilon)x + d). \qquad (6.3.30)$$

We now pick the Lyapunov function

$$V = x'P(\varepsilon)x \qquad (6.3.31)$$

and let $c > 0$ be such that

$$c \geq \sup_{x \in \mathcal{X}, \varepsilon \in (0,1]} x'P(\varepsilon)x. \qquad (6.3.32)$$

Such a c exists since $\lim_{\varepsilon \to 0} P(\varepsilon) = 0$ by Lemma 2.2.6 and $\mathcal{X}$ is bounded. Let $\varepsilon^* \in (0,1]$ be such that, for each $\varepsilon \in (0,\varepsilon^*]$, $x \in L_V(c)$ implies that $|B'P(\varepsilon)x|_\infty \leq \Delta$, where the level set $L_V(c)$ is defined as $L_V(c) = \{x \in \mathbf{R}^n : V(x) \leq c\}$. The existence of such an ε^* is again due to the fact that $\lim_{\varepsilon \to 0} P(\varepsilon) = 0$.

The evaluation of the derivative of V along the trajectories of the closed-loop system in the absence of d, using Remark 4.2.1, shows that for all $x \in L_V(c)$,

$$\begin{aligned}
\dot{V} &= -\varepsilon x'Qx + 2x'P(\varepsilon)B[\sigma(-(1+\rho)B'P(\varepsilon)x) + B'P(\varepsilon)x] \\
&= -\varepsilon x'Qx - 2\sum_{i=1}^{m} v_i[\sigma_i((1+\rho)v_i) - \text{sat}_\Delta(v_i)] \\
&\leq -\varepsilon x'Qx, \qquad (6.3.33)
\end{aligned}$$

where we have defined $v \in \mathbf{R}^m$ by $v = -B'P(\varepsilon)x$.

Inequality (6.3.33) shows that, for all $\varepsilon \in (0,\varepsilon^*]$ and $\rho \geq 0$ the equilibrium $x = 0$ of the closed-loop system (6.3.30) with $d \equiv 0$ is locally asymptotically stable with $\mathcal{X} \subset L_V(c)$ contained in its basin of attraction.

It remains to show that for each $\varepsilon \in (0,\varepsilon^*]$, there exists a $\rho^*(\varepsilon) > 0$ such that, for all $\rho \geq \rho^*(\varepsilon)$, $\varepsilon \in (0,\varepsilon^*]$, the closed-loop system has property (6.2.4). This can be shown in a similar way as we did in the proof of Theorem 6.3.1. ⊠

Proof of Theorem 6.3.3. Again, we prove this theorem by first explicitly constructing a family of state feedback laws $u = F(\rho)x$, parameterized in ρ, and then showing that, for any given $\eta > 0$, there exists a ρ^* such that, for all $\rho > \rho^*$ and for each $p \in (1, \infty]$ and each $D > 0$, the two properties in the statement of Problem 6.2.3 hold. This family of state feedback laws takes the form of

$$u = -\rho B' P x, \quad \rho \geq 0, \tag{6.3.34}$$

where P is the unique positive definite solution of the following Lyapunov equation,

$$A'P + PA + Q = 0, \tag{6.3.35}$$

and where Q is any positive definite matrix.

With this family of state feedback laws, the closed-loop system takes the form of

$$\dot{x} = Ax + B\sigma(-\rho B' P x + d). \tag{6.3.36}$$

We now pick the Lyapunov function

$$V = x' P x. \tag{6.3.37}$$

The evaluation of this Lyapunov function along the trajectories of the closed-loop system in the absence of the disturbance d shows that

$$
\begin{aligned}
\dot{V} &= -x'Qx + 2x'PB\sigma(-\rho B'Px) \\
&= -x'Qx - 2\sum_{i=1}^{m} v_i \sigma_i(\rho v_i) \\
&\leq -x'Qx,
\end{aligned}
\tag{6.3.38}
$$

where we have defined $v \in \mathbb{R}^m$ by $v = -B'Px$. This shows that the equilibrium $x = 0$ of the closed-loop system (6.3.36) with $d \equiv 0$ is globally asymptotically stable.

It remains to show that there exists a $\rho^* > 0$ such that, for all $\rho \geq \rho^*$, the closed-loop system has property (6.2.5). To this end, we evaluate the derivative of this V along the trajectories of the system (6.3.36) yielding that,

$$
\begin{aligned}
\dot{V} &= -x'Qx + 2x'PB\sigma(-\rho B'Px + d) \\
&= -x'Qx - 2\sum_{i=1}^{m} v_i \sigma_i(\rho v_i + d_i).
\end{aligned}
\tag{6.3.39}
$$

Recalling Remark 4.2.1, we have that,

$$|\rho v_i| \geq |d_i| \implies -v_i \sigma_i(\rho v_i + d_i) \leq 0,$$

and

$$|\rho v_i| \leq |d_i| \implies -v_i\sigma_i(\rho v_i + d_i) = -v_i[\sigma_i(\rho v_i + d_i) - \sigma_i(0)]$$

$$\leq \frac{|d_i|^\theta}{\rho^\theta}|v_i|^{1-\theta}[\sigma_i(\rho v_i + d_i) - \sigma_i(0)]$$

$$\leq \frac{2\delta}{\rho^\delta}|B'P|^{1-\theta}|x|^{1-\theta}|d_i|^{1+\theta},$$

where $\delta \in (0,1)$ is such that $1 + \theta \leq p$.

Hence, we conclude that

$$\dot{V} \leq -\alpha_1 V + \frac{1}{\rho^\theta}\beta_1 V^{\frac{1-\theta}{2}}|d|^{1+\theta}, \tag{6.3.40}$$

for some positive constants α_1 and β_1 independent from ρ.

Noting that (6.3.40) is identical to (6.3.17) in the proof of Theorem 6.3.1, the rest of the proof follows the same way as in the proof of Theorem 6.3.1. ⊠

Finally, we conclude this section with a remark regarding Theorem 6.3.3.

Remark 6.3.3. *It was shown in [71] that when the open loop system has only simple $j\omega$ poles and the matrix A is, without loss of generality, skew-symmetric, the state feedback $u = -\rho B'x$ achieves global asymptotic stabilization in the absence of the disturbance d and and finite gain L_p-stabilization for any $\rho > 0$. One might naturally wonder if the same class of state feedback laws would achieve semi-global ADDP/GAS global asymptotic stability as $\rho \to \infty$. The following example shows that, in general, this is not the case.*   ℝ

Example 6.3.1. Consider the following linear system subject to actuator saturation,

$$\dot{x} = \begin{bmatrix} 0 & 1 \\ -1 & 0 \end{bmatrix} x + \begin{bmatrix} 1 \\ 1 \end{bmatrix} \text{sat}(u + d), \tag{6.3.41}$$

where $\text{sat}(u) = \text{sign}(t)\min\{|u|, 1\}$ and $|d| \leq 1/5$. The open loop system has two poles at $\pm j$. Pick the family of state feedback laws as

$$u = -\rho B'x = -\rho[1 \quad 1]x, \quad \rho > 1. \tag{6.3.42}$$

Assuming that the saturation element is nonexistent in the closed-loop system, we calculate the impulse response from d to u as

$$h(t) = -\frac{\rho\left(\rho + \sqrt{\rho^2 - 1}\right)}{\sqrt{\rho^2 - 1}}e^{-\left(\rho + \sqrt{\rho^2 - 1}\right)t} - \frac{\rho\left(-\rho + \sqrt{\rho^2 - 1}\right)}{\sqrt{\rho^2 - 1}}e^{-\left(\rho - \sqrt{\rho^2 - 1}\right)t}.$$

It can be shown that

$$\int_0^\infty |h(t)|dt \leq 4,$$

which, in turn, shows that $\|u+d\|_{L_\infty} \leq 1$ and hence the closed-loop system will operate linearly even in the presence of the saturation element. For the linear closed-loop system, the transfer function from d to x is given by

$$H(s) = \frac{1}{s^2 + 2\rho s + 1} \begin{bmatrix} s+1 \\ s-1 \end{bmatrix}.$$

Hence

$$H(0) = \begin{bmatrix} 1 \\ -1 \end{bmatrix},$$

which shows that for a constant disturbance d, $|d| \leq 1/5$, the steady state of the state will remain a constant of the same magnitude. $\boxdot$

6.4. Concluding Remarks

We have considered the problem of semi-global almost disturbance decoupling with stability (semi-global ADDPS) for linear systems subject to actuator saturation and input-additive disturbance via linear static state feedback. Using the low-and-high gain design technique of Chapter 4, we established that semi-global disturbance decoupling problem with local asymptotic stability (semi-global ADDP/LAS) is always solvable as long as the system in the absence of actuator saturation is stabilizable, no matter where the poles of the open-loop system are, and the locations of these poles play a role only in the solution of semi-global ADDP/SGAS, and global ADDP/GAS, where semi-global or global asymptotic stability is required. In view of the results on global and semi-global stabilization, the fact that semi-global ADDP/LAS can always be solved by linear feedback no matter where the open loop poles are is rather surprising. In the next chapter when we introduce some techniques for scheduling the low gain and high gain parameters, we will show how the high gain parameter can be scheduled as a function of the state to solve the global ADDPS.

Chapter 7

Scheduling Low and High Gain Parameters — Turning Semi-Global Results into Global Ones

7.1. Introduction

In the previous Chapters we presented two types of feedback laws, low gain feedback and low-and-high gain feedback laws. These feedback laws are parameterized in a low gain parameter ε or/and a high gain parameter ρ. They were utilized to solve various control problems for linear systems subject to actuator saturation in Chapters 3-6. All these problems are of semi-global nature, in the sense that the values of low gain and high gain parameters are determined according to the *a priori* given design specifications, for example, the size of the basin of attraction, the degree of disturbance rejection and the bound on the allowable disturbances and uncertainties.

As in any parameterization, the determination of the values for the low gain and high gain parameters is of great practical importance and often difficult. One way around this difficulty is to schedule these parameters as functions of the state of the system. In doing so, we expect the above mentioned semi-global results to be upgraded into global ones.

The idea of scheduling the low gain parameter in an ARE based low gain feedback law has been exploited in several recent papers [77,104,112]. In this chapter we present two other scheduling techniques. The first of these two techniques is for scheduling the high gain parameter of the low-and-high gain feedback laws of Chapter 6 so that global, instead of semi-global, almost disturbance decoupling problems with stability for linear systems subject to actuator saturation can be solved. The second technique involves the scheduling of both low gain and high gain parameters of the low-and-high gain feedback laws of Chapter 4 so that robust global, instead of semi-global, stabilization can be achieved for linear systems subject to actuation saturation. These two scheduling techniques are presented in Sections 7.2 and 7.3 respectively. A brief concluding remark is made in Section 7.3.

7.2. Solutions of Global ADDPS for Linear Systems with Saturating Actuators

In this section we present a gain scheduling technique for the high gain parameter in the low-and-high gain feedback laws of Chapter 6. The resulting scheduled low-and-high gain feedback laws solve the global almost disturbance decoupling problems with stability for linear systems subject to actuator saturation, as given by (6.2.1) of Chapter 6. In the following two subsections, we will developed two families of scheduled low-and-high gain feedback laws to solve the global ADDP/LAS and global ADDP/SGAS respectively.

A precise definition of global and semi-global ADDPS can be found in Section 6.2. For convenience of presentation, we recall the system equation (6.2.1) as follows.

$$\dot{x} = Ax + B\sigma(u + d), \quad x \in \mathbb{R}^n, u \in \mathbb{R}^m, d \in \mathbb{R}^m, \qquad (7.2.1)$$

where (A, B) is stabilizable, and $\sigma : \mathbb{R}^m \to \mathbb{R}^m$ is a class $\mathcal{S}_5$ saturation function as defined in Section 4.2, i.e., $\sigma \in \mathcal{S}_5(\Delta, 0, \delta)$ for some known constants $\Delta > 0$ and $\delta > 0$. For a detailed description of class $\mathcal{S}_5$ saturation functions, refer to Section 4.2.

7.2.1. Solution of Global ADDP/LAS

To solve the global ADDP/LAS without any assumption on the locations of the open loop poles, we construct a family of scheduled low-and-high gain feedback laws in the following three steps.

Step 1: Low Gain State Feedback Design. The low gain state feedback law takes the form of

$$u_{\text{L}} = -B'Px, \qquad (7.2.2)$$

where P is the unique positive definite solution to the following algebraic Ricatti equation (ARE),

$$PA + A'P - PBB'P = -Q, \tag{7.2.3}$$

and where Q is any positive definite matrix.

Step 2: Scheduled High Gain State Feedback Design. We form the scheduled high gain feedback law as,

$$u_{\text{H}} = -\frac{\rho}{1 - (\kappa x'Px)^2}B'Px, \tag{7.2.4}$$

where $\kappa > 0$ and $\rho \geq 0$ are design parameters whose values are to be tuned according to the system parameters and the desired degree of disturbance rejection γ. We note here that, the above feedback gain is dependent on the state x, and as $\rho \to \infty$ or $\kappa x'Px \to 1$, the feedback gain increases to infinity. For this reason, the feedback law (7.2.4) is referred to as the scheduled high gain state feedback law. In contrast, we refer to the feedback law (7.2.2) as a low gain state feedback.

Step 3: Scheduled Low-and-High Gain State Feedback Design. We add the low gain state feedback and the scheduled high gain state feedback as obtained in the previous two steps to form the scheduled low-and-high gain state feedback law.

$$u = -B'Px - \frac{\rho}{1 - (\kappa x'Px)^2}B'Px, \quad \kappa > 0, \ \rho \geq 0. \tag{7.2.5}$$

△

The following theorem then shows that the scheduled low-and-high gain state feedback law we just constructed solves the global ADDP/LAS for the system (7.2.1) no matter where the open-loop poles are.

Theorem 7.2.1. *Consider the system (7.2.1). Then, there exists a $\kappa > 0$ such that the scheduled low-and-high gain state feedback law (7.2.5) solves the global L_2-almost disturbance decoupling problem with local asymptotic stability (global L_2-ADDP/LAS). More specifically, under this feedback law,*

1. *in the absence of the disturbance d, the equilibrium $x = 0$ of the closed-loop system is locally asymptotically stable for any $\rho \geq 0$;*

2. *for any given $\gamma > 0$, there exists a $\rho^*(\gamma) > 0$ such that, for each $\rho \geq \rho^*$, the closed-loop system is finite gain L_2-stable and its L_2-gain from d to x is less than or equal to γ, i.e.,*

$$\|x\|_{L_2} \leq \gamma\|d\|_{L_2}, \quad \forall d \in L_2^m.$$

$\boxed{\text{T}}$

Remark 7.2.1. *As will become clear from the proof below, the scheduled low-and-high gain state feedback law also guarantees a fixed basin of attraction of the locally asymptotically stable equilibrium $x = 0$ that will not shrink as the high gain parameter is increased to achieve almost disturbance decoupling required in Item 2 of the theorem.* $\boxed{\text{R}}$

Proof of Theorem 7.2.1. Under the scheduled low-and-high gain state feedback law (7.2.5), the closed-loop system takes the following form,

$$\dot{x} = Ax + B\sigma\left(-\left(1 + \frac{\rho}{1 - (\kappa x'Px)^2}\right)B'Px + d\right). \qquad (7.2.6)$$

We choose a Lyapunov function candidate, $V(x) = x'Px$ and let $\kappa > 0$ be such that $x \in L_V^o(1/\kappa)$ implies that $|B'Px|_\infty \leq \Delta$. Clearly, the right hand side of the above closed-loop system is locally Lipschitz for $x \in L_V^o(1/\kappa)$.

We now prove part 1 of the theorem: local asymptotic stability of the equilibrium $x = 0$ of the closed loop system when $d \equiv 0$. In this case, the evaluation of the derivatives of V along the trajectories of the closed loop system, using Remark 4.2.1, gives that, for all $x \in L_V^o(1/\kappa)$,

$$\dot{V} = -x'Qx + 2x'PB\left[\sigma\left(-\left(1 + \frac{\rho}{1 - (\kappa x'Px)^2}\right)B'Px\right) + B'Px\right]$$
$$\quad - x'PBB'Px$$
$$\leq -x'Qx - 2\sum_{i=1}^{m} v_i\left[\sigma_i\left(\left(1 + \frac{\rho}{1 - (\kappa x'Px)^2}\right)v_i\right) - \text{sat}_\Delta(v_i)\right]$$
$$\leq -x'Qx, \qquad (7.2.7)$$

where we have defined $v \in \mathbb{R}^m$ by $v = -B'Px$ and used v_i to denote the ith element of v.

The above shows that, for all $\rho \geq 0$, the equilibrium $x = 0$ of the closed-loop system (7.2.6) with $d \equiv 0$ is locally asymptotically stable with $L_V^o(1/\kappa)$ contained in its basin of attraction. It remains to show that there exists a $\rho^* > 0$ such that, for all $\rho \geq \rho^*$, the closed-loop system (7.2.6) is also finite gain L_2-stable with its L_2-gain from d to x less than or equal to γ. To this end, we evaluate the derivatives of V along the trajectories of the system (7.2.6) and yield that, for $x \in L_V^o(1/\kappa)$,

$$\dot{V} = -x'Qx + 2x'PB \left[\sigma \left(-\left(1 + \frac{\rho}{1 - (\kappa x'Px)^2} \right) B'Px + d \right) + B'Px \right]$$
$$-xPBB'Px$$
$$\leq -x'Qx - 2\sum_{i=1}^{m} v_i \left[\sigma_i \left(\left(1 + \frac{\rho}{1 - (\kappa x'Px)^2} \right) v_i + d_i \right) - \operatorname{sat}_\Delta(v_i) \right].$$

$$(7.2.8)$$

Recalling Remarks 4.2.1, we have,

$$\text{if } \left| \frac{\rho}{1 - (\kappa x'Px)^2} v_i \right| \geq |d_i| \Longrightarrow$$
$$-v_i \left[\sigma_i \left(\left(1 + \frac{\rho}{1 - (\kappa x'Px)^2} \right) v_i + d_i \right) - \operatorname{sat}_\Delta(v_i) \right] \leq 0,$$

and

$$\text{if } \left| \frac{\rho}{1 - (\kappa x'Px)^2} v_i \right| < |d_i| \Longrightarrow$$
$$-v_i \left[\sigma_i \left(\left(1 + \frac{\rho}{1 - (\kappa x'Px)^2} \right) v_i + d_i \right) - \operatorname{sat}_\Delta(v_i) \right]$$
$$= -v_i \left[\sigma_i \left(\left(1 + \frac{\rho}{1 - (\kappa x'Px)^2} \right) v_i + d_i \right) - \sigma(v_i) + \sigma(v_i) - \operatorname{sat}_\Delta(v_i) \right]$$
$$\leq \frac{2\delta(1 - (\kappa x'Px)^2)}{\rho} |d_i|^2,$$

where $\delta > 0$ is a Lipschitz constant of σ.

Hence, we conclude that

$$\dot{V} \leq -\lambda_{\min}(Q)x'x + \frac{4\delta(1 - (\kappa x'Px)^2)}{\rho} |d|^2, \quad d \in L_2^m, \qquad (7.2.9)$$

which implies that

$$\frac{1}{1 - (\kappa V(x))^2} \frac{\delta(\kappa V(x))}{\delta t} \leq \frac{4\delta\kappa}{\rho} |d|^2. \qquad (7.2.10)$$

Integrating both sides of the above inequality from 0 to t yields,

$$\frac{1}{2} \ln \frac{1 + \kappa V(x)}{1 - \kappa V(x)} \leq \frac{4\delta\kappa}{\rho} \|d\|_{L_2}^2 < \infty, \quad \forall d \in L_2^m \text{ and } x(0) = 0, \qquad (7.2.11)$$

which implies that trajectories starting from $x(0) = 0$ will remain in $L_V^o(1/\kappa)$ forever.

In conclusion, we have shown that

$$V(x) < 1/\kappa, \quad \forall d \in L_2^m \text{ and } x(0) = 0, \tag{7.2.12}$$

and hence (7.2.9) holds for all $d \in L_2^m$ and $x(0) = 0$. It follows from (7.2.9) that

$$\dot{V} \leq -\lambda_{\min}(Q)x'x + \frac{4\delta}{\rho}|d|^2, \quad \forall d \in L_2^m \text{ and } x(0) = 0. \tag{7.2.13}$$

By integrating both sides of the above inequality from 0 to ∞ and noting that $V(x)$ is non-negative and bounded with $V(0) = 0$, we obtain,

$$\|x\|_{L_2} \leq \sqrt{\frac{4\delta}{\rho\lambda_{\min}(Q)}}\|d\|_{L_2}, \quad \forall d \in L_2^m, \tag{7.2.14}$$

which shows the finite gain L_2-stability of the closed-loop system.

Finally, choosing

$$\rho^* = \frac{4\delta}{\gamma^2\lambda_{\min}(Q)},$$

we conclude our proof.   ⊠

7.2.2. Solution of Global ADDP/SGAS

We now construct another family of scheduled low-and-high gain feedback laws, in which the low gain feedback is also parameterized. We show that this family of feedback laws solves the global L_2-ADDP/SGAS, as long as the system in the absence of saturation is stabilizable with all its open loop poles located in the closed left-half plane.

Step 1: Low Gain State Feedback Design. The low gain state feedback law takes the form of

$$u_L = -B'P(\varepsilon)x, \tag{7.2.15}$$

where $P(\varepsilon)$ is the unique positive solution to the algebraic Ricatti equation (ARE),

$$PA + A'P - PBB'P = -\varepsilon I \quad \varepsilon \in (0, 1]. \tag{7.2.16}$$

We recall that the above feedback laws are the ARE based low gain feedback laws of Section 2.2.2

Step 2: Scheduled High Gain State Feedback Design. We form the scheduled high gain feedback law as,

$$u_H = -\frac{\rho}{1 - (\kappa x'P(\varepsilon)x)^2}B'P(\varepsilon)x, \tag{7.2.17}$$

where $\kappa > 0$ and $\rho \geq 0$ are design parameters whose values are to be tuned according to the system parameters, the *a priori* given set of initial conditions and the desired degree of disturbance rejection γ.

Step 3: Scheduled Low-and-High Gain State Feedback Design. We add the low gain state feedback and the scheduled high gain state feedback as obtained in the previous two steps to form the scheduled low-and-high gain state feedback law. More specifically, this feedback law takes the following form,

$$u = -B'Px - \frac{\rho}{1 - (\kappa x' P(\varepsilon)x)^2} B'P(\varepsilon)x, \quad \kappa > 0, \ \rho \geq 0. \qquad (7.2.18)$$

A

The following theorem then shows that the scheduled low-and-high gain state feedback law we just constructed solves the global L_2-ADDP/SGAS as long as the system in the absence of saturation is stabilizable with all its open loop poles located in the closed left-half plane.

Theorem 7.2.2. *Consider the system (7.2.1) and assume that all the eigenvalues of A are in the closed left-half plane. Then, the scheduled low-and-high gain state feedback law (7.2.18) solves the global L_2-ADDP/SGAS. More specifically, for any given (arbitrarily large) bounded set $\mathcal{X}$, there exist a $\kappa(\mathcal{X}) > 0$ and an $\varepsilon^* \in (0,1]$, such that for all $\varepsilon \in (0, \varepsilon^*]$, the closed-loop satisfies,*

1. *in the absence of the disturbance d, the equilibrium $x = 0$ of the closed-loop system is locally asymptotically stable with $\mathcal{X}$ contained in its basin of attraction for any $\rho \geq 0$;*

2. *for any given $\gamma > 0$, there exists a $\rho^*(\gamma, \varepsilon) > 0$ such that, for each $\rho \geq \rho^*$, the closed-loop system is finite gain L_2-stable and its L_2-gain from d to x is less than or equal to γ, i.e.,*

$$\|x\|_{L_2} \leq \gamma \|d\|_{L_2}, \quad \forall d \in L_2^m.$$

T

Proof of Theorem 7.2.2. Under the scheduled low-and-high gain state feedback law (7.2.18), the closed-loop system takes the following form,

$$\dot{x} = Ax + B\sigma \left(- \left(1 + \frac{\rho}{1 - (\kappa x' P(\varepsilon)x)^2} \right) B'P(\varepsilon)x + d \right). \qquad (7.2.19)$$

We choose a Lyapunov function candidate $V(x) = x'P(\varepsilon)x$ and let $\kappa > 0$ be such that

$$1/\kappa > \sup_{x \in W, \varepsilon \in (0,1]} x'P(\varepsilon)x. \tag{7.2.20}$$

Such a κ exists since $\lim_{\varepsilon \to 0} P(\varepsilon) = 0$ by Lemma 2.2.6 and $\mathcal{X}$ is bounded. Let $\varepsilon^* \in (0,1]$ be such that, for each $\varepsilon \in (0,\varepsilon^*]$, $x \in L_V^o(1/k)$ implies that $|B'P(\varepsilon)x|_\infty \leq \Delta$. The existence of such an ε^* is again due to the fact that $\lim_{\varepsilon \to 0} P(\varepsilon) = 0$. Clearly, the right hand side of the above closed-loop system is locally Lipschitz for $x \in L_V^o(1/\kappa)$.

Using the same arguments as used in the proof of Item 1 of Theorem 7.2.1, we can show that, for each $\varepsilon \in (0,\varepsilon^*]$, the equilibrium $x = 0$ of the closed-loop system in the absence of the disturbances is locally asymptotically with $L_V^o(1/\kappa)(\supset \mathcal{X})$ contained in its basin of attraction for all $\rho \geq 0$. Item 2 of the theorem can also be shown in the same way as that of Theorem 7.2.1.   ⊠

7.3. Robust Global Stabilization of Linear Systems with Saturating Actuators

In the previous section, we presented a gain scheduling technique for the high gain component of the low-and-high gain feedback laws of Chapter 6 so that global, instead of semi-global, ADDPS for linear systems subject to actuator saturation can be solved. In this section, we present another technique for scheduling both low gain and high gain components so that robust global stabilization for linear systems subject to actuator saturation, instead of semi-global stabilization as considered in Chapter 4, can be achieved by nonlinear feedback.

This section is organized as follows. In Section 7.3.1, we pose the problems to be solved in this paper. In Section 7.3.2, we present a scheduled low-and-high-gain state feedback design technique which leads to our state feedback results in Section 7.3.3. In Section 7.3.4 we give the scheduled low-and-high gain output feedback design which leads to our output feedback results of Section 7.3.5. An example is included in Section 7.3.6 to demonstrate the obtained results.

7.3.1. Problem Statement

We consider a linear system in the presence of actuator saturation and input additive disturbance and uncertainties,

$$\begin{cases} \dot{x} = Ax + B\sigma(u + g(x,t) + d(t)), \\ y = Cx, \end{cases} \tag{7.3.1}$$

where $x \in \mathbb{R}^n$ is the state, $u \in \mathbb{R}^m$ is the control input, $y \in \mathbb{R}^p$ is the measurement output, $g : \mathbb{R}^n \times \mathbb{R}_+ \to \mathbb{R}^m$ represents disturbance and (possibly time-varying) uncertainties and $d : \mathbb{R}_+ \to \mathbb{R}$ the disturbances, and finally $\sigma : \mathbb{R}^m \to \mathbb{R}^m$ is a class $\mathcal{S}_5$ saturation function as defined in Section 4.2, i.e., $\sigma \in \mathcal{S}_5(\Delta, b, \delta)$ for some known constants $\Delta > 0$ and $b \geq 0$ and some known function $\delta : \mathbb{R}^+ \to \mathbb{R}^+$. For simplicity in presentation, we assume that $b = 0$. The techniques of Chapter 4 can be used to treat the case that $b > 0$.

Regarding the triple (A, B, C) that represents the nominal linear system, we make the following assumptions.

Assumption 7.3.1. *The pair (A, B) is asymptotically null controllable with bounded controls (ANCBC), i.e.,*

1. *The eigenvalues of A are all located in the closed left-half plane;*

2. *The pair (A, B) is stabilizable.* ⊞

Regarding the uncertain element g and the disturbance d, we only require knowing an upper bound on the norm of g, and an ultimate bound on d. More specifically, we make the following assumptions.

Assumption 7.3.2. *The uncertain element $g(x, t)$ is piecewise continuous in t, locally Lipschitz in x and its norm is bounded by a known function,*

$$|g(x, t)| \leq g_0(|x|), \quad \forall (x, t) \in \mathbb{R}^n \times \mathbb{R}_+, \tag{7.3.2}$$

where the known function $g_0 : \mathbb{R}_+ \to \mathbb{R}_+$ is locally Lipschitz and satisfies $g_0(0) = 0$. We also assume, without loss of generality, that g_0 is non-decreasing. ⊞

Assumption 7.3.3. *The disturbance d is ultimately bounded by a known bound $D_0 > 0$, i.e., there exists a finite time $T_d \geq 0$ such that,*

$$|d(t)| \leq D_0, \quad \forall t \geq T_d, \tag{7.3.3}$$

where D_0 is independent of the disturbances and T_d is dependent on the disturbances. ⊞

We note here that, in comparison with Chapter 4, where d is required to be *uniformly* bounded by a known bound, Assumption 7.3.3 requires only ultimate boundedness and is much weaker.

We will be interested in finding controllers that achieve global results independent of the precise $\sigma \in \mathcal{S}_5(\Delta, 0, \delta)$ and independent of the precise g and d that satisfy Assumptions 7.3.2 and 7.3.3. To state the problems we will solve, we make the following preliminary definition.

Definition 7.3.1. *The data* $(\Delta, \delta, g_0, D_0, W_0)$ *is said to be admissible for state [output] feedback if* Δ *is a strictly positive real number,* $\delta : \mathbb{R}_+ \to \mathbb{R}_+$ *is strictly positive, locally Lipschitz and non-decreasing,* $g_0 : \mathbb{R}_+ \to \mathbb{R}_+$ *is locally Lipschitz and non-decreasing with* $g_0(0) = 0$, $D_0 \geq 0$, *and* $W_0 \subset \mathbb{R}^n \, [\mathbb{R}^{2n-p}]$ *which contains the origin as an interior point.* ▫

We will deal with both state feedback and output feedback.

Problem 7.3.1. *Given the data* $(\Delta, \delta, g_0, D_0, W_0)$, *admissible for state feedback, find a state feedback law* $u = F(x)$, *such that, for all* $\sigma \in \mathcal{S}_5(\Delta, 0, \delta)$ *and all* $g(x,t)$ *and* $d(t)$ *satisfying Assumptions 7.3.2 and 7.3.3 with* g_0 *and* D_0 *respectively, as long as either* σ *is bounded or* $T_d = 0$, *the closed-loop system satisfies*

1. *if* $D_0 = 0$, *the point* $x = 0$ *is globally asymptotically stable;*

2. *if* $D_0 > 0$, *every trajectory enters and remains in* W_0 *after some finite time.* ▣

Problem 7.3.2. *Given the data* $(\Delta, \delta, g_0, D_0, W_0)$, *admissible for output feedback, find an output feedback law of the form,*

$$\begin{cases} \dot{z} = \phi(z,y), & z \in \mathbb{R}^{n-p}, \\ u = F(z,y), \end{cases}$$

such that, for all $\sigma \in \mathcal{S}_5(\Delta, 0, \delta)$ *which are uniformly bounded over* $\mathcal{S}_5(\Delta, 0, \delta)$ *and for all* $g(x,t)$ *and* $d(t)$ *satisfying Assumptions 7.3.2 and 7.3.3 with* g_0 *and* D_0 *respectively, the closed loop system satisfies*

1. *if* $D_0 = 0$, *the point* $(x,z) = (0,0)$ *is globally asymptotically stable;*

2. *if* $D_0 > 0$, *every trajectory enters and remains in* W_0 *after some finite time.* ▣

Remark 7.3.1. *These two problems are the global counterparts of the semi-global problems formulated and solved in Chapter 4.*

Following the terminology used in Chapter 4, corresponding to specific values for g_0 *and* D_0, *the above problems can be given special names. For the case when* $g_0 \equiv 0$ *and* $D_0 = 0$, *this is called the global asymptotic stabilization problem. When* $g_0 \not\equiv 0$ *but* $D_0 = 0$, *this is called the robust global asymptotic stabilization problem. When* $g_0 \equiv 0$ *but* $D_0 > 0$, *this is called the global disturbance rejection problem. When* $g_0 \not\equiv 0$ *and* $D_0 > 0$, *this is called the robust global disturbance rejection problem. Since the choice of* $F(x)$ *depends on* g_0 *and* D_0, *the solutions to Problems 7.3.1 and 7.3.2 are automatically adapted to the appropriate special problems.* ▣

7.3.2. Scheduled Low-and-High Gain State Feedback

We construct a family of scheduled low-and-high gain state feedback laws that solves Problem 7.3.1 (the state feedback problem) as formulated in the previous subsection. The construction of the scheduled low-and-high gain state feedback is sequential. First a scheduled low gain control law [77] is designed and then a scheduled high gain control law is constructed. The construction will be carried out in the following three steps.

Step 1: Scheduled Low Gain State Feedback. We start by choosing a continuously differentiable function $Q : (0, 1] \to \mathbb{R}^{n \times n}$ such that $Q(r)$ is positive definite with $dQ(r)/dr > 0$ for each $r \in (0, 1]$ and $\lim_{r \to 0} Q(r) = 0$. A simple choice is $Q(r) = rI$. We next form the H_2 algebraic Riccati equation (ARE),

$$PA + A'P - PBB'P = -Q(r), \quad r \in (0, 1] \tag{7.3.4}$$

We have the following lemma regarding the above ARE.

Lemma 7.3.1. *Assume that (A, B) is stabilizable and A has all its eigenvalues in the closed left-half plane. Then,*

1. *for all $r \in (0, 1]$ there exists a unique matrix $P(r) > 0$ which solves the ARE (7.3.4) and is such that $A - BB'P(r)$ is an asymptotically stable matrix;*

2. $\lim_{r \to 0} P(r) = 0$;

3. *$P(r)$ is continuously differentiable and strictly increasing with respect to r, i.e., $dP(r)/dr > 0$.* ◻

Proof of Lemma 7.3.1. Items 1 and 2 were proven as part of Lemma 2.2.6 of Chapter 2. Item 3 was proven in [112] for H_∞-ARE. The continuous differentiability of $P(r)$ follows from that of the (generalized) eigenvectors of the continuously differentiable Hamilton matrix associated with the ARE (7.3.4). To show that $dP(r)/dr > 0$, we differentiate both sides of the ARE (7.3.4) and yield the following Lyapunov equation,

$$\frac{dP}{dr}(A - BB'P) + (A - BB'P)'\frac{dP}{dr} = -\frac{dQ(r)}{dr}. \tag{7.3.5}$$

The fact that $dP(r)/dr > 0$ then follows from the fact that $A - BB'P(r)$ is an asymptotically stable matrix and $dQ(r)/dr > 0$. ⊠

With the solution of the ARE (7.3.4), we construct the scheduled low gain feedback as,

$$u_{\mathrm{L}} = -B'P(R(x))x, \tag{7.3.6}$$

where

$$R(x) = \max\{r \in (0,1] : (x'P(r)x)\mathrm{tr}(B'P(r)B) \le \Delta^2\}. \tag{7.3.7}$$

For later use, we note here that,

$$\begin{aligned}
|u_{\mathrm{L}}|^2 &\le |B'P^{\frac{1}{2}}(R(x))|^2|P^{\frac{1}{2}}(R(x))x|^2 \\
&= \lambda_{\max}(B'P(R(x))B)(x'P(R(x))x) \\
&\le \mathrm{tr}(B'P(R(x))B)(x'P(R(x))x) \\
&\le \Delta^2,
\end{aligned} \tag{7.3.8}$$

and hence $|u_{\mathrm{L}}| \le \Delta$.

Regarding the function $R(x)$, the following lemma was proven in [77].

Lemma 7.3.2. *$R(x)$ is a continuous function of $x \in \mathbb{R}^n$ and is continuously differentiable in a neighborhood of any point x such that $0 < R(x) < \Delta$. Moreover, the function $B'P(R(x))x$ is globally Lipschitz.* ⊡

Finally, we note that the above feedback gain is dependent on the state x and as x increases to infinity, the gain decreases to zero. We thus refer to this feedback law as the scheduled low gain feedback law. As shown in [77], the scheduled low gain feedback renders the system (7.3.1) in the absence of uncertainties disturbance to state L_2-stable. It, however, is unable to solve our state feedback problem (Problem 7.3.1). To see this, consider the following simple system,

$$\dot{x} = 0x + \mathrm{sat}_\Delta(u + d(t)), \tag{7.3.9}$$

where $\mathrm{sat}_\Delta(\cdot)$ is standard saturation with a saturation level of Δ. The definition of $R(x)$, (7.3.7), implies that the scheduled low gain feedback law satisfies $|u_{\mathrm{L}}(x)| \le \Delta$ (see (7.3.8)). Hence any constant disturbance $d > \Delta$ will lead the closed-loop state to diverge to infinity.

Step 2: Scheduled High Gain State Feedback. We form the scheduled high gain state feedback law as,

$$u_{\mathrm{H}} = -\rho(x)B'P(R(x))x, \tag{7.3.10}$$

where

$$\rho(x) = \frac{\rho_0 \left(g_1(|x|) + D_0 + 1\right)^2 \delta(2g_0(|x|) + 2D_0 + 1)}{\mu(x)}, \qquad (7.3.11)$$

and where

- $\rho_0 \geq 0$ is a design parameter whose value is to be tuned according to the set of data admissible for state feedback;

- $g_1 : \mathbb{R}_+ \to \mathbb{R}_+$ is any locally Lipschitz function that satisfies

$$g_1(s) \geq g_0(s)/s, \quad \forall s > 0; \qquad (7.3.12)$$

- $\mu : \mathbb{R}^n \to \mathbb{R}_+$ is any strictly positive and locally Lipschitz function that satisfies

$$\mu(x) \leq \lambda_{\min}(Q(R(x))). \qquad (7.3.13)$$

We note here that the feedback gain in the above feedback law is dependent on the state x and as either ρ_0 or x increases to infinity so does the feedback gain. For this reason, the above feedback law is referred to as the scheduled high gain state feedback law.

Step 3: Scheduled Low-and-High Gain State Feedback. We add the scheduled low gain state feedback and the scheduled high gain state feedback to form the scheduled low-and-high gain state feedback law as follows,

$$u = -(1 + \rho(x))B'P(R(x))x, \qquad (7.3.14)$$

where $\rho(x)$ is as given by (7.3.11). Ⓐ

7.3.3. State Feedback Results

Theorem 7.3.1. *Let Assumption 7.3.1 hold. Given the data* $(\Delta, \delta, g_0, D_0, W_0)$, *admissible for state feedback, there exists a* $\rho_0^* > 0$ *such that, for all* $\rho_0 \geq \rho_0^*$, *the scheduled low-and-high gain feedback (7.3.14) solves Problem 7.3.1.* Ⓣ

Proof of Theorem 7.3.1. Under the scheduled low-and-high gain state feedback law, the closed-loop system takes the following form,

$$\dot{x} = Ax + B\sigma(-(1 + \rho(x))B'P(R(x))x + g(x,t) + d(t)), \qquad (7.3.15)$$

whose right hand side, in view of Lemma 7.3.2 is locally Lipschitz in x and piecewise continuous in t. We will show that the above closed-loop system satisfies both Items 1 and 2 of Problem 7.3.1. We will assume that $T_d = 0$.

This is without loss of generality since if $T_d \neq 0$, the assumption that σ is bounded guarantees that the state x will remain bounded in any finite time. For this closed-loop system, we pick the Lyapunov function,

$$V = x'P(R(x))x. \tag{7.3.16}$$

By Lemma 7.3.1 and the definition of $P(r)$, i.e., (7.3.7), it is clear that the function $V : \mathbb{R}^n \to \mathbb{R}_+$ as defined above is positive definite and radially unbounded.

The evaluation of the derivative of V along the trajectories of the closed loop system, using Remark 4.2.1 and the fact that for all $x \in \mathbb{R}^n$, $|B'P(R(x))x|_\infty \leq \Delta$, yields,

$$
\begin{aligned}
\dot{V} &= -x'[Q(R(x)) + P(R(x))BB'P(R(x))]x \\
&\quad + 2x'P(R(x))B[\sigma(-(1+\rho(x))B'P(R(x))x + g(x,t) + d(t)) \\
&\quad + B'P(R(x))x] + x'\frac{dP(R(x))}{dt}x \\
&\leq -\lambda_{\min}(Q(R(x)))x'x \\
&\quad - 2\sum_{i=1}^m v_i[\sigma_i((1+\rho(x))v_i + g_i(x,t) + d_i(t)) - \mathrm{sat}_\Delta(v_i)] \\
&\quad + x'\frac{dP(R(x))}{dt}x, \tag{7.3.17}
\end{aligned}
$$

where we have defined $v \in \mathbb{R}^m$ by $v = -B'P(R(x))x$ and used v_i, g_i and d_i to denote respectively the ith element of v, $g(x,t)$ and $d(t)$.

Following the arguments used in [77], in view of the definition of $R(x)$, (7.3.7), $(x'P(R(x))x)\mathrm{tr}(B'P(R(x))B) = \Delta^2$ whenever $R(x) \neq 1$, i.e., whenever $P(R(x))$ is not a constant locally. Therefore, $x'(dP(R(x))/dt)x$ and $dV(x)/dt$ are either both zero or have opposite signs. Hence, in the case that $g_0 \equiv D_0 = 0$, we have,

$$\dot{V} < 0, \quad \forall x \neq 0, \ \forall \rho_0 \geq 0, \tag{7.3.18}$$

which in turn shows that the equilibrium $x = 0$ of the closed-loop system (7.3.15) is globally asymptotically stable with $\rho_0^* = 0$.

We next consider the case that either $g_0 \not\equiv 0$ or $D_0 \neq 0$. In this case, we need $\rho_0 > 0$.

Using the same procedure as used in Chapter 4 in deriving (4.4.19) and (4.4.20), it can be verified that,

$$-v_i[\sigma_i((1+\rho(x))v_i+g_i+d_i)-\mathrm{sat}_\Delta(v_i)] \leq \frac{2(|g_i| + D_0)^2\delta(2|g_i| + 2D_0)}{\rho(x)}. \tag{7.3.19}$$

Hence, we conclude that

$$\dot{V} \leq -\lambda_{\min}(Q(R(x)))x'x + \frac{8m[g_0^2(|x|) + D_0^2]\delta(2g_0(|x|) + 2D_0)}{\rho(x)}$$

$$+x'\frac{dP(R(x))}{dt}x$$

$$\leq \begin{cases} -\lambda_{\min}(Q(R(x)))\left(1 - \dfrac{8m}{\rho_0}\right)|x|^2 + x'\dfrac{\delta P(R(x))}{\delta t}x & \text{if } D_0 = 0, \\[3mm] -\lambda_{\min}(Q(R(x)))\left[\left(1 - \dfrac{8m}{\rho_0}\right)|x|^2 - \dfrac{8m}{\rho_0}\right] + x'\dfrac{dP(R(x))}{dt}x & \text{if } D_0 > 0. \end{cases}$$

$$(7.3.20)$$

In the case that $D_0 = 0$, by recalling the fact that $x'(dP(R(x))/dt)x$ and $dV(x)/dt$ either are both zero or have the opposite signs whenever $R(x)$ is not a constant locally, we have,

$$\dot{V} < 0, \quad \forall x \neq 0 \text{ and } \forall \rho_0 \geq \rho_0^* = 9m, \tag{7.3.21}$$

which implies that $x = 0$ is a globally asymptotically stable equilibrium point of the closed-loop system whenever $D_0 = 0$ and $\rho_0 \geq \rho_0^*$. This completes the proof of Item 1 of Problem 7.3.1.

We next proceed with the proof of Item 2 of Problem 7.3.1. To this end, let $c > 0$ be such that $L_V(c) \subset \mathcal{W}_0$. Such a $c > 0$ exists since $\mathcal{W}_0$ contains the origin as an interior point. We next pick a $\rho_0^* > 0$ such that,

$$\rho_0^* > \max\left\{16m, \frac{16m\lambda_{\max}(P(1))}{c}\right\}. \tag{7.3.22}$$

Then, we have,

$$\dot{V} < x'\frac{dP(R(x))}{dt}x, \quad \forall x \notin L_V^o(c) \text{ and } \forall \rho_0 \geq \rho_0^*. \tag{7.3.23}$$

Again resorting to the fact that $x'(dP(R(x))/dt)x$ and $dV(x)/dt$ either are both zero or have the opposite signs whenever $R(x)$ is not a constant locally, we have,

$$\dot{V} < 0, \quad \forall x \notin L_V^o(c) = \{x : V(x) < c\} \text{ and } \forall \rho_0 \geq \rho_0^*, \tag{7.3.24}$$

which shows that every trajectory will enter the set $L_V(c) \subset \mathcal{W}_0$ in a finite time and remain in it thereafter. ☒

7.3.4. Scheduled Low-and-High Gain Output Feedback

We construct a family of scheduled low-and-high gain output feedback control laws that would solve our Problem 7.3.2. This family of output feedback control laws are of reduced order observer-based type, in that a reduced order observer is used to implement the scheduled low-and-high gain state feedback laws constructed previously. In order to construct such an observer, we make the following assumption.

Assumption 7.3.4. *The linear system represented by (A,B,C) is of minimum-phase, left invertible and all the elements in the lists $\mathcal{I}_2$ and $\mathcal{I}_4$ of Morse structural invariants are 1's.*   ⊞

Remark 7.3.2. *For the definition of Morse structural invariants, see [79]. In view of Property A.1.3 of the SCB, the fact that all the elements in the list $\mathcal{I}_4$ are 1's implies that all the infinite zeros are of order one. If the system is right invertible, then $\mathcal{I}_2$ is an empty set. Hence, Assumption 7.3.4 is automatically satisfied by a system that is square invertible with all its infinite zeros being of order one. On the special coordinate basis, Assumption 7.3.4 implies that x_c is nonexistent and all integers q_i and r_i equal to one (and hence, $\bar{y} = [x_b', \, x_d']'$).*

In comparison with the state feedback result, this assumption reduces the number of required measurements by the number of invariant zeros. For a minimum-phase single input single output system with relative degree one, only one output needs to be measured.

We recall that the assumption on lists $\mathcal{I}_2$ and $\mathcal{I}_4$ of Morse structural invariants is not needed for the semi-global output feedback results of Chapter 4, where plant initial conditions are assumed to be in an a priori given bounded set and the state feedback gain is constant. As will become apparent shortly, this assumption, however, not only is essential in arriving at our global output feedback results, but also results in an infinite gain margin of our output feedback law.   ⊞

The reduced order observer based low-and-high gain output feedback law is constructed in the following three steps.

Step 1. By Assumption 7.3.4, the linear system

$$\begin{cases} \dot{x} = Ax + Bu, \\ y = Cx \end{cases} \tag{7.3.25}$$

is left invertible with all the elements in the lists $\mathcal{I}_2$ and $\mathcal{I}_4$ being all ones. By Theorem A.1.1 (SCB), there exist nonsingular state transformation and output transformation,

$$x = \Gamma_s \bar{x}, \ \ y = \Gamma_o \bar{y},$$

that put the system into the following SCB form,

$$\bar{x} = [x'_a, x'_b, x'_d]', \ x_b = [x_{b1}, x_{b2}, \cdots, x_{bp-m}]',$$

$$x_d = [x_1, x_2, \cdots, x_m]', \ x_{bi}, x_i \in \mathbf{R},$$

$$\bar{y} = [y'_b, y'_d]', \ y_b = [y_{b1}, y_{b2}, \cdots, y_{bp-m}]', \ \bar{y}_d = [y_1, y_2, \cdots, y_m]',$$

$$u = [u_1, u_2, \cdots, u_m]',$$

$$\dot{x}_a = A_{aa}x_a + L_{ab}y_b + L_{ad}y_d, \ x_a \in \mathbf{R}^{n-p}, \tag{7.3.26}$$

$$\dot{x}_{bi} = L_{bib}y_b + L_{bid}y_d, \ y_{bi} = x_{bi}, \ i = 1, 2, \cdots, p - m, \tag{7.3.27}$$

$$\dot{x}_i = (L_i + E_{id})y_d + E_{ia}x_a + E_{ib}y_b + u_i, \ y_i = x_i, \ i = 1, 2, \cdots, n. \tag{7.3.28}$$

Step 2. Let $f : \mathbf{R}_+ \to \mathbf{R}_+$ be the locally Lipschitz function defined by,

$$f(s) = [g_0^2(s) + D_0^2]\delta(2g_0(s) + 2D_0), \tag{7.3.29}$$

and let $\beta_0 : \mathbf{R}_+ \to \mathbf{R}_+$ be any locally Lipschitz function such that

$$|f(s + s_0) - f(s)| + 1 \le \beta_0(s), \ \ \forall s \in \mathbf{R}_+, \tag{7.3.30}$$

for any fixed positive scalar s_0.

Step 3. The family of the reduced observer based scheduled low-and-high output feedback control laws is then given by,

$$\begin{cases} \dot{\hat{x}}_a = A_{aa}\hat{x}_a + [L_{ab}, L_{af}]\Gamma_o^{-1}y, \\ u = -(1 + \beta(\breve{x})\rho(\breve{x}))B'P(R(\breve{x}))\breve{x}, \end{cases} \tag{7.3.31}$$

where the functions ρ and R are as defined in Section 7.3.2, $\beta(\breve{x}) = \beta_0(|\breve{x}|)$, and $\breve{x}$ is given by,

$$\breve{x} = \Gamma_s \begin{bmatrix} \hat{x}_a \\ \Gamma_o^{-1}y \end{bmatrix}. \tag{7.3.32}$$

Ⓐ

7.3.5. Output Feedback Results

Theorem 7.3.2. *Let Assumptions 1 and 7.3.4 hold. Given the data* $(\Delta, \delta, g_0, D_0, \mathcal{W}_0)$, *admissible for the output feedback problem, there exists a* $\rho_0^* > 0$ *such that, for any* $\rho_0 \ge \rho_0^*$, *the reduced order observer based low-and-high gain output feedback control law (7.3.31) solves Problem 7.3.2.* Ⓣ

Remark 7.3.3. *We note here that, in view of Theorem 7.3.2 and the definition of function* β_0 *in (7.3.30), the output feedback law (7.3.31) has an infinite gain margin. The semi-global output feedback results of Chapter 4 do not possess this property.* Ⓡ

Proof of Theorem 7.3.2. For the system (7.3.1) under the reduced order observer based scheduled low-and-high gain output feedback law (7.3.31), the closed-loop system takes the form of,

$$\dot{x} = Ax + B\sigma(u + g(x,t) + d(t)), \tag{7.3.33}$$

$$\dot{\hat{x}}_a = A_{aa}\hat{x}_a + [L_{ab}, L_{af}]\Gamma_o^{-1}y, \tag{7.3.34}$$

$$u = -(1 + \beta(\check{x})\rho(\check{x}))B'P(R(\check{x}))\check{x}. \tag{7.3.35}$$

Recall that Γ_s and Γ_o are the state and output transformations that take the system (A,B,C) into its SCB form. Partition the state $\bar{x} = \Gamma_s^{-1}x$ as,

$$\bar{x} = [x_a', x_b', x_d']', \quad x_a \in \mathbf{R}^{n-p}, \ x_b \in \mathbf{R}^{p-m}, \ \text{and} \ x_d \in \mathbf{R}^m. \tag{7.3.36}$$

We then perform a state transformation,

$$\tilde{x} = \Gamma_s[\hat{x}_a', x_b', x_d']' = \check{x}, \ e_a = x_a - \hat{x}_a. \tag{7.3.37}$$

In the new states $\tilde{x}$ and e_a, the closed-loop system (7.3.33)-(7.3.35) can be written as,

$$\dot{\tilde{x}} = A\tilde{x} + B[\sigma(u + g(\tilde{x} + \Gamma_{sa}e_a, t) + d(t))) + E_a e_a], \tag{7.3.38}$$

$$\dot{e}_a = A_{aa}e_a, \tag{7.3.39}$$

$$u = -(1 + \beta(\tilde{x})\rho(\tilde{x}))B'P(R(\tilde{x}))\tilde{x}, \tag{7.3.40}$$

where $E_a = [\, E_{1a}' \quad E_{2a}' \quad \cdots \quad E_{ma}' \,]'$ and Γ_{sa} is defined through the following partitioning,

$$\Gamma_s = [\Gamma_{sa} \quad \Gamma_{sb} \quad \Gamma_{sd}], \ \Gamma_{sa} \in \mathbf{R}^{n\times(n-p)}, \Gamma_{sb} \in \mathbf{R}^{n\times(p-m)}, \Gamma_{sd} \in \mathbf{R}^{n\times m}. \tag{7.3.41}$$

Remark 7.3.4. *Unlike its semi-global counterpart of Chapter 4, where plant initial conditions are in an a priori given bounded set and the state feedback gain is independent of the state, the state feedback gain here is a highly nonlinear function of the state itself. Assumption 7.3.4 is essential in representing the output feedback law (7.3.40) free of the state e_a of the error dynamics (7.3.39).*

®

The fact that (A, B, C) is of minimum-phase implies, by the SCB Property A.1.2, that A_{aa} is asymptotically stable. Hence, there exists a $T_a(e_a(0)) \geq 0$ such that $|e_a(t)| \leq s_0/|\Gamma_{sa}|$, for all $t \geq T_a(e_a(0))$. We recall that for any disturbance d, there exists a T_d such that $|d(t)| \leq D_0$ for all $t \geq T_d$. Here the constant β_0 is as specified in (7.3.30). In what follows, we will examine the behavior of $\tilde{x}$ for $t \geq T(e_a(0), d) = \max\{T_a(e_a(0)), T_d\}$. Since the state $\tilde{x}$ remains bounded in any finite time due to the boundedness assumption on σ, we, without loss of generality, assume throughout this proof that $T(e_a(0), d) = 0$. Consider the Lyapunov function,

$$V(\tilde{x}, e_a) = \tilde{x}' P(R(\tilde{x}))\tilde{x} + (1 + \tau^2)e_a' P_a e_a, \qquad (7.3.42)$$

where $\tau = \sqrt{\lambda_{\max}(E_a' E_a)}$ and $P_a > 0$ is such that $A_{aa}' P_a + P_a A_{aa}' = -I$.

By Lemma 7.3.1 and the definition of $R(x)$, i.e., (7.3.7), it is clear that the function $V : \mathbb{R}^n \to \mathbb{R}_+$ as defined above is positive definite and radially unbounded.

The evaluation of the derivative of V along the trajectories of (7.3.38)-(7.3.39), using Remark 4.2.1, yields,

$$
\begin{aligned}
\dot{V} =\; & -\tilde{x}'[Q(R(\tilde{x})) + P(R(\tilde{x}))BB'P(R(\tilde{x}))]\tilde{x} + 2\tilde{x}'P(R(\tilde{x}))B \\
& \times[\sigma(-(1 + \beta(\tilde{x})\rho(\tilde{x}))B'P(R(\tilde{x}))\tilde{x} + g(\tilde{x} + \Gamma_{sa}e_a, t) + d(t)) + B'P(R(\tilde{x}))\tilde{x}] \\
& + 2\tilde{x}'P(R(\tilde{x}))BE_a e_a + \tilde{x}'\frac{\delta P(R(\tilde{x}))}{\delta t}\tilde{x} - (1 + \tau^2)e_a' e_a \\
\leq\; & -\tilde{x}'Q(R(\tilde{x}))\tilde{x} + 2\tilde{x}'P(R(\tilde{x}))B \\
& \times[\sigma(-(1 + \beta(\tilde{x})\rho(\tilde{x}))B'P(R(\tilde{x}))\tilde{x} + g(\tilde{x} + \Gamma_{sa}e_a, t) + d(t)) + B'P(R(\tilde{x}))\tilde{x}] \\
& + \tilde{x}'\frac{\delta P(R(\tilde{x}))}{\delta t}\tilde{x} - e_a' e_a \\
\leq\; & -\lambda_{\min}(Q(R(x)))\tilde{x}'\tilde{x} \\
& - 2\sum_{i=1}^{m} v_i[\sigma_i((1 + \beta(\tilde{x})\rho(\tilde{x}))v_i + g_i(\tilde{x} + \Gamma_{sa}e_a, t) + d_i(t)) - \mathrm{sat}_\Delta(v_i)] \\
& + \tilde{x}'\frac{\delta P(R(\tilde{x}))}{\delta t}\tilde{x} - e_a' e_a, \qquad (7.3.43)
\end{aligned}
$$

where we have defined $v \in \mathbb{R}^m$ by $v = -B'P(R(\tilde{x}))\tilde{x}$, and used v_i, g_i and d_i to denote respectively the ith element of v, $g(\tilde{x} + \Gamma_{sa}e_a, t)$ and $d(t)$.

Since $(\delta \tilde{x}' P(R(\tilde{x}))\tilde{x})/\delta t$ and $\tilde{x}'(\delta P(R(\tilde{x}))/\delta t)\tilde{x}$ either are both zero or have opposite signs whenever $R(\tilde{x})$ is not a constant locally and $\delta(e_a' P_a e_a)/\delta t$ is always negative whenever $e_a \neq 0$, in the case that $g_0 \equiv D_0 = 0$, we have,

$$\dot{V} < 0, \quad \forall(\tilde{x}, e_a) \neq (0,0), \ \forall \rho_0 \geq 0, \qquad (7.3.44)$$

which in turn shows that the equilibrium $(\tilde{x}, e_a) = (0,0)$ of (7.3.38)-(7.3.39) is globally asymptotically stable.

We next consider the case that either $g_0 \not\equiv 0$ or $D_0 \neq 0$. In this case, we need $\rho_0 > 0$.

Using the same procedure as used in Chapter 4 in deriving (4.4.19) and (4.4.20), it can be verified that

$$-v_i[\sigma_i((1+\beta(\tilde{x})\rho(\tilde{x}))v_i + g_i + d_i) - \text{sat}_\Delta(v_i)] \le \frac{2(|g_i| + D_0)^2\delta(2|g_i| + 2D_0)}{\beta(\tilde{x})\rho(\tilde{x})}.$$

$$(7.3.45)$$

Hence, we conclude that,

$$\begin{aligned}
\dot{V} &\le -\lambda_{\min}(Q(R(\tilde{x})))\tilde{x}'\tilde{x} \\
&\quad + \frac{8m[g_0^2(|\tilde{x}| + |\Gamma_{sa}||e_a|) + D_0^2]\delta(2g_0(|\tilde{x}| + |\Gamma_{sa}||e_a|) + 2D_0)}{\beta(\tilde{x})\rho(\tilde{x})} \\
&\quad + \tilde{x}'\frac{dP(R(\tilde{x}))}{dt}\tilde{x} - e_a'e_a \\
&\le -\lambda_{\min}(Q(R(\tilde{x})))\tilde{x}'\tilde{x} + \frac{8m[g_0^2(|\tilde{x}| + s_0) + D_0^2]\delta(2g_0(|\tilde{x}| + s_0) + 2D_0)}{\beta(\tilde{x})\rho(\tilde{x})} \\
&\quad + \tilde{x}'\frac{dP(R(\tilde{x}))}{dt}\tilde{x} - e_a'e_a \\
&\le -\lambda_{\min}(Q(R(\tilde{x})))\tilde{x}'\tilde{x} + \frac{8m[g_0^2(|\tilde{x}|) + D_0^2]\delta(2g_0(|\tilde{x}| + 2D_0) + 8m\beta(\tilde{x})}{\beta(\tilde{x})\rho(\tilde{x})} \\
&\quad + \tilde{x}'\frac{dP(R(\tilde{x}))}{dt}\tilde{x} - e_a'e_a \\
&\le -\lambda_{\min}(Q(R(\tilde{x})))\left[\left(1 - \frac{8m}{\rho_0}\right)|\tilde{x}|^2 - \frac{8m\delta(1) + 8m}{\rho_0\delta(1)}\right] + \tilde{x}'\frac{\delta P(R(\tilde{x}))}{\delta t}\tilde{x} - e_a'e_a.
\end{aligned}$$

$$(7.3.46)$$

For $\rho_0 \ge 16m$, the inequality (7.3.46) can be continued as

$$\dot{V} \le -\lambda_{\min}(Q(R(\tilde{x})))\gamma\left[V - \frac{8m(\delta(1) + 1)}{\gamma\rho_0\delta(1)}\right] + \tilde{x}'\frac{\delta P(R(\tilde{x}))}{\delta t}\tilde{x}, \qquad (7.3.47)$$

where

$$\gamma = \min\left\{\frac{1}{2\lambda_{\max}(P(1))}, \frac{1}{(1+\tau^2)\lambda_{\min}(Q(R(1)))\lambda_{\max}(P_a)}\right\}. \qquad (7.3.48)$$

Let $c_0 > 0$ be such that $L_V(c_0) = \{(\tilde{x}, e_a) \in \mathbb{R}^n \times \mathbb{R}^{n-p} : V(\tilde{x}, e_a) \le c_0\} \subset \mathcal{W}_0$. Such a $c_0 > 0$ exists since $\mathcal{W}_0$ contains the origin as an interior point. We next choose a $\rho_0^* > 0$ such that,

$$\rho_0^* > \left\{16m, \frac{8m(\delta(1) + 1)}{\gamma c_0\delta(1)}\right\}. \qquad (7.3.49)$$

Then, we have,

$$\dot{V} < \tilde{x}'\frac{\delta P(R(\tilde{x}))}{\delta t}\tilde{x}, \quad \forall(\tilde{x}, e_a) \notin L_V^o(c_0) \text{ and } \forall\rho_0 \ge \rho_0^*. \qquad (7.3.50)$$

Again resorting to the facts that $\tilde{x}'(\delta P(R(\tilde{x}))/\delta t)\tilde{x}$ and $\delta\tilde{x}'P(R(\tilde{x}))\tilde{x})/\delta t$ either are both zero or have the opposite signs whenever $R(x)$ is not a constant locally and that $\delta(e_a'P_a e_a)/\delta t$ is always negative whenever $e_a \neq 0$, we have,

$$\dot{V} < 0, \quad \forall(\tilde{x}, e_a) \notin L_V^o(c_0) \text{ and } \forall \rho_0 \geq \rho_0^*, \tag{7.3.51}$$

which shows that every trajectory starting from $\mathbf{R}^n \times \{e_a : |e_a| \leq s_0/|\Gamma_{sa}|\}$ will enter the set $L_V(c_0) \subset \mathcal{W}_0$ in a finite time and remain in it thereafter. This is Item 2 of Problem 7.3.2.

Item 1 of Problem 2, i.e., the case that $D_0 = 0$, would follow if we could show that there exists a neighborhood $\mathcal{A}$ of the origin and a $\rho_0^* > 0$ such that, for each $\rho_0 \geq \rho_0^*$, the origin is a locally uniformly asymptotically stable equilibrium of the closed loop system (7.3.38)-(7.3.40) and $\mathcal{A}$ is contained in its basin of attraction. This can be done as follows. We begin with the same Lyapunov function,

$$V = \tilde{x}'P(R(\tilde{x}))\tilde{x} + (1+\tau^2)e_a'P_a e_a, \tag{7.3.52}$$

and let $c > 0$ be such that $(\tilde{x}, e_a) \in L_V(c)$ implies $R(\tilde{x}) = 1$. Such a c exists since there exists a neighborhood of the origin in which $R(\tilde{x}) \equiv 1$. Then $\mathcal{A}$ can be chosen as any subset of $L_V(c)$. Inside the set $L_V(c)$, we have,

$$\beta(\tilde{x})\rho(\tilde{x}) \geq \frac{\rho_0 \delta(1)}{\lambda_{\max}(Q(1))}, \tag{7.3.53}$$

$$|g_0(|\tilde{x}| + |\Gamma_{sa}||e_a|)| \leq \alpha_1|\tilde{x}| + \alpha_2|e_a|, \tag{7.3.54}$$

$$\delta(2g_0(|\tilde{x}| + |\Gamma_{sa}||e_a|)) \leq \alpha_3, \tag{7.3.55}$$

where $\alpha_1 > 0$, $\alpha_2 > 0$ and $\alpha_3 > 0$ are independent of ρ_0.

Using (7.3.53)-(7.3.55), $\dot{V}$ along the trajectories of (7.3.38)-(7.3.40) inside $L_V(c)$ can be evaluated as,

$$\begin{aligned}
\dot{V} &= -\tilde{x}'[Q(1) + P(1)BB'P(1)]\tilde{x} + 2\tilde{x}'P(1)B[\sigma(-(1+\beta(\tilde{x})\rho(\tilde{x}))B'P(1)\tilde{x} \\
&\quad +g(\tilde{x} + \Gamma_{sa}e_a, t) + B'P(1)\tilde{x}] + 2\tilde{x}'P(R(1)BE_a e_a - (1+\tau^2)e_a'e_a \\
&\leq -\tilde{x}'Q(1)\tilde{x} + 2\tilde{x}'P(1)B[\sigma(-(1+\beta(\tilde{x})\rho(\tilde{x}))B'P(1)\tilde{x} + g(\tilde{x} + \Gamma_{sa}e_a, t)) \\
&\quad +B'P(1)\tilde{x}] - e_a'e_a, \tag{7.3.56}
\end{aligned}$$

where we have defined $v \in \mathbf{R}^m$ by $v = -B'P(1)\tilde{x}$, and used v_i and g_i to denote respectively the ith element of v and $g(\cdot, \cdot)$. Using the same arguments as used in deriving (7.3.46), the above inequality can be continued as,

$$\dot{V} \leq -\lambda_{\min}(Q(1))|\tilde{x}|^2 + \frac{8\lambda_{\min}(Q(1))m\alpha_1^2\alpha_3}{\delta(1)\rho_0}|\tilde{x}|^2 + \frac{8\lambda_{\min}(Q(1))m\alpha_2^2\alpha_3}{\delta(1)\rho_0}|e_a|^2$$

$$-|e_a|^2$$

$$\leq -\left[\lambda_{\min}(Q(1)) - \frac{8\lambda_{\min}(Q(1))m\alpha_1^2\alpha_3}{\delta(1)\rho_0}\right]|\tilde{x}|^2$$

$$-\left[1 - \frac{8\lambda_{\min}(Q(1))m\alpha_2^2\alpha_3}{\delta(1)\rho_0}\right]|e_a|^2. \tag{7.3.57}$$

It is now clear that there exists a $\rho_0^* > 0$ such that, for each $\rho_0 \geq \rho_0^*$,

$$\dot{V} < 0, \quad \forall(\tilde{x}, e_a) \in L_V(c) \setminus \{0, 0\}, \tag{7.3.58}$$

and hence the closed-loop system (7.3.38)-(7.3.40) is locally uniformly asymptotically stable at the origin with $\mathcal{A} \subset L_V(c)$ contained in the basin of attraction. ⊠

7.3.6. An Example

Example 7.3.1. Consider the system (7.3.1) with

$$A = \begin{bmatrix} -1 & 1 \\ -1 & 1 \end{bmatrix}, \quad B = \begin{bmatrix} 0 \\ 1 \end{bmatrix}, \quad C = [0 \quad 1]. \tag{7.3.59}$$

Let the saturation function $\sigma \in \mathcal{S}_5(1, 0, 2)$, the uncertainties $g(x, t)$ be such that $|g(x, t)| \leq 0.1|x| + 0.1|x|^2$ and the ultimate bound on the disturbance $d(t)$ be $D_0 = 4$. It is easy to verify that A has two repeated eigenvalues at the origin and (A, B) is controllable. Hence Assumption 7.3.1 is satisfied. Following scheduled low-and-high gain state feedback design procedure, we pick $Q(r) = rI$ and solve the ARE (7.3.4) for the triple (A, B, C) as given by (7.3.59), obtaining

$$P(r) = \begin{bmatrix} \sqrt{r + 2\sqrt{2r}}\left(1 + \sqrt{2r} - \sqrt{r + 2\sqrt{2r}}\right) & \sqrt{2r} - \sqrt{r + 2\sqrt{2r}} \\ \sqrt{2r} - \sqrt{r + 2\sqrt{2r}} & \sqrt{r + 2\sqrt{2r}} \end{bmatrix},$$

and hence,

$$R(x) = \max\left\{r \in (0, 1] : \left(r + 2\sqrt{2r}\right)\left(1 + \sqrt{2r} - \sqrt{r + 2\sqrt{2r}}\right)x_1^2 \right.$$
$$\left. + 2\left(\sqrt{2r^2 + 4r} - r - 2\sqrt{2r}\right)x_1 x_2 + \left(r + 2\sqrt{2r}\right)x_2^2 \leq 1\right\}.$$

Choosing $g_1(s) = 0.1s + 0.1$ and $\mu(x) = R(x)$, the function $\rho(x)$ of (7.3.11) is given by

$$\rho(x) = \frac{2\rho_0(0.1|x| + 5.1)^2}{R(x)},$$

and the scheduled low-and-high gain state feedback law by

$$u = -(1 + \rho(x))B'P(R(x))x.$$

Some simulation results for $\sigma(s) = 2\tanh(s)$, $g(x,t) = 0.1x_1 + 0.1x_2^2$ and $d(t) = 4\cos(10t)$ are shown in Fig. 7.3.1, from which the robust disturbance rejection is clear.

We next consider the case when only the output is available for feedback. In this case, we note that the triple (A, B, C) is already in the SCB with $x_a = x_1$ and $x_d = x_2$, and it is easily seen that it is of minimum-phase with one invariant zero at $s = -1$ and one infinite zero of order 1. Thus Assumption 7.3.4 is satisfied. Following the scheduled low-and-high gain output feedback design procedure, we obtain the following reduced order observer based controller.

$$\begin{cases} \dot{\hat{x}}_1 = -\hat{x}_1 + y, \\ u = -(1 + \beta(\check{x})\rho(\check{x}))B'P(R(\check{x}))\check{x}, \quad \check{x} = [\hat{x}_1, y]', \end{cases}$$

where $\beta(\check{x}) = 0.08(|\check{x}|+1)^3 + 1$ is such that (7.3.30) holds for $s_0 = 1$. Simulations for the same saturation function $\sigma(\cdot)$, uncertainties $g(x,t)$, disturbance $d(t)$, and plant initial conditions are shown in Fig. 7.3.2. ▣

7.4. Concluding Remarks

Two gain scheduling techniques were developed in this chapter to demonstrate how the low gain and high gain parameters in the low-and-high gain feedback laws can be appropriately scheduled as functions of the state so that the semi-global results, typical products of low gain and low-and-high gain feedback laws, can be turned into global ones. The two scheduling techniques presented led to the solution of global ADDPS for and global robust stabilization of linear systems subject to actuator saturation.

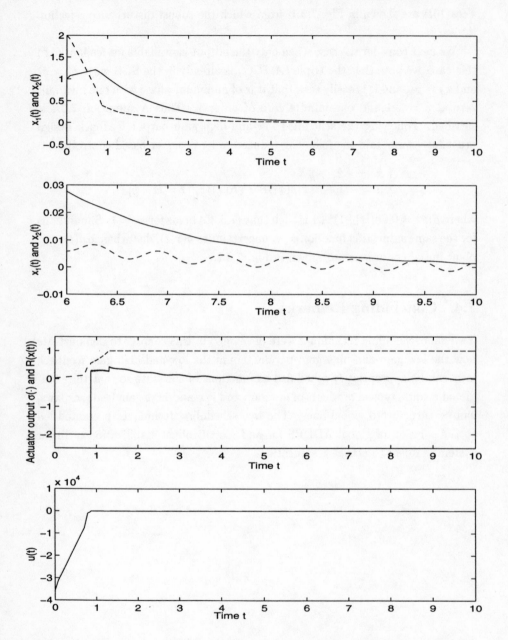

Figure 7.3.1: $\sigma(s) = 2\tanh(s)$, $g(x,t) = 0.1x_1 + 0.1x_2^2$, and $d(t) = 4\cos(10t)$; $\rho_0 = 16$.

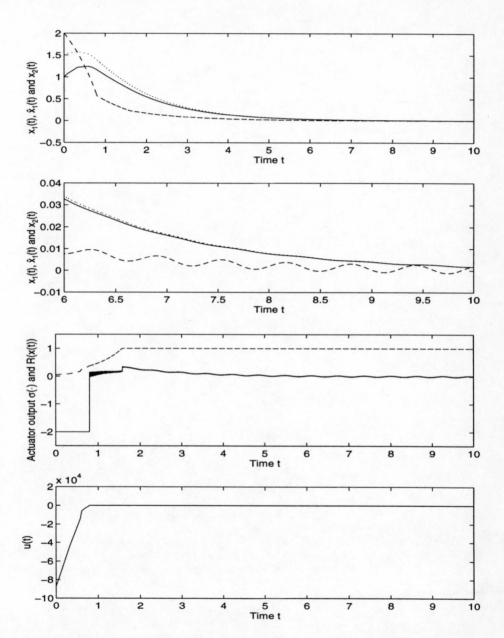

Figure 7.3.2: $\sigma(s) = 2\tanh(s)$, $g(x,t) = 0.1x_1 + 0.1x_2^2$, and $d(t) = 4\cos(10t)$; $\rho_0 = 16$

Chapter 8

Semi-Global Stabilization of Linear Systems with Magnitude and Rate Saturating Actuators

8.1. Introduction

There exists a vast literature on the stabilization of linear systems using magnitude and/or rate saturating actuators. The majority of this literature, however, address only, actuator magnitude saturation (see, for example, [3]). In the case of magnitude saturating actuators, a crucial result on the subject appeared in [103], where it was established that a linear system subject to actuator magnitude saturation can be globally asymptotically stabilized if and only if the system in the absence of the saturation is asymptotically null controllable with bounded controls. It is well known that asymptotic null controllability with bounded controls is equivalent to the usual notion of linear stabilizability plus the added condition that all the open loop poles be in the closed left-half plane ([98,102]). Several globally asymptotically stabilizing feedback laws were constructed ([19,104,108,110]).

Another crucial result related to global asymptotic stabilization of such systems using magnitude saturating actuators is that, in general, *linear* feedback cannot achieve global asymptotic stabilization [18,109]. Being aware of this result, we introduced the notion of semi-global stabilization for both continuous-time and discrete-time linear systems subject to actuator magnitude saturation (see Chapter 3). The semi-global framework for stabilization requires feedback laws that yield a closed-loop system which has an asymptotically stable equilibrium whose basin of attraction includes an *a priori* given (arbitrarily large) bounded set. More specifically, in Chapter 3 it was shown via explicit construction of feedback laws that, under the condition of asymptotic null controllability with bounded controls, one can achieve semi-global stabilization of linear systems subject to actuator magnitude saturation using *linear* feedback laws.

On the other hand, actuator rate saturation also presents a serious challenge to control engineers. The phase lag associated with rate saturation has a destabilizing effect. For example, investigators have identified rate saturation as a contributing factor to the recent mishaps of YF-22 [14] and Gripen [39] prototypes and the first production Gripen [101]. For further discussion on the destabilizing effect of actuator rate saturation, see, for example, [2]. It can be expected that the problem of actuator rate saturation is more severe when the actuator is also subject to magnitude saturation since small actuator output results in small stability margin even in the absence of rate saturation.

In spite of the importance of the problem of actuator rate saturation, it has received far less attention. Beside [2], some other references are [7,28,100,118]. The purpose of this chapter is to show how the idea of low gain feedback design can be utilized to establish some new results on the control of linear systems subject to both magnitude and rate saturation. In particular, we will show that, if a continuous-time or a discrete-time linear system is asymptotically null-controllable with bounded controls, then when subject to both actuator magnitude and rate saturation, it is semi-globally asymptotically stabilizable by *linear* state feedback. If, in addition, the system is also detectable, then it is semi-globally asymptotically stabilizable via *linear* output feedback. This was proven by explicit construction of feedback laws.

Although the results for continuous-time systems and their discrete-time counterpart bear resemblance, the feedback laws we construct to establish these results represent completely different ways of application of low gain feedback design techniques. For example, the feedback laws constructed for continuous-time systems are of singular perturbation type and, in the absence of actuator saturation, would result in two time scales in the closed-loop system. The slow subsystem is induced from the plant dynamics by a low gain feedback (constructed basing on the solution of a family of parameterized AREs) and the fast subsystem from the actuator dynamics by a high gain feedback of the actuator state. It is clear from the singular perturbation theory, such high gain feedback is necessary even in the absence of actuator saturation. These feedback laws, while providing high degree of robustness with respect to the uncertainties in the actuator nonlinearities, do not extend to discrete-time systems due to the presence of the necessary high gain component.

The remainder of this chapter is organized as follows. Sections 8.2 and 8.3 present the continuous-time and the discrete-time results respectively. Section 8.4 contains some brief concluding remarks.

8.2. Continuous-Time Systems

8.2.1. Problem Statement

We consider a system of the form,

$$\begin{cases} \dot{x} = Ax + B\sigma_p(v), \\ \dot{v} = \sigma_r(-v + u), \\ y = Cx, \end{cases} \tag{8.2.1}$$

where $x \in \mathbb{R}^n$ is the plant state, $v \in \mathbb{R}^m$ is the actuator state, $u \in \mathbb{R}^m$ is the control input to the actuators, $y \in \mathbb{R}^p$ is the measurement output, and the functions $\sigma_p, \sigma_r : \mathbb{R}^m \to \mathbb{R}^m$ are saturation functions to be defined next. They represent respectively the actuator magnitude and rate saturation. Here we have also without loss of generality assumed that all actuators have unity time constant.

Definition 8.2.1. *A function $\sigma : \mathbb{R}^m \to \mathbb{R}^m$ is said to be a saturation function if,*

1. *$\sigma(s)$ is decentralized, i.e., $\sigma(s) = [\sigma_1(s_1), \sigma_2(s_2), \cdots, \sigma_m(s_m)]$; and for each $i = 1$ to m,*

2. *σ_i is locally Lipschitz;*

3. *There exists a $\Delta_i > 0$, b_i and c_i, $b_i \geq c_i > 0$, such that,*

$$\begin{cases} c_i s_i^2 \leq s_i \sigma_i(s_i) \leq b_i s_i^2 & \text{if } |s_i| \leq \Delta_i, \\ c_i \Delta_i \leq |\sigma_i(s_i)| \leq b_i |s_i| & \text{if } |s_i| > \Delta_i. \end{cases}$$

Remark 8.2.1.   *1. Graphically, each element of the vector valued saturation function resides in the shaded area form some constants $\Delta > 0$, and $b_i \geq c_i > 0$. For notational simplicity, but without loss of generality, we will assume throughout this paper that for each i, $\Delta_i = \Delta$, $b_i = b \geq 1$ and $c_i = 1$;*

2. *We recall that the saturation function as defined here is a class S_2 saturation function with the extra property that it has a linear growth rate*

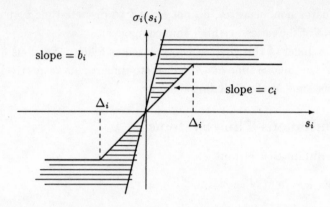

Figure 8.2.1: Qualitative description of the saturation function

outside a neighborhood of the origin. For this reason, we will call such a saturation function as a class S_2^{LG} saturation function. It follows from the above definition, $\sigma(s) = s$, $\arctan(s)$, $\tanh(s)$ and the standard saturation function $\sigma(s) = \text{sign}(s) \min\{|s|, 1\}$ are all class S_2^{LG} saturation functions.

®

Definition 8.2.2. *The set of all class S_2^{LG} saturation functions with constant Δ and b is denoted by $S_2^{LG}(\Delta, b)$.* Ⓓ

The problems to be solved in this section are the following.

Problem 8.2.1. *Consider the system (8.2.1), where $\sigma_p, \sigma_r \in S_2^{LG}(\Delta, b)$ for some $\Delta > 0$ and $b \geq 1$. The problem of semi-global stabilization via linear state feedback is defined as follows. For any given (arbitrarily large) bounded set $W_0 \subset \mathbb{R}^{n+m}$, find a linear state feedback law $u = Fx + Gv$, such that, for all $\sigma_p, \sigma_r \in S_2^{LG}(\Delta, b)$, the equilibrium of the closed-loop system $(x, v) = (0, 0)$ is locally asymptotically stable with W_0 contained in its basin of attraction.* Ⓟ

Problem 8.2.2. *Consider the system (8.2.1), where $\sigma_p, \sigma_r \in S_2^{LG}(\Delta, b)$ for some $\Delta > 0$ and $b \geq 1$. The problem of semi-global stabilization via linear dynamic output feedback of dynamical order n is defined as follows. For any given (arbitrarily large) bounded set $W_0 \subset \mathbb{R}^{2n+m}$, find a linear dynamic output feedback law of the form,*

$$\begin{cases} \dot{\zeta} = M\zeta + Ny, \quad \zeta \in \mathbb{R}^n, \\ u = F\zeta + Gv, \end{cases} \tag{8.2.2}$$

such that, for all $\sigma_p, \sigma_r \in S_2^{\mathrm{LG}}(\Delta, b)$, the equilibrium of the closed-loop system $(x, v, \zeta) = (0, 0, 0)$ is locally asymptotically stable with $\mathcal{W}_0$ contained in its basin of attraction. ⊡

We make the following assumptions.

Assumption 8.2.1. *The pair (A, B) is asymptotically null controllable with bounded controls (ANCBC), i.e.,*

1. *(A, B) is stabilizable;*

2. *All eigenvalues of A are located in the closed left-half plane.* ⊞

Assumption 8.2.2. *The pair (A, C) is stabilizable.* ⊞

8.2.2. Semi-Global Stabilization by Linear Feedback

Concerning semi-global stabilization of the system (8.2.1), we have the following result.

Theorem 8.2.1. *Consider system (8.2.1). If Assumption 8.2.1 is satisfied, then Problem 8.2.1 is solvable. If Assumptions 8.2.1 and 8.2.2 are satisfied, then Problem 8.2.2 is solvable.* ⊡

Proof of Theorem 8.2.1. We will separate the proof into two parts, one for the state feedback result and the other for the output feedback.

State Feedback Result:

We will show that the following family of linear state feedback laws solves Problem 8.2.1,

$$u = -\frac{1}{\varepsilon^2} B' P(\varepsilon) x - \left(\frac{1}{\varepsilon^2} - 1 \right) v, \qquad (8.2.3)$$

where $P(\varepsilon)$ is the solution to the following ARE,

$$A'P + PA - PBB'P + \varepsilon I = 0, \quad \varepsilon \in (0, 1]. \qquad (8.2.4)$$

We recall that this ARE is the same as the ARE (2.2.51) and hence it has the properties of Lemma 2.2.6.

More specifically, we will show that, for any given bounded set $\mathcal{W}_0$, there exists an $\varepsilon^* \in (0, 1]$ such that, for all $\varepsilon \in (0, \varepsilon^*]$, the equilibrium $(x, v) = (0, 0)$ of the closed-loop system is locally asymptotically stable with $\mathcal{W}_0$ contained in its basin of attraction.

We note here that the feedback law is parameterized in a single parameter ε. As is clear from the proof below, as $\varepsilon \to 0$, $B'P(\varepsilon) \to 0$ to enlarge the basin of attraction, while the overall gain approaches to ∞ to fully utilize the available actuator rate.

Under this feedback law, the closed-loop system is given by,

$$\begin{cases} \dot{x} = Ax + B\sigma_p(v), \\ \dot{v} = \sigma_r(-v/\varepsilon^2 - B'Px/\varepsilon^2). \end{cases} \tag{8.2.5}$$

Pick the Lyapunov function,

$$V = x'Px + (v + B'Px)'(v + B'Px), \tag{8.2.6}$$

and let $c > 0$ be such that

$$c \geq \sup_{\varepsilon \in (0,1], (x,v) \in W_0} \{x'Px + (v + B'Px)'(v + B'Px)\}. \tag{8.2.7}$$

Such a c exists since W_0 is bounded and $\lim_{\varepsilon \to 0} P = 0$. Let $\varepsilon_1^* \in (0,1]$ be such that for all $\varepsilon \in (0, \varepsilon_1^*]$,

$$(x,v) \in L_V(c) \Longrightarrow |B'Px| < \frac{\Delta}{3b}, \; |B'PAx| < \frac{\Delta}{3m}, \; |B'PB\sigma_p(v)| < \frac{\Delta}{3m}. \tag{8.2.8}$$

The existence of such an ε_1^* is again due to Lemma 2.2.6 and the fact that $|\sigma_{pi}(s)| \leq b|s|$.

The evaluation of $\dot{V}$ along the trajectories of the closed-loop system then gives that, for $(x,v) \in L_V(c)$, $\varepsilon \in (0, \varepsilon_1^*]$,

$$\dot{V} = -\varepsilon x'x - x'PBB'Px + 2x'PB[\sigma_p(v) + B'Px]$$
$$+2(v + B'Px)'[\sigma_r(-v/\varepsilon^2 - B'Px/\varepsilon^2) + B'PAx + B'PB\sigma_p(v)]$$
$$\leq -\varepsilon x'x + 2\sum_{i=1}^{m} \left[-\mu_i[\sigma_{pi}(v_i) - \mu_i)] + (v_i - \mu_i)[\sigma_{ri}(-(v_i - \mu_i)/\varepsilon^2) \right.$$
$$\left. +B_i'PAx + B_i'PB\sigma_p(v)] \right], \tag{8.2.9}$$

where B_i is the ith column of the matrix B and μ_i is the ith element of $\mu = -B'Px$.

In the case that $v_i - \mu_i > \varepsilon^2 \Delta$, it follows from Definition 8.2.1 and (8.2.8) that, if $\mu_i \geq 0$, then $v_i > \mu_i$ and

$$-\mu_i[\sigma_{pi}(v_i) - \mu_i)] + (v_i - \mu_i)[\sigma_{ri}(-(v_i - \mu_i)/\varepsilon^2) + B_i'PAx + B_i'PB\sigma_p(v)]$$
$$\leq -\frac{\Delta}{3}(v_i - \mu_i) < 0,$$

moreover,

$$-\mu_i[\sigma_{pi}(v_i) - \mu_i)] + (v_i - \mu_i)[\sigma_{ri}(-(v_i - \mu_i)/\varepsilon^2) + B_i'PAx + B_i'PB\sigma_p(v)]$$

$$+ \sum_{|v_j - \mu_j| \le \varepsilon^2 \Delta} (v_j - \mu_j)[B_j' PAx + B_j' PB\sigma_p(v)]$$

$$\le -\frac{\Delta}{3}(v_i - \mu_i) < 0,$$

and if $\mu_i < 0$,

$$-\mu_i[\sigma_{pi}(v_i) - \mu_i)] + (v_i - \mu_i)[\sigma_{ri}(-(v_i - \mu_i)/\varepsilon^2) + B_i' PAx + B_i' PB\sigma_p(v)]$$

$$\le \begin{cases} -b\mu_i(v_i - \mu_i) - \frac{\Delta}{3}(v_i - \mu_i), & \text{if } v_i \ge 0 \\ -\mu_i(v_i - \mu_i) - \frac{\Delta}{3}(v_i - \mu_i), & \text{if } v_i < 0 \text{ (and hence } |v_i| \le |\mu_i| \le \Delta) \end{cases}$$

$$< 0,$$

moreover,

$$-\mu_i[\sigma_{pi}(v_i) - \mu_i)] + (v_i - \mu_i)[\sigma_{ri}(-(v_i - \mu_i)/\varepsilon^2) + B_i' PAx + B_i' PB\sigma_p(v)]$$

$$+ \sum_{|v_j - \mu_j| \le \varepsilon^2 \Delta} (v_j - \mu_j)[B_j' PAx + B_j' PB\sigma_p(v)]$$

$$\le \begin{cases} -b\mu_i(v_i - \mu_i) - \frac{\Delta}{3}(v_i - \mu_i), & \text{if } v_i \ge 0 \\ -\mu_i(v_i - \mu_i) - \frac{\Delta}{3}(v_i - \mu_i), & \text{if } v_i < 0 \text{ (and hence } |v_i| \le |\mu_i| \le \Delta) \end{cases}$$

$$< 0.$$

Similarly, for the case that $v_i - \mu_i < -\varepsilon^2 \Delta$, we also have,

$$-\mu_i[\sigma_{pi}(v_i) - \mu_i)] + (v_i - \mu_i)[\sigma_{ri}(-(v_i - \mu_i)/\varepsilon^2) + B_i' PAx + B_i' PB\sigma_p(v)] < 0.$$

Hence, if $|v_i - \mu_i| > \varepsilon^2 \Delta$ for all $i = 1$ to m, we have,

$$\dot{V} < 0, \quad \forall \varepsilon \in (0, \varepsilon_1^*]. \tag{8.2.10}$$

If $|v_i - \mu_i| > \varepsilon^2 \Delta$ for at least one, but not all, i, then for any i such that $|v_i - \mu_i| \le \varepsilon^2 \Delta$, we have that, for $\varepsilon \in (0, \sqrt{2/3}]$, $|v_i| \le \varepsilon^2 \Delta + \Delta/3 \le \Delta$ and,

$$-\mu_i[\sigma_{pi}(v_i) - \mu_i)] + (v_i - \mu_i)\sigma_{ri}(-(v_i - \mu_i)/\varepsilon^2)$$

$$\le \begin{cases} -\mu_i(v_i - \mu_i) - \frac{1}{\varepsilon^2}(v_i - \mu_i)^2, & \text{if } \mu_i v_i \ge 0 \\ -b\mu_i(v_i - \mu_i) - \frac{1}{\varepsilon^2}(v_i - \mu_i)^2, & \text{if } \mu_i v_i < 0 \end{cases}$$

$$\le \frac{\varepsilon}{2m} x' x - \left(\frac{1}{\varepsilon^2} - \frac{mb^2 |B_i' P|^2}{2\varepsilon}\right)(v_i - \mu_i)^2.$$

Letting $\varepsilon_2^* \in (0, \min\{\varepsilon_1^*, \sqrt{2/3}\}]$ be such that,

$$mb^2 |B' P|^2 \varepsilon < 1, \quad \forall \varepsilon \in (0, \varepsilon_2^*],$$

we have,

$$\dot{V} < -\frac{\varepsilon}{2} x' x - \frac{1}{2\varepsilon^2} \sum_{|v_i - \mu_i| \le \varepsilon^2 \Delta} (v_i - \mu_i)^2 < 0, \quad \forall (x, v) \neq (0, 0), \varepsilon \in (0, \varepsilon_2^*].$$

$$\tag{8.2.11}$$

Finally, in the case that $|v_i - \mu_i| \le \varepsilon^2 \Delta$ holds for all $i = 1$ to m. In this case, we note that, for $\varepsilon \in (0, \sqrt{2/3}]$, $|v_i| \le \varepsilon^2 \Delta + \Delta/3 \le \Delta$ and,

$$-\mu_i[\sigma_{pi}(v_i) - \mu_i)] + (v_i - \mu_i)[\sigma_{ri}(-(v_i - \mu_i)/\varepsilon^2) + B_i'PAx + B_i'PB\sigma_p(v)]$$

$$\le \begin{cases} -\mu_i(v_i - \mu_i) - \frac{1}{\varepsilon^2}(v_i - \mu_i)^2 + |v_i - \mu_i|(|B_i'PA||x| + b|B_i'PB||v|), \\ \qquad\qquad\qquad\qquad\qquad\qquad\qquad\qquad \text{if } \mu_i v_i \ge 0 \\ -b\mu_i(v_i - \mu_i) - \frac{1}{\varepsilon^2}(v_i - \mu_i)^2 + |v_i - \mu_i|(|B_i'PA||x| + b|B_i'PB||v|), \\ \qquad\qquad\qquad\qquad\qquad\qquad\qquad\qquad \text{if } \mu_i v_i < 0 \end{cases}$$

$$\le (b|B'P| + |B'PA| + b|B'PB||B'P|)|v - \mu||x| - \frac{1}{\varepsilon^2}(v_i - \mu_i)^2$$
$$+ b|B'PB||v - \mu|^2,$$

and hence,

$$\dot{V} \le -\varepsilon|x|^2 - \frac{2}{\varepsilon^2}|v - \mu|^2 + 2m(b|B'P| + |B'PA| + b|B'PB||B'P|)|v - \mu||x|$$
$$+ 2mb|B'PB||v - \mu|^2$$

$$\le -\frac{\varepsilon}{2}|x|^2 - \frac{1}{\varepsilon^2}|v + B'Px|^2 - \frac{1}{\varepsilon^2}[1 - 2m^2(b|B'P| + |B'PA| \qquad (8.2.12)$$
$$+ b|B'PB||B'P|)^2\varepsilon - 2mb|B'PB|\varepsilon^2]|v - \mu|^2. \qquad (8.2.13)$$

Let $\varepsilon_3^* \in (0, \min\{\varepsilon_1^*, \sqrt{2/3}\}]$ be such that for all $\varepsilon \in (0, \varepsilon_3^*]$,

$$2m^2(b|B'P| + |B'PA| + b|B'PB||B'P|)^2\varepsilon + 2mb|B'PB|\varepsilon^2 \le 1.$$

Such an ε_3^* exists since $\lim_{\varepsilon \to 0} P(\varepsilon) = 0$. With this choice of ε_3^*, we have,

$$\dot{V} \le -\frac{\varepsilon}{2}|x|^2 - \frac{1}{\varepsilon^2}|v + B'Px|^2, \quad \forall (x, v) \in L_V(c), \ \varepsilon \in (0, \varepsilon_3^*]. \qquad (8.2.14)$$

In conclusion, we have shown that,

$$\dot{V} < 0, \quad \forall (x, v) \in L_V(c) \setminus (0, 0), \ \varepsilon \in (0, \varepsilon^*], \qquad (8.2.15)$$

where $\varepsilon^* = \min\{\varepsilon_2^*, \varepsilon_3^*\}$. This implies that the equilibrium $(x, v) = (0, 0)$ of the closed-loop system is locally asymptotically stable with $\mathcal{W}_0 \subset L_V(c)$ contained in its basin of attraction. This ends the proof of the state feedback result.

Output Feedback Result:

We will show that the following family of observer based liner output feedback laws solves Problem 8.2.2,

$$\begin{cases} \dot{\hat{x}} = A\hat{x} - L(y - C\hat{x}), \\ u = -\frac{1}{\varepsilon^2}B'P\hat{x} - \left(\frac{1}{\varepsilon^2} - 1\right)v, \end{cases} \qquad (8.2.16)$$

where P is the solution to the ARE (8.2.4) and L is any matrix such that $A + LC$ is asymptotically stable. The existence of such an L is due to Assumption 8.2.2.

More specifically, we will show that, for any given bounded set $\mathcal{W}_0$, there exists an $\varepsilon^* \in (0, 1]$ such that, for all $\varepsilon \in (0, \varepsilon^*]$, the equilibrium $(x, \hat{x}, v) = (0, 0, 0)$ of the closed-loop system is locally asymptotically stable with $\mathcal{W}_0$ contained in its basin of attraction.

Under this feedback law, the closed-loop system is given by,

$$\begin{cases} \dot{x} = Ax + B\sigma_p(v), \\ \dot{\hat{x}} = A\hat{x} - L(y - C\hat{x}), \\ \dot{v} = \sigma_r(-v/\varepsilon^2 - B'P\hat{x}/\varepsilon^2). \end{cases} \tag{8.2.17}$$

Letting $e = x - \hat{x}$, the closed-loop system can be rewritten as,

$$\begin{cases} \dot{x} = Ax + B\sigma_p(v), \\ \dot{e} = (A + LC)e + B\sigma_p(v), \\ \dot{v} = \sigma_r(-v/\varepsilon^2 - B'Px/\varepsilon^2 + B'Pe/\varepsilon^2). \end{cases} \tag{8.2.18}$$

Pick the Lyapunov function,

$$V = x'Px + (v + B'Px - B'Pe)'(v + B'Px - B'Pe) + \lambda_{\max}^{\frac{1}{2}}(P)e'P_0e, \tag{8.2.19}$$

where $P_0 > 0$ is the unique solution to the following Lyapunov equation

$$(A + LC)'P_0 + P_0(A + LC) = -I. \tag{8.2.20}$$

Let $c > 0$ be such that

$$c \geq \sup_{\varepsilon \in (0,1], (x, \hat{x}, v) \in \mathcal{W}_0} V. \tag{8.2.21}$$

Such a c exists since $\mathcal{W}_0$ is bounded and $\lim_{\varepsilon \to 0} P = 0$. We note here that with this choice of c, $(x, \hat{x}, v) \in \mathcal{W}_0$ implies that $(x, e, v) \in L_V(c)$.

Let $\varepsilon_1^* \in (0, 1]$ be such that for all $\varepsilon \in (0, \varepsilon_1^*]$,

$$(x, e, v) \in L_V(c) \implies |B'Px| < \frac{\Delta}{8b}, \ |B'Pe| < \frac{\Delta}{8b}, \ |B'PAx| < \frac{\Delta}{8m},$$

$$|\lambda_{\max}^{\frac{1}{2}}(P)B'P_0e| < \frac{\Delta}{8b}, \ |B'P(A + LC)e| \leq \frac{\Delta}{8m}. \tag{8.2.22}$$

The existence of such an ε_1^* is again due to Lemma 2.2.6.

The evaluation of $\dot{V}$ along the trajectories of the closed-loop system then gives that, for $(x, e, v) \in L_V(c)$, $\varepsilon \in (0, \varepsilon_1^*]$,

$$\dot{V} = -\varepsilon x'x - x'PBB'Px + 2x'PB[\sigma_p(v) + B'Px] - \lambda_{\max}^{\frac{1}{2}}(P)e'e$$

$$+ 2\lambda_{\max}^{\frac{1}{2}}(P)e'P_0B\sigma_p(v) + 2(v + B'Px - B'Pe)'$$

$$\times [\sigma_r(-v/\varepsilon^2 - B'Px/\varepsilon^2 + B'Pe/\varepsilon^2) + B'PAx - B'P(A + LC)e]$$

$$= -\varepsilon x'x - \lambda_{\max}^{\frac{1}{2}}(P)e'e + 2\sum_{i=1}^{m} \left[-\lambda_{\max}^{\frac{1}{2}}(P)\eta_i\sigma_{pi}(v_i) - \frac{1}{4}\mu_i^2 \right]$$

$$+2 \sum_{i=1}^{m} \left[-\mu_i[\sigma_{pi}(v_i) - \mu_i] - \frac{1}{4}\mu_i^2 + (v_i - \mu_i + \omega_i)[\sigma_{ri}(-(v_i - \mu_i + \omega_i)/\varepsilon^2) \right.$$

$$\left. + B_i'PAx - B_i'P(A + LC)e] \right], \tag{8.2.23}$$

where B_i is the ith column of the matrix B, and η_i, μ_i and ω_i are respectively the ith elements of $\eta = -B'P_0 e$, $\mu = -B'Px$ and $\omega = -B'Pe$.

We first consider the summand under the first summation sign,

$$-\lambda_{\max}^{\frac{1}{2}}(P)\eta_i \sigma_{pi}(v_i) - \frac{1}{4}\mu_i^2 = \lambda_{\max}^{\frac{1}{2}}(P)\eta_i[-\sigma_{pi}(v_i - \mu_i + \omega_i)$$

$$+\sigma_{pi}(v_i - \mu_i + \omega_i) - \sigma_{pi}(v_i)] - \frac{1}{4}\mu_i^2. \tag{8.2.24}$$

We note that for $(x, e, v) \in L_V(c)$, $|v_i|$, $|\mu_i - \omega_i| \le D_i$ for some constant D_i independent of $\varepsilon \in (0, 1]$. Recalling that σ_{pi} is locally Lipschitz, it follows from [29, p. 93] that there exists a $\pi > 0$, independent of ε, such that, for all $i = 1$ to m,

$$|\sigma_{pi}(s_1) - \sigma_{pi}(s_2)| \le \pi|s_1 - s_2|, \quad \forall |s_1|, |s_2| \le D_i.$$

Hence, it follows from (8.2.24) that, in the case that $|v_i - \mu_i + \omega_i| > \varepsilon^2 \Delta$,

$$-\lambda_{\max}^{\frac{1}{2}}(P)\eta_i \sigma_{pi}(v_i) - \frac{1}{4}\mu_i^2 < \frac{\Delta}{8}|v_i - \mu_i + \omega_i|$$

$$+\lambda_{\max}^{\frac{1}{2}}(P)\pi|\eta_i\mu_i| + \lambda_{\max}^{\frac{1}{2}}(P)\pi|\eta_i\omega_i| - \frac{1}{4}\mu_i^2$$

$$\le \frac{\Delta}{8}|v_i - \mu_i + \omega_i| + \lambda_{\max}(P)$$

$$\times \left(\pi^2|B'P_0|^2 + \lambda_{\max}^{\frac{1}{2}}(P)\pi|P_0||B|^2\right)e'e, \tag{8.2.25}$$

and in the case that $|v_i - \mu_i + \omega_i| \le \varepsilon^2 \Delta$,

$$-\lambda_{\max}^{\frac{1}{2}}(P)\eta_i \sigma_{pi}(v_i) - \frac{1}{4}\mu_i^2$$

$$< \lambda_{\max}^{\frac{1}{2}}(P)b|\eta_i(v_i - \mu_i + \omega_i)| + \lambda_{\max}^{\frac{1}{2}}(P)\pi|\eta_i\mu_i| + \lambda_{\max}^{\frac{1}{2}}(P)\pi|\eta_i\omega_i| - \frac{1}{4}\mu_i^2$$

$$\le (v_i - \mu_i + \omega_i)^2 + \lambda_{\max}(P)\left(\pi^2|B'P_0|^2 + b^2|B'P_0|^2 + \lambda_{\max}^{\frac{1}{2}}(P)\pi|P_0||B|^2\right)e'e. \tag{8.2.26}$$

We now proceed to consider the summand under the second summation sign. To this end, we first consider the case that $v_i - \mu_i + \omega_i > \varepsilon^2 \Delta$, it follows from Definition 8.2.1 and (8.2.22) that, if $v_i \ge 0, -\mu_i + \omega_i \ge 0$, then,

$$-\mu_i[\sigma_{pi}(v_i) - \mu_i] - \frac{1}{4}\mu_i^2 \le -\mu_i[\sigma_{pi}(v_i) - \mu_i + \omega_i] + \omega_i^2$$

$$\leq b|\mu_i|(v_i - \mu_i + \omega_i) + \omega_i^2$$
$$< \frac{\Delta}{8}(v_i - \mu_i + \omega_i) + \omega_i^2,$$

if $v_i \leq 0, -\mu_i + \omega_i > 0$, then $|v_i| \leq |-\mu_i + \omega_i| \leq \Delta$ and hence,

$$-\mu_i[\sigma_{pi}(v_i) - \mu_i] - \frac{1}{4}\mu_i^2 \leq -\mu_i[\sigma_{pi}(v_i) - \mu_i + \omega_i] + \omega_i^2$$
$$\leq |\mu_i|(v_i - \mu_i + \omega_i) + \omega_i^2$$
$$< \frac{\Delta}{8}(v_i - \mu_i + \omega_i) + \omega_i^2,$$

and if $v_i > 0, -\mu_i + \omega_i \leq 0$, then,

$$-\mu_i[\sigma_{pi}(v_i) - \mu_i] - \frac{1}{4}\mu_i^2$$
$$\leq \begin{cases} -b\mu_i(v_i - \mu_i + \omega_i) + b^2\omega_i^2, & \text{if } \mu_i \leq 0 \\ -\mu_i(v_i - \mu_i + \omega_i) + \omega_i^2, & \text{if } \mu_i > 0, v_i \leq \Delta \\ -\frac{1}{4}\mu_i^2, & \text{if } \mu_i > 0, v_i > \Delta \end{cases}$$
$$< \frac{\Delta}{4}(v_i - \mu_i + \omega_i) + b^2\omega_i^2.$$

The above shows that in the case of $v_i - \mu_i + \omega_i > \varepsilon^2\Delta$, we have,

$$-\mu_i[\sigma_{pi}(v_i) - \mu_i] - \frac{1}{4}\mu_i^2 + (v_i - \mu_i + \omega_i)[\sigma_{ri}(-(v_i - \mu_i + \omega_i)/\varepsilon^2)$$
$$+B_i'PAx - B_i'P(A + LC)e]$$
$$< -\frac{\Delta}{2}|v_i - \mu_i + \omega_i| + b^2\omega_i^2. \tag{8.2.27}$$

Moreover,

$$-\mu_i[\sigma_{pi}(v_i) - \mu_i] - \frac{1}{4}\mu_i^2 + (v_i - \mu_i + \omega_i)[\sigma_{ri}(-(v_i - \mu_i + \omega_i)/\varepsilon^2)$$
$$+B_i'PAx - B_i'P(A + LC)e]$$
$$+ \sum_{|v_j - \mu_j + \omega_j| \leq \varepsilon^2\Delta} (v_j - \mu_j + \omega_j)[B_j'PAx + B_j'P(A + LC)e]$$
$$< -\frac{\Delta}{2}|v_i - \mu_i + \omega_i| + b^2\omega_i^2. \tag{8.2.28}$$

Similarly, we can show that (8.2.27) and (8.2.28) also hold for the case that, $v_i - \mu_i + \omega_i < -\varepsilon^2\Delta$.

Hence, if $|v_i - \mu_i + \omega_i| > \varepsilon^2\Delta$ for all $i = 1$ to m, it follows from (8.2.23), (8.2.25) and (8.2.27) that,

$$\dot{V} < -\varepsilon x'x - \lambda_{\max}^{\frac{1}{2}}(P)\left(1 - 2\lambda_{\max}^{\frac{1}{2}}(P)m\pi^2|B'P_0|^2 - 2\lambda_{\max}(P)m\pi|P_0||B|^2\right.$$
$$\left. -2b^2\lambda_{\max}^{\frac{3}{2}}(P)|B|^2\right)e'e - \frac{\Delta}{4}(v - \mu + \omega)'(v - \mu + \omega).$$

Letting $\varepsilon_2^* \in (0, \varepsilon_1^*]$ be such that, for all $\varepsilon \in (0, \varepsilon_2^*]$,

$$2\lambda_{\max}^{\frac{1}{2}}(P)m\pi^2|B'P_0|^2 + 2\lambda_{\max}(P)m\pi|P_0||B|^2 + 2b^2\lambda_{\max}^{\frac{3}{2}}(P)|B|^2 \leq 1,$$

we have that,

$$\dot{V} < 0, \quad \forall \varepsilon \in (0, \varepsilon_2^*]. \tag{8.2.29}$$

If $|v_i - \mu_i + \omega_i| > \varepsilon^2\Delta$ for at least one, but not all, i, then for any i such that $|v_i - \mu_i + \omega_i| \leq \varepsilon^2\Delta$, we have that, for $\varepsilon \in (0, \sqrt{3/4}]$, $|v_i| \leq \varepsilon^2\Delta + \Delta/8 + \Delta/8 \leq \Delta$, and,

$$-\mu_i[\sigma_{pi}(v_i) - \mu_i] - \frac{1}{4}\mu_i^2 + (v_i - \mu_i + \omega_i)\sigma_{ri}(-(v_i - \mu_i + \omega_i)/\varepsilon^2)$$

$$\leq \begin{cases} -\mu_i(v_i - \mu_i + \omega_i) + \omega_i^2 - \frac{1}{\varepsilon^2}(v_i - \mu_i + \omega_i)^2 & \text{if } \mu_i v_i \geq 0 \\ -b\mu_i(v_i - \mu_i + \omega_i) + b^2\omega_i^2 - \frac{1}{\varepsilon^2}(v_i - \mu_i + \omega_i)^2 & \text{if } \mu_i v_i < 0 \end{cases}$$

$$\leq \frac{\varepsilon}{2m}x'x + \lambda_{\max}^2(P)b^2|B|^2e'e$$

$$- \left(\frac{1}{\varepsilon^2} - \frac{\lambda_{\max}^2(P)mb^2|B|^2}{2\varepsilon}\right)(v_i - \mu_i + \omega_i)^2. \tag{8.2.30}$$

It then follows from (8.2.23), (8.2.25), (8.2.26), (8.2.28) and (8.2.30) that,

$$\dot{V} < -\frac{\varepsilon}{2}x'x - \lambda_{\max}^{\frac{1}{2}}(P)\left(1 - 2\lambda_{\max}^{\frac{1}{2}}(P)m\pi^2|B'P_0|^2 - 2\lambda_{\max}^{\frac{1}{2}}(P)mb^2|B'P_0|^2\right.$$

$$\left. -2\lambda_{\max}(P)m\pi|P_0||B|^2 - 2\lambda_{\max}^{\frac{3}{2}}(P)mb^2|B|^2\right)e'e$$

$$-\frac{\Delta}{2}\sum_{|v_j-\mu_j+\omega_j|>\varepsilon^2\Delta}|v_i - \mu_j + \omega_j|$$

$$-2\sum_{|v_j-\mu_j+\omega_j|\leq\varepsilon^2\Delta}\left(\frac{1}{\varepsilon^2} - \frac{\lambda_{\max}^2(P)mb^2|B|^2}{2\varepsilon} - m\right)(v_j - \mu_j + \omega_j)^2. \tag{8.2.31}$$

Let $\varepsilon_3^* \in (0, \min\{\varepsilon_1^*, \sqrt{3/4}\}]$ be such that, for all $\varepsilon \in (0, \varepsilon_3^*]$,

$$\lambda_{\max}^{\frac{1}{2}}(P)m\pi^2|B'P_0|^2 + \lambda_{\max}^{\frac{1}{2}}(P)mb^2|B'P_0|^2 + \lambda_{\max}(P)m\pi|P_0||B|^2$$

$$+ \lambda_{\max}^{\frac{3}{2}}(P)mb^2|B|^2 \leq 1/2,$$

and

$$\varepsilon\lambda_{\max}^2(P)mb^2|B|^2 + 2m\varepsilon^2 \leq 2.$$

Such an ε_3^* exists since $\lim_{\varepsilon\to 0} P = 0$. With this choice of ε_3^*, we have,

$$\dot{V} < 0, \quad \forall(x, e, v) \in L_V(c) \setminus (0, 0, 0), \varepsilon \in (0, \varepsilon_3^*]. \tag{8.2.32}$$

Finally, in the case that $|v_i - \mu_i + \omega_i| \le \varepsilon^2\Delta$ for all $i = 1$ to m. In this case, we note that, for $\varepsilon \in (0, \sqrt{3/4}]$, $|v_i| \le \varepsilon^2\Delta + \Delta/8 + \Delta/8 \le \Delta$ and,

$$
\begin{aligned}
&-\mu_i\left[\sigma_{pi}(v_i) - \mu_i\right] - \frac{1}{4}\mu_i^2 + (v_i - \mu_i + \omega_i) \\
&\quad \times [\sigma_{ri}(-(v_i - \mu_i + \omega_i)/\varepsilon^2) + B_i'PAx - B_i'P(A + LC)e] \\
&\le \begin{cases} -\mu_i(v_i - \mu_i + \omega_i) + \omega_i^2 & \text{if } \mu_i v_i \ge 0 \\ -b\mu_i(v_i - \mu_i + \omega_i) + b^2\omega_i^2 & \text{if } \mu_i v_i < 0 \end{cases} \\
&\quad -\frac{1}{\varepsilon^2}(v_i - \mu_i + \omega_i)^2 + |v_i - \mu_i + \omega_i|(|B_i'PA||x| + |B_i'P(A + LC)||e|) \\
&\le (b|B'P| + |B'PA|)|v - \mu + \omega||x| + |B'P(A + LC)||v - \mu - \omega||e| \\
&\quad -\frac{1}{\varepsilon^2}(v_i - \mu_i + \omega_i)^2 + b^2\omega_i^2.
\end{aligned} \tag{8.2.33}
$$

Hence it follows from (8.2.23), (8.2.26) and (8.2.33) that

$$
\begin{aligned}
\dot{V} &\le -\varepsilon|x|^2 - \lambda_{\max}^{\frac{1}{2}}(P)|e|^2 + 2|v - \mu + \omega|^2 \\
&\quad + 2m\lambda_{\max}(P)[\pi^2|B'P_0|^2 + b^2|B'P_0|^2 + \lambda_{\max}^{\frac{1}{2}}(P)\pi|P_0||B|^2]|e|^2 \\
&\quad + 2m(b|B'P| + |B'PA|)|v - \mu + \omega||x| \\
&\quad + 2m\lambda_{\max}(P)|B||(A + LC)||v - \mu - \omega||e| - \frac{2}{\varepsilon^2}|v - \mu + \omega|^2 + 2b^2|\omega|^2 \\
&\le -\frac{\varepsilon}{2}|x|^2 - \lambda_{\max}^{\frac{1}{2}}(P)\left[1 - 2m\lambda_{\max}^{\frac{1}{2}}(P)\pi^2|B'P_0|^2 - \lambda_{\max}^{\frac{1}{2}}(P)b^2|B'P_0|^2\right. \\
&\quad \left. -\lambda_{\max}(P)\pi|P_0||B|^2 - \lambda_{\max}^{\frac{1}{2}}(P) - 2\lambda_{\max}^{\frac{3}{2}}(P)b^2|B|^2\right]|e|^2 \\
&\quad -\frac{2}{\varepsilon^2}\left[1 - \varepsilon^2 - \varepsilon m^2(b|B'P| + |B'PA|)^2 - \frac{\varepsilon^2}{2}\lambda_{\max}(P)m^2|B|^2|A + LC|^2\right] \\
&\quad \times |v - \mu + \omega|^2.
\end{aligned} \tag{8.2.34}
$$

Let $\varepsilon_4^* \in (0, \min\{\varepsilon_1^*, \sqrt{1/2}\}]$ be such that, for all $\varepsilon \in (0, \varepsilon_4^*]$,

$$
\lambda_{\max}^{\frac{1}{2}}(P)\left[2m\pi^2|B'P_0|^2 + b^2|B'P_0|^2 + \lambda_{\max}^{\frac{1}{2}}(P)\pi|P_0||B|^2 + 1\right.
$$
$$
\left. + 2\lambda_{\max}(P)b^2|B|^2\right] \le \frac{1}{2},
$$

and

$$
\varepsilon^2 + \varepsilon m^2(b|B'P| + |B'PA|)^2 + \frac{\varepsilon^2}{2}\lambda_{\max}(P)m^2|B|^2|A + LC|^2 \le \frac{1}{2}.
$$

Such an ε_4^* exists since $\lim_{\varepsilon \to 0} P = 0$. With this choice of ε_4^*, we have,

$$
\dot{V} \le -\frac{\varepsilon}{2}|x|^2 - \frac{1}{2}\lambda_{\max}^{\frac{1}{2}}(P)|e|^2 - \frac{1}{\varepsilon^2}|x + B'Px - B'Pe|^2,
$$
$$
\forall(x, e, v) \in L_V(c), \varepsilon \in (0, \varepsilon_4^*]. \tag{8.2.35}
$$

In conclusion, (8.2.29), (8.2.32) and (8.2.35) together show that,

$$\dot{V} < 0, \quad \forall (x, e, v) \in L_V(c) \setminus (0, 0, 0), \ \varepsilon \in (0, \varepsilon^*], \tag{8.2.36}$$

where $\varepsilon^* = \min\{\varepsilon_2^*, \varepsilon_3^*, \varepsilon_4^*\}$. This implies that the equilibrium $(x, e, v) = (0, 0, 0)$ of the closed-loop system is locally asymptotically stable with $L_V(c)$ contained in its basin of attraction. Recalling that $(x, \hat{x}, v) \in \mathcal{W}_0$ implies $(x, e, v) \in L_V(c)$, we complete the proof of the output feedback result. ⊠

8.3. Discrete-Time Systems

8.3.1. Problem Statement

We consider a discrete-time system of the form,

$$\begin{cases} x^+ = Ax + B\sigma_p(v), \\ v^+ = v + \sigma_r((\alpha - 1)v + u), \ |\alpha| < 1, \\ y = Cx, \end{cases} \tag{8.3.1}$$

where $x \in \mathbb{R}^n$ is the plant state, $v \in \mathbb{R}^m$ is the actuator state, $u \in \mathbb{R}^m$ is the control input to the actuators, $y \in \mathbb{R}^p$ is the measurement output, and the functions σ_p and $\sigma_r : \mathbb{R}^m \to \mathbb{R}^m$ are saturation functions to be defined next. They represent respectively the actuator magnitude and rate saturation. Here we have also without loss of generality assumed that all actuators have the same "time constant" α.

Definition 8.3.1. *A function $\sigma : \mathbb{R}^m \to \mathbb{R}^m$ is said to be a saturation function if,*

1. *$\sigma(s)$ is decentralized, i.e., $\sigma(s) = [\sigma_1(s_1), \sigma_2(s_2), \cdots, \sigma_m(s_m)]$; and for each $i = 1$ to m,*

2. *σ_i is continuous;*

3. *σ_i is linear in a neighborhood of the origin and is bounded away from the vertical axis outside this neighborhood. Without loss of generality, assume that within this linear neighborhood the slope is unity, i.e.,*

$$\begin{cases} \sigma_i(s_i) = s_i & \text{if } |s_i| \leq \Delta, \\ \Delta \leq |\sigma_i(s_i)| \leq b|s_i| & \text{if } |s_i| > \Delta, \end{cases} \tag{8.3.2}$$

for some (arbitrarily small) $\Delta > 0$ and some (arbitrarily large) $b \geq 1$. ▣

Remark 8.3.1. 1. *The characteristics of the saturation σ outside the neighborhood where it is linear can be quite arbitrary and unknown. Hence, our results hold essentially for any actuator characteristics that are linear in an (arbitrarily small) neighborhood of the origin.*

2. We recall that the saturation function as defined here is a class $\mathcal{S}_1$ saturation function with the extra property that it has a linear growth rate outside a neighborhood of the origin. For this reason, we will call such a saturation function as a class $\mathcal{S}_1^{LG}$ saturation function. ℝ

The problems to be solved in this section are:

Problem 8.3.1. *Consider the system (8.3.1). The problem of semi-global stabilization via linear state feedback is defined as follows. For any given (arbitrarily large) bounded set $\mathcal{W}_0 \subset \mathbb{R}^{n+m}$, find a linear state feedback law $u = Fx + Gv$ such that the equilibrium of the closed-loop system $(x, v) = (0, 0)$ is locally asymptotically stable with $\mathcal{W}_0$ contained in its basin of attraction.* ℙ

Problem 8.3.2. *Consider the system (8.3.1). The problem of semi-global stabilization via linear dynamic output feedback of dynamical order n is defined as follows. For any given (arbitrarily large) bounded set $\mathcal{W}_0 \subset \mathbb{R}^{2n+m}$, find a linear dynamic output feedback law of the form,*

$$\begin{cases} \zeta^+ = M\zeta + Ny, & \zeta \in \mathbb{R}^n, \\ u = F\zeta + Gv, \end{cases} \tag{8.3.3}$$

such that the equilibrium of the closed-loop system $(x, v, \zeta) = (0, 0, 0)$ is locally asymptotically stable with $\mathcal{W}_0$ contained in its basin of attraction. ℙ

We make the following assumptions:

Assumption 8.3.1. *The pair (A, B) is asymptotically null controllable with bounded controls (ANCBC), i.e.,*

1. *(A, B) is stabilizable;*

2. *All eigenvalues of A are located inside or on the unit circle.* ⊞

Assumption 8.3.2. *The pair (A, C) is detectable.* ⊞

8.3.2. Semi-Global Stabilization by Linear Feedback

Concerning semi-global stabilization of the system (8.3.1), we have the following results.

Theorem 8.3.1. *Consider system (8.3.1). If Assumption 8.3.1 is satisfied, then Problem 8.3.1 is solvable. If Assumptions 8.3.1 and 8.3.2 are satisfied, then Problem 8.3.2 is solvable.* ⊤

Proof of Theorem 8.3.1. We will separate the proof into two parts, one for the state feedback result and the other for the output feedback.

State Feedback Result:

We will show that the following family of linear state feedback laws solves Problem 8.3.1,

$$
\begin{aligned}
u &= -(I + B'P(\varepsilon)B)^{-1}B'P(\varepsilon)AAx + (1 - 1/b)(I + B'PB)^{-1}B'P(\varepsilon)Ax \\
&\quad -[(\alpha + 1/b - 1)I + (I + B'P(\varepsilon)B)^{-1}B'P(\varepsilon)AB]v \\
&= -H(\varepsilon)Ax + (1 - 1/b)H(\varepsilon)x - [(\alpha + 1/b - 1)I + H(\varepsilon)B]v, \qquad (8.3.4)
\end{aligned}
$$

where $H(\varepsilon) = (I + B'P(\varepsilon)B)^{-1}B'P(\varepsilon)A$ and $P(\varepsilon)$ is the solution to the following ARE,

$$
P = A'PA + \varepsilon I - A'PB(B'PB + I)^{-1}B'PA. \qquad (8.3.5)
$$

We recall that this ARE is the same as the ARE of (2.3.69) and hence possesses the properties of Lemma 2.3.6.

More specifically, we will show that, for any given bounded set $\mathcal{W}_0$, there exists an $\varepsilon^* \in (0, 1]$ such that, for all $\varepsilon \in (0, \varepsilon^*]$, the equilibrium $(x, v) = (0, 0)$ of the closed-loop system is locally asymptotically stable with $\mathcal{W}_0$ contained in its basin of attraction.

We note here that, in the absence of actuator saturation, $b = 1$ and the above state feedback law simplifies to

$$
u = -(I + B'P(\varepsilon)B)^{-1}B'P(\varepsilon)AAx - [\alpha I + (I + B'P(\varepsilon)B)^{-1}B'P(\varepsilon)AB]v,
$$

which is a composite control law that induces an LQR type low gain feedback $-(I + B'P(\varepsilon)B)^{-1}B'P(\varepsilon)Ax$ on the plant dynamics. The terms with b in the feedback law (8.3.4) are included to cope with the uncertainties in the actuator nonlinearities outside the linear neighborhood. The mechanism behind the choice of these terms will be clear as we proceed with the proof. We recall that in the continuous-time case (Section 8.2), the induction of the low gain feedback for the plant is through the high gain action on the actuator dynamics. This high gain action itself possesses the ability to counteract various uncertainties.

We also note that the feedback law is parameterized in a single parameter ε. As $\varepsilon \to 0$, the feedback gain for the plant state goes to zero. For this reason, the feedback law is referred to as a low gain feedback. It will be clear shortly that low gain feedback, although crucial to the establishment of our results, does not help prevent saturation from occurring.

Under this feedback law, the closed-loop system is given by,

$$
\begin{cases}
x^+ = Ax + B\sigma_p(v), \\
v^+ = v + \sigma_r(-v/b + (1 - 1/b)Hx - HAx - HBv),
\end{cases} \qquad (8.3.6)
$$

where, for notational convenience, we have suppressed the dependency on ε of P and H. Pick the Lyapunov function,

$$V(x,v) = x'Px + 4b^2[v + Hx]'[v + Hx], \tag{8.3.7}$$

and let $c > 0$ be such that

$$c \geq \sup_{\varepsilon \in (0,1], (x,v) \in \mathcal{W}_0} \left\{ x'Px + 4b^2[v + Hx]'[v + Hx] \right\}. \tag{8.3.8}$$

Such a c exists since $\mathcal{W}_0$ is bounded and $\lim_{\varepsilon \to 0} P = 0$. Let $\varepsilon^* \in (0,1]$ be such that for all $\varepsilon \in (0, \varepsilon^*]$,

$$(x,v) \in L_V(c) = \{(x,v) : V(x,v) \leq c\} \Longrightarrow$$

$$B'PB \leq \frac{\Delta^2}{2[b(\sqrt{c} + \Delta) + \Delta]^2} I, \ |Hx| \leq \frac{\Delta}{32b}, \ |HAx| \leq \frac{\Delta}{32b},$$

$$|HBv| \leq \frac{\Delta}{32b\sqrt{4(b-1)^2 + 1}}, \ |HB\sigma_p(v)| \leq \frac{\Delta}{32b\sqrt{4(b-1)^2 + 1}}. \tag{8.3.9}$$

The existence of such an ε^* is due to Lemma 2.3.6 and the fact that $|\sigma_{pi}(s_i)| \leq b|s_i|$.

Also, it follows from (8.3.5) that

$$[A - BH]'P[A - BH] - P = -\varepsilon I - H'H. \tag{8.3.10}$$

The evaluation of the difference of V along the trajectories of the closed-loop system (8.3.6) for $(x,v) \in L_V(c)$, $\varepsilon \in (0, \varepsilon^*]$, can then be carried out as follows,

$$\begin{aligned}
\Delta V &= -\varepsilon x'x - x'H'Hx + [\sigma_p(v) + Hx]'B'PB[\sigma_p(v) + Hx] \\
&\quad + 2x'[A - BH]'PB[\sigma_p(v) + Hx] \\
&\quad + 4b^2[v + \sigma_r(-(v + Hx)/b + Hx - HAx - HBv) + HAx + HB\sigma_p(v)]' \\
&\quad \times [v + \sigma_r(-(v + Hx)/b + Hx - HAx - HBv) + HAx + HB\sigma_p(v)] \\
&\quad - 4b^2[v + Hx]'[v + Hx] \\
&= -\varepsilon x'x - x'H'Hx + [\sigma_p(v) + Hx]'B'PB[\sigma_p(v) + Hx] \\
&\quad + 2x'H'[\sigma_p(v) + Hx] \\
&\quad + 4b^2[v + \sigma_r(-(v + Hx)/b + Hx - HAx - HBv) + HAx + HB\sigma_p(v)]' \\
&\quad \times [v + \sigma_r(-(v + Hx)/b + Hx - HAx - HBv) + HAx + HB\sigma_p(v)] \\
&\quad - 4b^2[v + Hx))]'[v + Hx] \\
&\leq -\varepsilon x'x
\end{aligned}$$

$$+ \sum_{i=1}^{m} \left[-h_i^2 + 2h_i[\sigma_{pi}(v_i) + h_i] + \frac{\Delta^2}{2[b(\sqrt{c} + \Delta) + \Delta]^2}[\sigma_{pi}(v_i) + h_i]^2 \right.$$

$$\left. + 4b^2[v_i + \sigma_{ri}(-(v_i + h_i)/b + h_i - q_i - r_i) + q_i + t_i]^2 - 4b^2(v_i + h_i)^2 \right],$$

$$(8.3.11)$$

where h_i, q_i, r_i and t_i are respectively the ith element of Hx, HAx, HBv and $HB\sigma_p(v)$.

We now examine the summand under the summation sign. First consider the case that $|v_i| \geq \frac{7\Delta}{8}$. In this case, it follows from (8.3.8), (8.3.9) and the definition of the saturation function that,

$$[\sigma_{pi}(v_i) + h_i]^2 \leq [b(\sqrt{c} + \Delta) + \Delta]^2, \qquad (8.3.12)$$

and hence

$$-h_i^2 + 2h_i[\sigma_{pi}(v_i) + h_i] + \frac{\Delta^2}{2[b(\sqrt{c} + \Delta) + \Delta]^2}[\sigma_{pi}(v_i) + h_i]^2$$

$$+4b^2[v_i + \sigma_{ri}(-(v_i + h_i)/b + h_i - q_i - r_i) + q_i + t_i]^2 - 4b^2(v_i + h_i)^2$$

$$\leq \frac{\Delta^2}{32^2} + \frac{\Delta}{16}|v_i| + \frac{\Delta^2}{2} + 4b^2 \left[|v_i| - \frac{1}{b}\left(\frac{7\Delta}{8} - \frac{\Delta}{8}\right) + \frac{\Delta}{16b} \right]^2$$

$$-4b^2 \left[|v_i| - \frac{\Delta}{32b} \right]^2$$

$$= \frac{\Delta^2}{32^2} + \frac{\Delta^2}{2} + \frac{\Delta}{16}|v_i| - \frac{21\Delta}{4}|v_i| + \frac{483\Delta^2}{256}$$

$$< -\Delta^2. \qquad (8.3.13)$$

We next consider the case that $|v_i| < \frac{7\Delta}{8}$. In this case, it follows from (8.3.9) that,

$$\sigma_{pi}(v_i) = v_i, \quad \sigma_{ri}(-(v_i + h_i)/b + h_i - q_i - r_i) = -(v_i + h_i)/b + h_i - q_i - r_i, \quad (8.3.14)$$

and hence

$$-h_i^2 + 2h_i[\sigma_{pi}(v_i) + h_i] + \frac{\Delta^2}{2[b(\sqrt{c} + \Delta) + \Delta]^2}[\sigma_{pi}(v_i) + h_i]^2$$

$$+4b^2[v_i + \sigma_{ri}(-(v_i + h_i)/b + h_i - q_i - r_i) + q_i + t_i]^2 - 4b^2(v_i + h_i)^2$$

$$\leq -\frac{1}{2}h_i^2 + \frac{5}{2}[v_i + h_i]^2 + 4b^2[(1 - 1/b)(v_i + h_i) - r_i + t_i]^2 - 4b^2(v_i + h_i)^2$$

$$\leq \frac{5}{2}[v_i + h_i]^2 - 4(2b - 1)[v_i + h_i]^2 + [v_i + h_i]^2$$

$$+4b^2 \left[4(b - 1)^2 + 1 \right] (r_i - t_i)^2$$

$$\leq -\frac{1}{2}[v_i + h_i]^2 + 4b^2 \left[4(b - 1)^2 + 1 \right] (r_i - t_i)^2. \qquad (8.3.15)$$

Now, if $|v_i| < \frac{7\Delta}{8}$ for all $i = 1$ to m, then $HB\sigma_p(v) = HBv$ and $t_i = r_i$ for all $i = 1$ to m. It then follows from (8.3.11) and (8.3.15) that, $\forall (x, v) \in L_V(c)$ and $\forall \varepsilon \in (0, \varepsilon^*]$,

$$\Delta V \leq -\varepsilon x'x - \sum_{i=1}^{m} \frac{1}{2}[v_i + h_i]^2 = -\varepsilon x'x - \frac{1}{2}(v + Hx)'(v + Hx). \qquad (8.3.16)$$

On the other hand, if there exists at least one $i = i_0$ such that $|v_{i_0}| \geq \frac{7\Delta}{8}$, it follows from (8.3.11), (8.3.13) and (8.3.15) that, $\forall (x, v) \in L_V(c)$ and $\forall \varepsilon \in (0, \varepsilon^*]$,

$$\Delta V \leq -\varepsilon x'x - \Delta^2 + \sum_{|v_i| < \frac{7\Delta}{8}} \left[-\frac{1}{2}[v_i + h_i]^2 + 4b^2[4(b-1)^2 + 1](r_i - t_i)^2 \right]$$

$$\leq -\Delta^2 + \sum_{i=1}^{m} 4b^2[4(b-1)^2 + 1](r_i - t_i)^2$$

$$\leq -\Delta^2 + 4b^2[4(b-1)^2 + 1]|HBv - HB\sigma_p(v)|^2$$

$$\leq -\Delta^2 + \frac{\Delta^2}{64}$$

$$< 0, \qquad (8.3.17)$$

where we have used the following fact due to (8.3.9),

$$|HBv - HB\sigma_p(v)| \leq \frac{\Delta}{16b\sqrt{4(b-1)^2 + 1}}. \qquad (8.3.18)$$

In conclusion, we have shown that

$$\Delta V < 0, \quad \forall (x, v) \in L_V(c) \setminus (0, 0), \ \varepsilon \in (0, \varepsilon^*]. \qquad (8.3.19)$$

This implies that the equilibrium $(x, v) = (0, 0)$ of the closed-loop system is locally asymptotically stable with $\mathcal{W}_0 \subset L_V(c)$ contained in its basin of attraction. This ends the proof of the state feedback result.

Output Feedback Result:

We will show that the following family of observer based liner output feedback laws solves Problem 8.3.2,

$$\begin{cases} \hat{x}^+ = A\hat{x} - L(y - C\hat{x}) + Bv, \\ u = -HA\hat{x} + (1 - 1/b)H\hat{x} - [(\alpha + 1/b - 1)I + HB]v, \end{cases} \qquad (8.3.20)$$

where $H = (I + B'PB)^{-1}B'PA$ with P being the solution to the ARE (8.3.5) and L is any matrix such that $A + LC$ is asymptotically stable. The existence of such an L is due to Assumption 8.3.2. More specifically, we will show that, for any given bounded set W_0, there exists an $\varepsilon^* \in (0,1]$ such that, for all $\varepsilon \in (0,\varepsilon^*]$, the equilibrium $(x,\hat{x},v) = (0,0,0)$ of the closed-loop system is locally asymptotically stable with W_0 contained in its basin of attraction.

Under this feedback law, the closed-loop system is given by,

$$\begin{cases} x^+ = Ax + B\sigma_p(v), \\ \hat{x}^+ = A\hat{x} - L(y - C\hat{x}) + Bv, \\ v^+ = \sigma_r(-v/b + (1 - 1/b)H\hat{x} - HA\hat{x} - HBv). \end{cases} \tag{8.3.21}$$

Letting $e = x - \hat{x}$, the closed-loop system can be rewritten as,

$$\begin{cases} x^+ = Ax + B\sigma_p(v), \\ e^+ = (A + LC)e + B[\sigma_p(v) - v], \\ v^+ = \sigma_r(-v/b + (1 - 1/b)Hx - HAx - HBv - (1 - 1/b)He + HAe). \end{cases} \tag{8.3.22}$$

Pick the Lyapunov function,

$$V(x,e,v) = x'Px + 4b^2(v + Hx)'(v + Hx) + \lambda_{\max}^{\frac{1}{2}}(P)e'P_0e, \tag{8.3.23}$$

where $P_0 > 0$ is the unique solution to the following Lyapunov equation

$$(A + LC)'P_0(A + LC) - P_0 = -I. \tag{8.3.24}$$

Let $c > 0$ be such that

$$c \geq \sup_{\varepsilon \in (0,1], (x,\hat{x},v) \in W_0} \left\{ x'Px + (v + Hx)'(v + Hx) + \lambda_{\max}^{\frac{1}{2}}(P)e'P_0e \right\}. \tag{8.3.25}$$

Such a c exists since W_0 is bounded and $\lim_{\varepsilon \to 0} P = 0$. We note here that with this choice of c, $(x,\hat{x},v) \in W_0$ implies that $(x,e,v) \in L_V(c)$.

Let $\varepsilon^* \in (0,1]$ be such that for all $\varepsilon \in (0,\varepsilon^*]$,

$$(x,e,v) \in L_V(c) \Longrightarrow B'PB \leq \frac{\Delta^2}{2[b(\sqrt{c} + \Delta) + \Delta]^2}I, \ |Hx| \leq \frac{\Delta}{64b}, \ |HAx| \leq \frac{\Delta}{64b},$$

$$|HBv| \leq \frac{\Delta}{64b\sqrt{4(b-1)^2 + 1}}, \ |HB\sigma_p(v)| \leq \frac{\Delta}{64\sqrt{4(b-1)^2 + 1}},$$

$$(1 - 1/b)|HAe| \leq \frac{\Delta}{64b}, \ |HAe| \leq \frac{\Delta}{64b},$$

$$|H[(1 - 1/b)I - A]| \leq \frac{\lambda_{\max}^{\frac{1}{4}}(P)}{4b\sqrt{4(b-1)^2 + 1}},$$

$$\lambda_{\max}^{\frac{1}{2}}(P)|e'(A + LC)'P_0B[\sigma_p(v) - v]| \leq \frac{\Delta^2}{32},$$

$$\lambda_{\max}^{\frac{1}{2}}(P)|[\sigma_p(v) - v]'B'P_0B[\sigma_p(v) - v] \leq \frac{\Delta^2}{32},$$

$$|H[(1 - 1/b)I - A]e| \leq \frac{\Delta}{4b\sqrt{4(b-1)^2 + 1}}. \tag{8.3.26}$$

The existence of such an ε^* is again due to Lemma 2.3.6 and the fact that $|\sigma_{pi}(s_i)| \leq b|s_i|$.

The evaluation of the difference of V along the trajectories of the closed-loop system (8.3.22) for $(x, e, v) \in L_V(c)$, $\varepsilon \in (0, \varepsilon^*]$, can then be carried out as follows,

$$
\begin{aligned}
\Delta V = &-\varepsilon x'x - x'H'Hx + [\sigma_p(v) + Hx]'B'PB[\sigma_p(v) + Hx] \\
&+2x'[A - BH]'PB[\sigma_p(v) + Hx] \\
&-\lambda_{\max}^{\frac{1}{2}}(P)e'e + 2\lambda_{\max}^{\frac{1}{2}}(P)e'(A + LC)'P_0B[\sigma_p(v) - v] \\
&+\lambda_{\max}^{\frac{1}{2}}(P)[\sigma_p(v) - v]'B'P_0B[\sigma_p(v) - v] \\
&+4b^2[v + \sigma_r(-(v + Hx)/b + Hx - HAx - HBv - (1 - 1/b)He + HAe) \\
&+HAx + HB\sigma_p(v)]'[v + \sigma_r(-(v + Hx)/b + Hx - HAx - HBv \\
&-(1 - 1/b)He + HAe) + HAx + HB\sigma_p(v)] - 4b^2[v + Hx]'[v + Hx] \\
= &-\varepsilon x'x - x'H'Hx + [\sigma_p(v) + Hx]'B'PB[\sigma_p(v) + Hx] \\
&+2x'H'[\sigma_p(v) + Hx] - \lambda_{\max}^{\frac{1}{2}}(P)e'e \\
&+2\lambda_{\max}^{\frac{1}{2}}(P)e'(A + LC)'P_0B[\sigma_p(v) - v] \\
&+\lambda_{\max}^{\frac{1}{2}}(P)[\sigma_p(v) - v]'B'P_0B[\sigma_p(v) - v] \\
&+4b^2[v + \sigma_r(-(v + Hx)/b + Hx - HAx - HBv - (1 - 1/b)He + HAe) \\
&+HAx + HB\sigma_p(v)]'[v + \sigma_r(-(v + Hx)/b + Hx - HAx - HBv \\
&-(1 - 1/b)He + HAe) + HAx + HB\sigma_p(v)] - 4b^2[v + Hx]'[v + Hx] \\
\leq &-\varepsilon x'x - \lambda_{\max}^{\frac{1}{2}}(P)e'e + 2\lambda_{\max}^{\frac{1}{2}}(P)e'(A + LC)'P_0B[\sigma_p(v) - v] \\
&+\lambda_{\max}^{\frac{1}{2}}(P)[\sigma_p(v) - v]'B'P_0B[\sigma_p(v) - v] \\
&+\sum_{i=1}^{m}\left[-h_i^2 + 2h_i[\sigma_{pi}(v_i) + h_i] + \frac{\Delta^2}{2[b(\sqrt{c} + \Delta) + \Delta]^2}[\sigma_{pi}(v_i) + h_i]^2 \right. \\
&+4b^2[v_i + \sigma_{ri}(-(v_i + h_i)/b + h_i - q_i - r_i - f_i + g_i) + q_i + t_i]^2 \\
&\left. - 4b^2(v_i + h_i)^2\right],
\end{aligned}
\tag{8.3.27}
$$

where h_i, q_i, r_i, t_i, f_i and g_i are respectively the ith element of Hx, HAx, HBv, $HB\sigma_p(v)$, $(1 - 1/b)He$ and HAe.

We now examine the summand under the summation sign. First consider the case that $|v_i| \geq \frac{7\Delta}{8}$. In this case, it follows from (8.3.25), (8.3.26) and the definition of the saturation function that,

$$
[\sigma_{pi}(v_i) + h_i]^2 \leq [b(\sqrt{c} + \Delta) + \Delta]^2,
\tag{8.3.28}
$$

and hence

$$
-h_i^2 + 2h_i[\sigma_{pi}(v_i) + h_i] + \frac{\Delta^2}{2[b(\sqrt{c} + \Delta) + \Delta]^2}[\sigma_{pi}(v_i) + h_i]^2
$$

$$
+4b^2[v_i + \sigma_{ri}(-(v_i + h_i)/b + h_i - q_i - r_i - f_i + g_i) + q_i + t_i]^2
$$

$$
-4b^2(v_i + h_i)^2
$$

$$
\leq \frac{\Delta^2}{64^2} + \frac{\Delta}{32}|v_i| + \frac{\Delta^2}{2} + 4b^2\left[|v_i| - \frac{1}{b}\left(\frac{7\Delta}{8} - \frac{\Delta}{8}\right) + \frac{\Delta}{32b}\right]^2
$$

$$
-4b^2\left[|v_i| - \frac{1}{64b}\right]^2
$$

$$
= \frac{\Delta^2}{64^2} + \frac{\Delta^2}{2} + \frac{\Delta}{32}|v_i| - \frac{45\Delta}{8}|v_i| + \frac{2115\Delta^2}{1024}
$$

$$
\leq -2\Delta^2. \tag{8.3.29}
$$

We next consider the case that $|v_i| < \frac{7\Delta}{8}$. In this case, it follows from (8.3.26) that

$$
\sigma_{pi}(v_i) = v_i \tag{8.3.30}
$$

and

$$
\sigma_{ri}(-(v_i+h_i)/b+h_i-q_i-r_i-f_i+g_i) = -(v_i+h_i)/b+h_i-q_i-r_i-f_i+g_i, \tag{8.3.31}
$$

and hence

$$
-h_i^2 + 2h_i[\sigma_{pi}(v_i) + h_i] + \frac{\Delta^2}{2[b(\sqrt{c} + \Delta) + \Delta]^2}[\sigma_{pi}(v_i) + h_i]^2
$$

$$
+4b^2[v_i + \sigma_{ri}(-(v_i + h_i)/b + h_i - q_i - r_i - f_i + g_i) + q_i + t_i]^2
$$

$$
-4b^2(v_i + h_i)^2
$$

$$
\leq -\frac{1}{2}h_i^2 + \frac{5}{2}[v_i + h_i]^2 + 4b^2[(1 - 1/b)(v_i + h_i) - r_i + t_i - f_i + g_i]^2
$$

$$
-4b^2(v_i + h_i)^2
$$

$$
\leq \frac{5}{2}[v_i + h_i]^2 - 4(2b - 1)[v_i + h_i]^2 + [v_i + h_i]^2
$$

$$
+8b^2\left[4(b - 1)^2 + 1\right](r_i - t_i)^2 + 8b^2\left[4(b - 1)^2 + 1\right](f_i - g_i)^2
$$

$$
\leq -\frac{1}{2}[v_i + h_i]^2 + 8b^2\left[4(b - 1)^2 + 1\right](r_i - t_i)^2
$$

$$
+8b^2\left[4(b - 1)^2 + 1\right](f_i - g_i)^2. \tag{8.3.32}
$$

Now, if $|v_i| < \frac{7\Delta}{8}$ for all $i = 1$ to m, then $\sigma_p(v) = v$, $HB\sigma_p(v) = HBv$ and $t_i = r_i$ for all $i = 1$ to m. It then follows from (8.3.27), (8.3.32) and (8.3.26) that, $\forall(x, e, v) \in L_V(c)$ and $\forall \varepsilon \in (0, \varepsilon^*]$,

$$
\Delta V \leq -\varepsilon x'x - \lambda_{\max}^{\frac{1}{2}}(P)e'e
$$

$$+ \sum_{i=1}^{m} \left[-\frac{1}{2}[v_i + h_i]^2 + 8b^2 \left[4(b-1)^2 + 1 \right] (f_i - g_i)^2 \right]$$

$$= -\varepsilon x' x - \lambda_{\max}^{\frac{1}{2}}(P)e'e - \frac{1}{2}(v + Hx)'(v + Hx)$$

$$+ 8b^2 \left[4(b-1)^2 + 1 \right] e'[(1 - 1/b)I - A]'H'H[(1 - 1/b)I - A]e$$

$$\leq -\varepsilon x' x - \frac{1}{2}\lambda_{\max}^{\frac{1}{2}}(P)e'e - \frac{1}{2}(v + Hx)'(v + Hx). \qquad (8.3.33)$$

On the other hand, if there exists at least one $i = i_0$ such that $|v_{i_0}| \geq \frac{7\Delta}{8}$, it follows from (8.3.27), (8.3.29), (8.3.32), and (8.3.26) that,

$$\Delta V \leq -\varepsilon x' x - \lambda_{\max}^{\frac{1}{2}}(P)e'e + \frac{\Delta^2}{16} + \frac{\Delta^2}{32} - 2\Delta^2 + \sum_{|v_i| < \frac{7\Delta}{8}} \left[-\frac{1}{2}[v_i + h_i]^2 \right.$$

$$\left. + 8b^2[4(b-1)^2 + 1](r_i - t_i)^2 + 8b^2[4(b-1)^2 + 1](f_i - g_i)^2 \right]$$

$$\leq -\frac{61\Delta^2}{32}$$

$$+ \sum_{i=1}^{m} \left[8b^2[4(b-1)^2 + 1](r_i - t_i)^2 + 8b^2[4(b-1)^2 + 1](f_i - g_i)^2 \right]$$

$$= -\frac{61\Delta^2}{32} + 8b^2[4(b-1)^2 + 1]\|HB[v - \sigma_p(v)]\|^2$$

$$+ 8b^2[4(b-1)^2 + 1]\|H[(1 - 1/b)I - A]e\|^2$$

$$\leq -\frac{51\Delta^2}{128}. \qquad (8.3.34)$$

In conclusion, we have shown that,

$$\Delta V < 0, \quad \forall (x, e, v) \in L_V(c) \setminus (0, 0, 0), \ \varepsilon \in (0, \varepsilon^*]. \qquad (8.3.35)$$

This implies that the equilibrium $(x, e, v) = (0, 0, 0)$ of the closed-loop system is locally asymptotically stable with $L_V(c)$ contained in its basin of attraction. Recalling that $(x, \hat{x}, v) \in \mathcal{W}_0$ implies $(x, e, v) \in L_V(c)$, we complete the proof of the output feedback result.   ⊠

8.4. Concluding Remarks

Actuator rate saturation causes phase lag that could lead to instability of the closed-loop system. The problem is more sever when the actuator is also subject to magnitude saturation since small actuator output results in small stability margin even in the absence of rate saturation. In this chapter, we have utilized the low gain feedback design techniques of Chapter 2 to construct linear feedback laws that achieve semi-global asymptotic stabilization with magnitude and rate limited actuators. The assumptions made on the open loop system are the same as for semi-global asymptotic stabilization with actuators that are limited only in magnitude. The question of to what extent the low-and-high gain feedback design techniques as developed for linear systems subject to only actuator magnitude saturation (Chapter 4) can be applied to linear systems subject to both actuator magnitude and rate saturation remains to be answered.

Chapter 9

Robust Semi-Global Stabilization of Minimum Phase Input-Output Linearizable Systems

9.1. Introduction

We consider the class of multi-input multi-output input output linearizable systems,

$$\begin{cases} \dot{x} = f(x, C\xi), \quad x \in \mathbb{R}^l, \ C\xi \in \mathbb{R}^r, \\ \dot{\xi} = A\xi + B\left[G\left(x, \xi, d\left(t\right)\right)u + g\left(x, \xi, d\left(t\right)\right)\right], \quad \xi \in \mathbb{R}^n, \ u \in \mathbb{R}^m, \quad (9.1.1) \\ y = D\xi, \quad y \in \mathbb{R}^p, \end{cases}$$

where y is the only available measurement on the system, u is the control input, $d(t)$ is any continuous disturbance assumed to belong to a compact set $\mathcal{D} \subset \mathbb{R}^q$, f and g are $\mathcal{C}^1$ vector functions, G is a $\mathcal{C}^1$ matrix function which is invertible for all $(x, \xi, d) \in \mathbb{R}^l \times \mathbb{R}^n \times \mathcal{D}$, and finally, A, B, C and D are constant matrices of appropriate dimensions such that the pairs (A, B) and (A, D) are stabilizable and detectable respectively.

For such a class of systems, the dynamical system $\dot{x} = f(x, 0)$ is referred to as zero dynamics (e.g., [4–6]). Moreover, the system (9.1.1) is said to be of minimum-phase if its zero dynamics has $x = 0$ as a globally asymptotically stable equilibrium point.

Most of the early research activities on designing controllers for minimum-phase input output linearizable systems focused on *global* stabilization using

197

state and/or output feedback (e.g., [13,31,73,88]). The critical and very restrictive assumptions in these works were the *global growth conditions* (typically global Lipschitz-like conditions) on the mappings f and g. Recent research efforts on the stabilization of such systems concentrate on removing or relaxing these growth conditions on f and g.

The efforts of Kokotovic and his co-workers (e.g., [33,87,106,107]) showed that the removal of the global growth condition on f is possible if one restricts the dependency of f on the state ξ of the linearizable part of the system; namely, one can successfully remove the global growth condition on f and achieve global stabilization if a proper restriction is placed on the matrix C. Note that C represents the dependency of f on the state of the linearizable part of the system. They also showed that placing restrictions on C is "necessary" (see [87]). The peaking phenomenon that causes such an intricate problem was identified and studied in [107]. To capture the condition on C, we briefly recall the main results of [87] which deals with global stabilization of the minimum-phase system (9.1.1) in the absence of $d(t)$ using the measurement of the state of the system x and ξ. By redefining the input through pre-feedback one can exactly cancel G and g and the system (9.1.1) reduces to a simplified form,

$$\begin{cases} \dot{x} = f(x, C\xi), \\ \dot{\xi} = A\xi + Bu. \end{cases} \tag{9.1.2}$$

In [87] it was shown that under the minimum-phase assumption and with no growth condition on f, one can globally asymptotically stabilize such systems if C is such that (A, B, C) is right invertible and weakly minimum-phase[1].

Moreover, examples were given that indicate that this condition on C is "necessary" in the sense that they cannot be further relaxed unless a growth condition is imposed on f. We must emphasize that such restrictions on C were obtained under the assumption that all the state of the system is measured and available for feedback. A more recent line of research initiated by Teel (e.g., [55–57,111]) relaxed the restriction on C by considering semi-global stabilization of such systems. The relaxation of the condition on C is obtained obviously at the cost of replacing global stabilization by semi-global stabilization. In [57], we have shown that the minimum-phase system (9.1.2) can be semi-globally stabilizable via linear feedback laws using only the measurement of the state of the linearized part of the system, ξ, as long as (A, B, C) is right invertible and has all its invariant zeros located in the closed left-half s-plane. Note that the weakly minimum-phase assumption on (A, B, C) required in [87] for global

---

[1]A linear system (A, B, C) is said to be of weakly minimum phase if its zero dynamics is stable in the sense of Lyapunov.

stabilization is relaxed by allowing repeated invariant zeros on the $j\omega$-axis. This is a rather significant relaxation since some unstable zero dynamics of (A, B, C) would be allowed.

The global growth condition on g appears when one is dealing with either *global* output feedback stabilization problem or when the function g is not exactly known in the *global* state feedback stabilization problem. Obviously, for the case of *global* state feedback stabilization and when the function g is exactly known, one can always cancel out g by a simple feedback law. Recently, it was shown in [76] that such a global growth condition in the global output feedback stabilization problem in generally cannot be removed. This was shown by the following seemingly simple system

$$
\begin{cases}
\dot{\xi}_1 = \xi_2, \\
\dot{\xi}_2 = \xi_2^n + u, \\
y = \xi_1.
\end{cases}
\tag{9.1.3}
$$

Note that this system is an input output linearizable system which has no zero dynamics. Hence there is no problem regarding f. Moreover, for this system $g = \xi_2^n$. In [76] it is shown that for $n \geq 3$ there is no continuous finite dimensional dynamic output feedback law which globally stabilizes this system. Naturally, the next step is to examine the global growth requirements on g in the context of semi-global output feedback stabilization. Two very recent papers ([30,115]) have shown that in the case of semi-global output feedback stabilization and for some special cases of the system (9.1.1) one can relax these growth conditions on g to some local growth condition. These local growth conditions are, obviously, far less restrictive than the global ones and in fact they disappear when the zero dynamics of the system is globally exponentially stable or when g does not depend on x. Moreover, in [115] it was also shown that the global growth condition on g can be removed in the case of semi-global practical stabilization.

In this chapter we deal with the semi-global stabilization and/or semi-global practical stabilization of the minimum-phase system (9.1.1). It is related and can be considered as a continuation of the two recent works [30] and [115]. The work of [30] deals with the semi-global stabilization of the system (9.1.1) without zero dynamics. Here the high gain observer theory as developed in [93] is utilized to design stabilizing controllers. In order to avoid the destabilizing effect of the peaking phenomenon associated with the high gain observer, [30] proposed an efficient technique of saturating the control during a short time period of the excessively large overshoot in the error dynamics due to the peaking. The work of [115] utilized this technique of [30] and generalized the result of [30] to include the nonlinear zero dynamics. More specifically, in [115] it is assumed that (9.1.1) is a single input single output system with (A, B, C) possibly having invariant zeros at the origin. The goal of this chapter is to show how low gain feedback design techniques of Chapter 2 can be utilized to generalize the results of [115] to a general multi-input multi-output minimum-phase system (9.1.1). We assume that (A, B, C) is right invertible and has all its invariant zeros located in the closed left half plane. Note that this condition on (A, B, C) is "necessary" as was discussed earlier (see also [57,87]). We also require (A, B, D) to be left invertible with no invariant zeros.

Our approach to the problem of semi-global stabilization and/or practical stabilization of (9.1.1) consists of two parts. The first part involves the design of semi-globally stabilizing and/or practically stabilizing static partial state feedback laws. That is, we assume that the measurement of the state of the linearizable part of the system is available for feedback (i.e., $D = I$). In the second part, we utilize the design in the first part to develop observer-based-like controllers for the semi-global stabilization and/or practical stabilization of the system (9.1.1). Our proposed semi-global stabilizing and/or practically stabilizing controllers are robust in the sense that their construction does not require any knowledge of the mappings f, g and G. These controllers are designed with tuning parameters which allow the regulation of the closed-loop system state to the origin or to any *a priori* given (arbitrarily small) neighborhood of the origin for any initial conditions in any *a priori* given (arbitrarily large) bounded set by appropriate choice of the values of these parameters.

This chapter is organized as follows. In Section 9.2 we formulate the problems. The state feedback results and the output feedback results are presented in Sections 9.3 and 9.4 respectively. Finally, we draw some conclusions in Section 9.5.

9.2. Problem Statement

We are interested in the problems of semi-global asymptotic stabilization and/or semi-global practical stabilization of the minimum-phase system (9.1.1) via partial state and/or output feedback. To make our problems precise, we need the following definitions.

Definition 9.2.1. *The point $(x, \xi) = (0,0)$ (not necessarily an equilibrium) is said to be semi-globally practically stabilizable for the system (9.1.1) by linear static partial state feedback if, for any a priori given (arbitrarily large) bounded set $\mathcal{W} \subset \mathbb{R}^{l+n}$ and any a priori given (arbitrarily small) set $\mathcal{W}_0 \subset \mathbb{R}^{l+n}$, which* contains $(0,0)$ as an interior point, there exists a static state feedback control law using only the state ξ of the linearizable part of the system (9.1.1),

$$u = F_{w,w_0}\xi, \tag{9.2.1}$$

such that all the solutions of the closed-loop system with initial conditions in W enter the set W_0 in a finite time and remain in it thereafter. ▢

Definition 9.2.2. *The equilibrium $(x,\xi) = (0,0)$ of the system (9.1.1) is said to be semi-globally asymptotically stabilizable by linear static partial state feedback if, for any a priori given (arbitrarily large) bounded set $W \subset \mathbf{R}^{l+n}$, there exists a static state feedback control law using only the state ξ of the linearizable part of the system (9.1.1),*

$$u = F_w\xi, \tag{9.2.2}$$

such that the equilibrium $(0,0)$ of the closed-loop system is locally asymptotically stable and W is contained in its basin of attraction. ▢

Definition 9.2.3. *The point $(x,\xi) = (0,0)$ (not necessarily an equilibrium) is said to be semi-globally practically stabilizable for the system (9.1.1) by dynamic output feedback of dynamical order k, if for any a priori given (arbitrarily large) bounded set $W \subset \mathbf{R}^{l+n+k}$ and any a priori given (arbitrarily small) set $W_0 \subset \mathbf{R}^{l+n+k}$, which contains $(0,0,0)$ as an interior point, there exists a dynamic output feedback control law*

$$\begin{cases} \dot{z} = \varphi_{w,w_0}(z,y), \ z \in \mathbf{R}^k, \\ u = F_{w,w_0}(z,y), \end{cases} \tag{9.2.3}$$

such that all the solutions of the closed-loop system with initial conditions in W enter the set W_0 in a finite time and remain in it thereafter. ▢

Definition 9.2.4. *The equilibrium $(x,\xi) = (0,0)$ of the system Σ is said to be semi-globally asymptotically stabilizable by dynamic output feedback of dynamical order k if, for any a priori given (arbitrarily large) bounded set $W \subset \mathbf{R}^{l+n+k}$, there exists a dynamic output feedback control law*

$$\begin{cases} \dot{z} = \varphi_w(z,y), \ z \in \mathbf{R}^k, \\ u = F_w(z,y), \end{cases} \tag{9.2.4}$$

such that the equilibrium $(x,\xi,z) = (0,0,0)$ of the closed-loop system is locally asymptotically stable and W is contained in its basin of attraction. ▢

The goal of this chapter is to utilize the low gain feedback design technique and the classical high gain feedback to establish semi-global asymptotic stabilizablity and semi-global practical stabilizability of the system (9.1.1) as defined above. To establish our main results, we make the following assumptions on the system (9.1.1). Each of our results will requires a subset of these assumptions.

Assumption 9.2.1. *The zero dynamics*

$$\dot{x} = f(x, 0) \tag{9.2.5}$$

has 0 as a globally asymptotically stable equilibrium point, i.e., the system (9.1.1) is of minimum-phase. ⊞

Assumption 9.2.2. *The pair (A, B) is stabilizable.* ⊞

Assumption 9.2.3. *The triple (A, B, C) is right invertible with all its invariant zeros located in the closed left-half plane.* ⊞

Assumption 9.2.4. *There exist a constant matrix $M \in \mathbb{R}^{m \times m}$ and a positive definite matrix $P \in \mathbb{R}^{m \times m}$, such that*

$$PG(x, \xi, d(t))M + M'G'(x, \xi, d(t))P = -Q(x, \xi, d(t)), \tag{9.2.6}$$

where $Q(x, \xi, d(t))$ is a positive definite matrix uniformly in all its arguments, i.e., there exists a positive definite constant matrix Q_0 such that $Q(x, \xi, d(t)) \geq Q_0$ for all $(x, \xi, d(t)) \in \mathbb{R}^l \times \mathbb{R}^n \times \mathcal{D}$. ⊞

Assumption 9.2.5. *There exist a neighborhood of $x = 0$, $\mathcal{U}_0 \subset \mathbb{R}^l$, a positive definite Lyapunov function $V_0 : \mathbb{R}^l \to \mathbb{R}$, and a class $\mathcal{K}$ function ψ, such that for all $x \in \mathcal{U}_0$,*

$$\frac{\partial V_0(x)}{\partial x} f(x, 0) \leq -\psi^2(|x|), \tag{9.2.7}$$

$$\left| \frac{\partial V_0(x)}{\partial x} \right| \leq \gamma_f \psi(|x|), \tag{9.2.8}$$

and

$$|g(x, 0, d)| \leq \gamma_g \psi(|x|), \tag{9.2.9}$$

for some positive constant numbers γ_f and γ_g. ⊞

Assumption 9.2.6. *The triple (A, B, D) is left invertible with no invariant zeros.* ⊞

Remark 9.2.1. *Assumption 9.2.4 holds whenever $G(x, \xi, d(t))$ is a matrix that is positive definite uniformly in all its arguments. In such a case M and P can be identity matrices. For single input single output systems, Assumption 9.2.4 reduces to requiring $G(x, \xi, d(t))$ to have a definite sign which is known. Thus Assumption 9.2.4 can be viewed as a generalization of the knowledge of the sign of high frequency gain of the linear single input single output system to the case of nonlinear multi-input multi-output systems.* ℝ

9.3. State Feedback Results

The main results concerning semi-global practical stabilizability and semi-global asymptotic stabilizability via linear partial state feedback of the system (9.1.1) is given in the following two theorems.

Theorem 9.3.1. *If Assumptions 9.2.1-9.2.4 hold, then the point* $(0,0)$ *is semi-globally practically stabilizable for the system (9.1.1) via linear static partial state feedback.* Ⓣ

Theorem 9.3.2. *If Assumptions 9.2.1-9.2.5 hold, then the equilibrium* $(0,0)$ *of the system* Σ *is semi-globally asymptotically stabilizable via linear static partial state feedback.* Ⓣ

Remark 9.3.1. *In the proof of Theorems 9.3.1 and 9.3.2, a family of state feedback laws is explicitly constructed based solely on the knowledge of the matrices* A, B *and* C, *and no knowledge of the mappings* f, g *and* G *is required. In this sense, our state feedback design is robust.* Ⓡ

To prove Theorem 9.3.1, we need to present some technical lemmas. These lemmas not only play critical roles in proving the main results of this chapter, but also are of interest on their own.

Lemma 9.3.1. (Squaring-Up) *Consider the linear time invariant system*

$$\begin{cases} \dot{\xi} = A\xi + Bu, \\ y = C\xi, \end{cases} \tag{9.3.1}$$

where the state vector $\xi \in \mathbb{R}^n$, *the input vector* $u \in \mathbb{R}^m$ *and the output vector* $y \in \mathbb{R}^p$. *Assume that the system* (A,B,C) *is right invertible. Then, there exists a* $\tilde{C} \in \mathbb{R}^{(m-p)\times n}$, *such that the system* $(A,B,\hat{C})$, $\hat{C} := [C', \tilde{C}']'$, *is square invertible and has the same invariant zeros as the system* (A,B,C). Ⓛ

Proof of Lemma 9.3.1. For the given system (9.3.1), choose nonsingular transformations Γ_s, $\Gamma_o = I$ and Γ_i as in the SCB theorem (Theorem A.1.1) such that, in the notations of Theorem A.1.1, the system (9.3.1) can be written as,

$$\dot{\xi}_a = A_{aa}\xi_a + L_{ad}y_d, \tag{9.3.2}$$

$$\dot{\xi}_c = A_{cc}\xi_c + L_{cd}y_d + B_c[E_{ca}\xi_a + u_c], \tag{9.3.3}$$

and for $i = 1$ to m_d,

$$\dot{\xi}_i = A_{q_i}\xi_i + L_{id}y_d + B_{q_i}[u_i + E_{ia}\xi_a + E_{ic}\xi_c + E_{id}\xi_d], \tag{9.3.4}$$

$$y_i = C_{q_i}\xi_i = \xi_{i1}. \tag{9.3.5}$$

We note that the nonexistence of ξ_b due to the right invertibility of the system (A, B, C). By Theorem A.1.1, the pair (A_{cc}, B_c) is controllable, hence, as shown in [85], there exist nonsingular state and input transformations Γ_c and Γ_v,

$$\xi_c = \Gamma_c \xi_{cc}, \ \xi_{cc} = [\xi'_{cc1}, \xi'_{cc2}, \cdots, \xi'_{ccm_c}]', \ u_c = \Gamma_v[v_1, v_2, \cdots, v_{m_c}]',$$

and integer indices l_i, $i = 1$ to m_c, such that the dynamics of ξ_c, (9.3.3), can be written as,

$$\dot{\xi}_{cci} = A_{l_i}\xi_{cci} + L_{cid}y_d + B_{l_i}[v_i + E_{cia}\xi_a + E_{cic}\xi_{cc}], \ i = 1, 2, \cdots, m_c, \quad (9.3.6)$$

where for an integer l_i, the matrices A_{l_i}, B_{l_i} and C_{l_i} are as defined in Theorem A.1.1. The system (9.3.1) can thus be rewritten as,

$$\dot{\xi}_a = A_{aa}\xi_a + L_{ad}y_d, \tag{9.3.7}$$

$$\dot{\xi}_{cci} = A_{l_i}\xi_{cci} + L_{cid}y_d + B_{l_i}[v_i + E_{cia}\xi_a + E_{cic}\xi_{cc}], \ i = 1, 2, \cdots, m_c, \tag{9.3.8}$$

and for $i = 1$ to m_f,

$$\dot{\xi}_i = A_{q_i}\xi_i + L_{id}y_d + B_{q_i}[u_i + E_{ia}\xi_a + E_{ic}\Gamma_c\xi_{cc} + E_{id}\xi_d], \tag{9.3.9}$$

$$\bar{y}_i = C_{q_i}\xi_i = \xi_{i1}, \tag{9.3.10}$$

which is clearly an SCB representation of the system $(A, B, \hat{C})$ with

$$\hat{C} = \begin{bmatrix} C \\ [0 \ \ C_c\Gamma_c^{-1} \ \ 0]\Gamma_S^{-1} \end{bmatrix}, \quad C_c = \text{blkdiag}\{C_{l_1}, C_{l_2}, \cdots, C_{l_{m_c}}\}.$$

Moreover, it is square invertible and has the same invariant zeros as (A, B, C). This completes the proof.   ⊠

Lemma 9.3.2. (Stability Lemma) *Let* $\beta : \mathbb{R}_+ \to \mathbb{R}_+$ *be a continuous non-increasing function, satisfying that* $\beta(s) \to 0$ *as* $s \to \infty$, $\gamma : \mathbb{R}_+ \to \mathbb{R}_+$ *be a class* $\mathcal{K}$ *function,* $\alpha : \mathbb{R}_+ \to \mathbb{R}_+$ *be a function positive definite on* $\mathbb{R}_+ \setminus \{0\}$, *and* $\varepsilon > 0$. *Define* $\mathcal{I}(m, \alpha, \beta, \gamma, \varepsilon)$ *as the set of all* $\mathbb{R}^m$-*valued measurable functions* $v(t) = (v_1(t), v_2(t), \cdots, v_m(t))$ *on* $[0, \infty)$ *that satisfy the bound,*

$$|v_i(t)| \leq \beta(t/\gamma(\varepsilon)) + \gamma(\varepsilon)\beta(\alpha(\varepsilon)t), \ i = 1, 2, \cdots, m. \tag{9.3.11}$$

Assume that the equilibrium $x = 0$ *of the system* $\dot{x} = f(x, 0)$ *is globally asymptotically stable. Then for any given* $R > 0$ *and any given triple* (α, β, γ), *there exists an* $\varepsilon^* > 0$ *such that for every* $\varepsilon \in (0, \varepsilon^*]$, *every solution* $t \to x(t)$ *of*

$$\dot{x} = f(x, v), \ v \in \mathcal{I}(m, \alpha, \beta, \gamma, \varepsilon), \tag{9.3.12}$$

with initial conditions in the set $\{x : |x| \leq R\}$ *satisfies* $x(t) \to 0$ *as* $t \to \infty$. *Moreover, the convergence is uniform with respect to* $x(0) \in \{x : |x| \leq R\}$ *and* $v \in \mathcal{I}(m, \alpha, \beta, \gamma, \varepsilon)$.   Ⓛ

Proof of Lemma 9.3.2. This lemma is a generalization of Theorem 4.1 of [107]. Its proof is a slight modification of that of Theorem 4.1 in [107].

Since the equilibrium $x = 0$ of the system $\dot{x} = f(x, 0)$ is globally asymptotically stable, the converse Lyapunov theorem (see [36]) shows that there exists a smooth Lyapunov function $V : \mathbb{R}^n \to \mathbb{R}_+$ such that $V(0) = 0$, $V(x) > 0$ for $x \neq 0$, $\nabla V(x) \cdot f(x, 0) < 0$ for $x \neq 0$, and V is proper (The properness of V implies that the level set $L_V(c) = \{x : V(x) \leq c\}$ is compact for all $c > 0$.).

Let $c = \max\{V(x) : |x| \leq R\}$. The function $(x, v) \to \nabla V(x) \cdot f(x, v)$ is continuous. So there is a constant $h > 0$ such that

$$\nabla V(x) \cdot f(x, v) \leq h, \ \forall x \in L_V(c+1) \text{ and } \forall |v_i| \leq (1 + \gamma(1))\beta(0), \ i = 1, 2, \cdots, m.$$

Now let $\tau = 1/h$. It then follows, that if $v \in_\mathbb{I} (m, \alpha, \beta, \gamma, \varepsilon)$ and $\varepsilon \in (0, 1]$, then $V(x(t)) \leq c + 1$ for all $0 \leq t \leq \tau$, for every solution of $\dot{x} = f(x, v)$ such that $V(x(0)) \leq c$.

Next, use the fact that the function $(x, v) \to \nabla V(x) \cdot f(x, v)$ is continuous, and its values are strictly negative for $c \leq V(x) \leq c + 1$, $v = 0$. It then follows that there is a $\delta > 0$ such that,

$$\nabla V(x) \cdot f(x, v) < 0, \ \forall x \in \{x : c \leq V(x) \leq c + 1\} \text{ and } |v_i| \leq \delta, \ i = 1, 2, \cdots, m.$$

Then choose an $\varepsilon^* > 0$ such that, for every $\varepsilon \in (0, \varepsilon^*]$,

$$\beta(t/\gamma(\varepsilon)) + \gamma(\varepsilon)\beta(\alpha(\varepsilon)t) \leq \delta, \ t \geq \tau.$$

This can be done as follows. Since $\beta(s) \to 0$ as $s \to \infty$, $s_\delta > 0$ can be chosen such that $\beta(s) \leq \delta/2$, $s \geq s_\delta$. Since γ is of class $\mathcal{K}$, $\varepsilon^* > 0$ can be chosen such that $\tau/\gamma(\varepsilon^*) \geq s_\delta$ and $\gamma(\varepsilon^*) \leq \delta/2\beta(0)$. Clearly, with this choice of ε^*, $\beta(\tau/\gamma(\varepsilon^*)) + \gamma(\varepsilon^*)\beta(\alpha(\varepsilon^*)\tau) \leq \delta$. Then, the non-increasing property of function β and the strictly increasing property of class $\mathcal{K}$ function γ show that $\beta(t/\gamma(\varepsilon)) + \gamma(\varepsilon)\beta(\alpha(\varepsilon)t) \leq \beta(\tau/\gamma(\varepsilon^*)) + \gamma(\varepsilon^*)\beta(\alpha(\varepsilon^*)\tau) \leq \delta$, $t \geq \tau$, $\varepsilon \in (0, \varepsilon^*]$.

Now for any $\varepsilon \in (0, \varepsilon^*]$, if v is an input belonging to $\mathcal{I}(m, \alpha, \beta, \gamma, \varepsilon)$ and $x(t)$ is a solution of $\dot{x} = f(x, v)$ with $|x(0)| \leq R$, defined on an interval $[0, T)$, then $V(x(0)) \leq c$, and hence $V(x(t)) \leq c + 1$ for $0 \leq t \leq \tau$. For $t \geq \tau$, $V(x(t)) \leq c + 1$, since $|v_i(t)| \leq \delta$ for $t \geq \tau$. (Indeed, let the set $I = \{t : \tau \leq t \leq T, V(x(t)) > c + 1\}$. Clearly, I is relatively open in $[\tau, \infty)$ and does not contain τ. Hence it is a union of a finite or countable sequence $\{I_j\}$ of open intervals. On each I_j, the function $t \to V(x(t))$ is absolutely continuous and has a negative derivative, so it is decreasing. Therefore, if a_j is the left endpoint of I_j, then $V(x(t)) \leq V(x(a_j))$ for $t \in I_j$. But $a_j \notin I_j$, and so $V(x(a_j)) \leq c + 1$. Therefore $V(x(t)) \leq c + 1$ for all $t \in I$. This shows that I is empty.)

We now know that $V(x(t)) \leq c + 1$ for all $t \in [0, T)$, if $x(t)$ is a trajectory as above. This implies, in particular, using standard continuation theorems for ordinary differential equations, that the maximal interval of definition of such a trajectory is $[0, +\infty)$ and therefore, $V(x(t)) \leq c + 1$ for all $t \in [0, \infty)$. We next show that for each fixed $\varepsilon \in (0, \varepsilon^*]$, $x(t) \to 0$ as $t \to \infty$. To this end, fix any $\varepsilon \in (0, \varepsilon^*]$. Now, for any $\eta > 0$, let $\tilde{c} > 0$ be such that $V(x) \leq \tilde{c}$ implies $|x| \leq \eta$. Noting that $\nabla V(x) \cdot f(x, 0) < 0$ for all $x \in \{\tilde{c} \leq V(x) \leq c + 1\}$, it follows from the continuity of the function $(x, v) \to \nabla V(x) \cdot f(x, v)$ that there exists a $\mu > 0$ and $\theta > 0$ such that,

$$\nabla V(x) \cdot f(x, v) \leq -\theta, \quad \forall x \in \{x : \tilde{c} \leq V(x) \leq c + 1\} \text{ and } |v_i| \leq \mu, \ i = 1, 2, \cdots, m.$$

Then let $T_1(\varepsilon) > 0$ be such that $\beta(t/\gamma(\varepsilon)) + \gamma(\varepsilon)\beta(\alpha(\varepsilon)t) \leq \mu, t \geq T_1(\varepsilon)$, and let $T_2(\varepsilon)$ be such that $\theta(T_2(\varepsilon) - T_1(\varepsilon)) > c + 1$. The existence of such a $T_1(\varepsilon)$ is due to the fact that $\beta(s) \to 0$ as $s \to \infty$. Then $V(x(t)) \leq c + 1$ for $0 \leq t \leq T_1(\varepsilon)$. For $T_1(\varepsilon) \leq t < \infty$, $\nabla V(x) \cdot f(x, v) \leq -\theta$ as long as $x \in \{x : \tilde{c} \leq V(x) \leq c + 1\}$. It then follows that there is a $\hat{t} \in [T_1(\varepsilon), T_2(\varepsilon)]$ such that $V(x(\hat{t})) \leq \tilde{c}$. (Indeed, if we had $V(x(t)) > \tilde{c}$ for all $t \in [T_1(\varepsilon), T_2(\varepsilon)]$, it would follow that $\nabla V(x) \cdot f(x, v) \leq -\theta$ for all $t \in [T_1(\varepsilon), T_2(\varepsilon)]$, and then $V(x(T_2(\varepsilon))) \leq V(x(T_1(\varepsilon))) - \theta(T_2(\varepsilon) - T_1(\varepsilon)) < 0$.) It is then clear that, if $V(x(\hat{t})) \leq \tilde{c}$ for some $\hat{t}$, it would follow that $V(x(t)) \leq \tilde{c}$ for all larger t.

In particular, we have shown that, for any fixed $\varepsilon \in (0, \varepsilon^*]$, $|x(t)| \leq \eta$ for all $t \geq T(\varepsilon)$, where $T(\varepsilon) = T_2(\varepsilon)$. This is true for all initial conditions such that $|x(0)| \leq R$ and all input v in $\mathcal{I}(m, \alpha, \beta, \gamma, \varepsilon)$. ⊠

Lemma 9.3.3. (Semi-Global Backstepping) *Consider the* $\mathcal{C}^1$ *nonlinear control system*

$$\begin{cases} \dot{x} = f(x, \zeta, d(t)), \ x \in \mathbb{R}^n, \\ \dot{\zeta} = g(x, \zeta, d(t)) + \frac{1}{\mu} G(x, \zeta, d(t))\zeta, \ \zeta \in \mathbb{R}^m, \end{cases} \tag{9.3.13}$$

where $\mu > 0$ *and* $d(t)$ *is any continuous time-varying signal assumed to belong to the compact set* $\mathcal{D} \subset \mathbb{R}^q$. *Assume that for the system*

$$\dot{x} = f(x, 0, d(t)), \tag{9.3.14}$$

there exists a neighborhood $\mathcal{W}_0$ *of the origin in* $\mathbb{R}^n$ *and a* $\mathcal{C}^1$ *function* $V_0 :$ $\mathcal{W}_0 \to \mathbb{R}_+$ *which is positive definite on* $\in \mathcal{W}_0 \setminus \{0\}$ *and proper on* $\mathcal{W}_0$ *and satisfies*

$$\frac{\partial V_0}{\partial x} f(x, 0, d(t)) \leq -\Psi_0(x), \tag{9.3.15}$$

where $\Psi_0(x)$ is continuous on $\mathcal{W}_0$ and positive definite on $\{x : \rho_0 < V_0(x) \leq c_0 + 1\}$ for some nonnegative real number $\rho_0 < 1$ and some real number $c_0 \geq 1$. Also assume that there exists a positive definite matrix $P \in \mathbb{R}^{m \times m}$, such that

$$PG(x, \zeta, d(t)) + G^{\mathrm{T}}(x, \zeta, d(t))P = -Q(x, \zeta, d(t)), \qquad (9.3.16)$$

where $Q(x, \zeta, d(t))$ is a positive definite matrix uniformly in all its arguments, i.e., there exists a positive definite constant matrix Q_0 such that $Q(x, \zeta, d(t)) \geq Q_0$ for all $(x, \zeta, d(t)) \in \mathbb{R}^n \times \mathbb{R}^m \times \mathcal{D}$. Given $c_1 \geq 1$, we define the function

$$V_1(x, \zeta) = c_0 \frac{V_0(x)}{c_0 + 1 - V_0(x)} + c_1 \frac{\zeta' P \zeta}{c_1 + 1 - \zeta' P \zeta} \qquad (9.3.17)$$

and the set

$$\mathcal{W}_1 = \{x : V_0(x) < c_0 + 1\} \times \{\zeta : \zeta' P \zeta < c_1 + 1\}.$$

Then $V_1 : \mathcal{W}_1 \to \mathbb{R}_+$ is positive definite on $\mathcal{W}_1 \setminus \{0\}$ and proper on $\mathcal{W}_1$. Furthermore, for each $\rho_1 > 0$, there exists a $\mu^*(\rho_1) > 0$, such that, for each $\mu \in (0, \mu^*(\rho_1)]$, the derivative of V_1 along the trajectories of (9.3.13) satisfies

$$\dot{V}_1 \leq -\Psi_1(x, \zeta), \qquad (9.3.18)$$

where $\Psi_1(x, \zeta)$ is positive definite on $\{(x, \zeta) : \rho_0 + \rho_1 \leq V_1(x, \zeta) \leq c_0^2 + c_1^2 + 1\}$.

□

Proof of Lemma 9.3.3. The lemma is a multivariable version of Lemma 2.2 of [115]. Its proof follows that of Lemma 2.2 of [115] with the only modification being that, instead of using ζ^2 in function $V_1(x, \zeta)$ and the assumption that $G(x, \zeta, d(t)) \geq b > 0$ for the case $m = 1$, we need to use $\zeta' P \zeta$ in the function $V_1(x, \zeta)$ with P being assumed to satisfy the Lyapunov equation type (9.3.16).

✠

Remark 9.3.2. Note that if the system $\dot{x} = f(x, 0, d(t))$ is autonomous and its equilibrium $x = 0$ is locally asymptotically stable with basin of attraction $\mathcal{W}_0$, then the existence of such a C^1 function V_0 is guaranteed by Theorem 7 of [36]. Furthermore, ρ_0 can be chosen equal to zero and c_0 can be chosen arbitrarily large.

R

We are now ready to prove Theorem 9.3.1.

Proof of Theorem 9.3.1. We will prove this theorem by first explicitly constructing a family of linear static partial state feedback laws $u = F(\varepsilon, \mu)\xi$, parameterized in ε and μ, and then showing that, under Assumptions 9.2.1-9.2.4, the family of feedback laws indeed achieves semi-global practical stabilization of the system (9.1.1). More specifically, we will show that for any *a priori* given (arbitrarily large) bounded set $W \subset \mathbb{R}^{l+n}$ and any *a priori* given (arbitrarily small) $W_0 \subset W$, which contains $(0,0)$ as an interior point, there exists an $\varepsilon^* > 0$ and for each $\varepsilon \in (0, \varepsilon^*]$ there exists a $\mu^*(\varepsilon) > 0$ such that, for all $\mu \in (0, \mu^*(\varepsilon)], \varepsilon \in (0, \varepsilon^*]$, all the solutions of the closed-loop system with initial conditions in W enter W_0 in a finite time and remain in it thereafter.

To start with, consider the linearizable part of the system (9.1.1)

$$\begin{cases} \dot{\xi} = A\xi + B[G(x, \xi, d(t))u + g(x, \xi, d(t))], \ \eta = C\xi, \\ y = D\xi, \end{cases} \qquad (9.3.19)$$

where η represent the part of the dynamics of this system which enters the zero dynamics of the system (9.1.1). In case that the triple (A, B, C) is not square invertible, find a $\tilde{C} \in \mathbb{R}^{(m-r) \times n}$ such that the system $(A, B, \hat{C})$, $\hat{C} = [C', \tilde{C}']'$, is square invertible and has the same invariant zeros as (A, B, C). An explicit algorithm for choosing such a $\tilde{C}$ is given in the proof of Lemma 9.3.1. We then perform a nonsingular state transformation for the system (9.3.19),

$$\xi = \Gamma_s \bar{\xi},$$

such that the linearizable system can be written as,

$$\begin{cases} \dot{\bar{\xi}} = \bar{A}\bar{\xi} + \bar{B}[\bar{G}(x, \bar{\xi}, d(t))u + \bar{g}(x, \bar{\xi}, d(t))], \ \hat{\eta} = \bar{C}\bar{\xi}, \\ y = \bar{D}\bar{\xi}, \end{cases} \qquad (9.3.20)$$

where

$$\bar{A} = \Gamma_s^{-1} A \Gamma_s, \ \bar{B} = \Gamma_s^{-1} B \Gamma_I, \ \bar{C} = \hat{C} \Gamma_s, \ \bar{D} = D \Gamma_s,$$

$$\bar{G}(x, \bar{\xi}, d(t)) = \Gamma_I^{-1} G(x, \Gamma_s \bar{\xi}, d(t)), \ \bar{g}(x, \bar{\xi}, d(t)) = \Gamma_I^{-1} g(x, \Gamma_s \bar{\xi}, d(t)),$$

and where Γ_s and Γ_I are such that $(\bar{A}, \bar{B}, \bar{C})$ is in the SCB form of $(A, B, \hat{C})$. The existence of such transformation matrices Γ_s and Γ_I is established in Theorem A.1.1. Now in view of Theorem A.1.1, the system (9.3.20) has the following dynamic equations,

$$\bar{\xi} = [\xi_a', \xi_d']', \ \xi_d = [\xi_1', \xi_2', \cdots, \xi_m']', \ \xi_i = [\xi_{i1}, \xi_{i2}, \cdots, \xi_{iq_i}]',$$

$$\hat{\eta} = \hat{C}\xi = [\hat{\eta}_1, \hat{\eta}_2, \cdots, \hat{\eta}_m]', \ \hat{\eta}_i = \xi_{i1},$$

$$\hat{\eta} = [\eta', \tilde{\eta}']', \ \eta = C\xi = [\hat{\eta}_1, \hat{\eta}_2, \cdots, \hat{\eta}_r]', \ \tilde{\eta} = \tilde{C}\xi = [\hat{\eta}_{r+1}, \hat{\eta}_{r+2}, \cdots, \hat{\eta}_m]',$$

and

$$\dot{\xi}_a = A_{aa}\xi_a + L_{ad}\hat{\eta}, \ \xi_a \in \mathbb{R}^{n_a}, \qquad (9.3.21)$$

$$\dot{\xi}_i = A_{q_i}\xi_i + L_{id}\hat{\eta} + B_{q_i}[E_{ia}\xi_a + E_{id}\xi_d + \omega_i(x, \bar{\xi}, u, d)], \ i = 1, 2, \cdots, m, \qquad (9.3.22)$$

where $\omega_i(x, \bar{\xi}, u, d)$ is the ith element of $\bar{G}(x, \bar{\xi}, d(t))u + \bar{g}(x, \bar{\xi}, d(t))$.

We proceed with the design of the state feedback laws. The design of these state feedback laws are to be completed in three steps.

Step 1 - Low Gain Design. We know from Property A.1.2 of the SCB that the eigenvalues of A_{aa} are the invariant zeros of the linear system $(A, B, \hat{C})$ and hence are all located in the closed left-half plane. Moreover, the pair (A_{aa}, L_{ad}) is stabilizable. Hence we can design a low gain feedback gain $F_a(\varepsilon)$ for the pair (A_{aa}, L_{ad}) as in Section 2.2.1. Moreover, without loss of generality, we assume that the pair (A_{aa}, L_{ad}) is in the block diagonal control canonical form of (2.2.2)-(2.2.3) with $l = l_a$, and, the gain matrix $F_{ai}(\varepsilon)$ corresponding to each diagonal pair (A_{aai}, L_{adi}) of (A_{aa}, L_{ad}) has all the properties of Lemmas 2.2.1-2.2.4. We correspondingly denote the matrices $Q(\varepsilon)$ and $S(\varepsilon)$ in Lemmas 2.2.1-2.2.4 as $Q_{ai}(\varepsilon)$ and $S_{ai}(\varepsilon)$ respectively.

For later use, we partition the matrix $F_a(\varepsilon)$ as,

$$F_a(\varepsilon) = [F'_{a1}(\varepsilon), F'_{a2}(\varepsilon), \cdots, F'_{am}(\varepsilon)]', \qquad (9.3.23)$$

where each $F_{ai}(\varepsilon) \in \mathbb{R}^{1 \times n_a}$.

Step 2 - Output Renaming. Change variables as follows,

$$\zeta_a = \xi_a, \qquad (9.3.24)$$

and for each $i = 1$ to m,

$$\bar{\hat{\eta}}_i = \zeta_{i1} = \xi_{i1} - F_{ai}(\varepsilon)\xi_a, \qquad (9.3.25)$$

$$\zeta_{ij} = \xi_{ij}, \quad j = 2, 3, \cdots, q_i - 1, \qquad (9.3.26)$$

$$\zeta_{iq_i} = \xi_{iq_i} + L_{iq_i-1}\hat{\eta} + \frac{c_{i1}}{\varepsilon^{q_i-1}}(\xi_{i1} - F_{ai}(\varepsilon)\xi_a) + \frac{c_{i2}}{\varepsilon^{q_i-2}}\xi_{i2}$$
$$+ \cdots + \frac{c_{iq_i-1}}{\varepsilon}\xi_{iq_i-1}, \qquad (9.3.27)$$

where c_{ij}'s are chosen in such a way that all the polynomials

$$s^{q_i-1} + c_{iq_i-1}s^{q_i-2} + \cdots + c_{i2}s + c_{i1} \qquad (9.3.28)$$

are Hurwitz, and $L_{iq_i-1} \in \mathbb{R}^{1 \times n_a}$ is the $q_i - 1$th row of L_{id}.

With this set of new state variables, the system (9.3.21)-(9.3.22) can be written as,

$$\dot{\zeta}_a = (A_{aa} + L_{ad}F_a(\varepsilon))\zeta_a + L_{ad}\bar{\hat{\eta}}, \quad \bar{\hat{\eta}} = \hat{\eta} - F_a(\varepsilon)\zeta_a, \qquad (9.3.29)$$

$$\dot{\zeta}_{i1} = \zeta_{i2} - F_{ai}(\varepsilon)A_{aa}\zeta_a + (L_{i1} - F_{ai}(\varepsilon)L_{ad})F_a(\varepsilon)\zeta_a$$
$$+ (L_{i1} - F_{ai}(\varepsilon)L_{ad})\bar{\tilde{\eta}}, \tag{9.3.30}$$

$$\dot{\zeta}_{ij} = \zeta_{ij+1} + L_{ij}(\bar{\tilde{\eta}} + F_a(\varepsilon)\zeta_a), \quad j = 2,3,\cdots,q_i - 2, \tag{9.3.31}$$

$$\dot{\zeta}_{iq_i-1} = -\frac{c_{i1}}{\varepsilon^{q_i-1}}\zeta_{i1} - \frac{c_{i2}}{\varepsilon^{q_i-2}}\zeta_{i2} - \cdots - \frac{c_{iq_i-1}}{\varepsilon}\zeta_{iq_i-1} + \zeta_{iq_i}, \tag{9.3.32}$$

$$\dot{\zeta}_{iq_i} = N_i(\varepsilon)\zeta + \bar{\omega}_i(x,\bar{\zeta},u,d), \quad \bar{\tilde{\eta}}_i = \zeta_{i1}, \quad i = 1,2,\cdots,m, \tag{9.3.33}$$

where

$$\zeta = [\zeta_a', \zeta_s', \zeta_q']', \quad \bar{\tilde{\eta}} = [\bar{\tilde{\eta}}_1, \bar{\tilde{\eta}}_2, \cdots, \bar{\tilde{\eta}}_m]',$$

$$\zeta_s = [\zeta_{s1}', \zeta_{s2}', \cdots, \zeta_{sm}']', \quad \zeta_{si} = [\zeta_{i1}, \zeta_{i2}, \cdots, \zeta_{iq_i-1}]',$$

$$\zeta_q = [\zeta_{1q_1}, \zeta_{2q_2}, \cdots, \zeta_{mq_m}]',$$

$N_i(\varepsilon) \in \mathbb{R}^{1\times n}$ is a vector whose elements are functions of ε, L_{ij} is the jth row of L_i, and $\bar{\omega}_i(x,\bar{\zeta},u,d)$ is the ith element of $\bar{G}(x,\bar{\xi},d(t))u + \bar{g}(x,\bar{\xi},d(t))$.

Note that in the above we have assumed that for all $i = 1$ to m, $q_i \geq 2$. This is without loss of generality since otherwise some of the ζ_{si}'s would have dropped out automatically.

Step 3 - High Gain Design. We finally obtain the state feedback laws

$$u = \frac{1}{\mu}M\Gamma_1\bar{\zeta}_q, \tag{9.3.34}$$

where $\mu \in (0,1]$ is a positive scalar whose value is to be chosen later. Ⓐ

We now proceed to show that, under Assumptions 9.2.1-9.2.4, the family of static partial state feedback laws (9.3.34) indeed achieves semi-global practical stabilization for the system (9.1.1). With the state feedback (9.3.34), the closed-loop system can be written as

$$\dot{x} = f(x,\eta), \tag{9.3.35}$$

$$\dot{\tilde{\zeta}}_a = A_{aa}^c(\varepsilon)\tilde{\zeta}_a + S_a(\varepsilon)Q_a^{-1}(\varepsilon)L_{ad}\bar{\tilde{\eta}}, \quad \bar{\tilde{\eta}}_i = \tilde{\zeta}_{i1}, \tag{9.3.36}$$

$$\dot{\tilde{\zeta}}_s^c = \frac{1}{\varepsilon}A_s^c\tilde{\zeta}_s + L_{sa}(\varepsilon)F_a(\varepsilon)Q_a(\varepsilon)S_a^{-1}(\varepsilon)\tilde{\zeta}_a$$
$$+ L_{saa}F_a(\varepsilon)A_{aa}Q_a(\varepsilon)S_a^{-1}(\varepsilon)\tilde{\zeta}_a + L_{ss}(\varepsilon)\tilde{\zeta}_s + L_{sq}\tilde{\zeta}_q, \tag{9.3.37}$$

$$\dot{\tilde{\zeta}}_q = \frac{1}{\mu}\tilde{G}(x,\tilde{\zeta},d(t);\varepsilon)M\Gamma_1\tilde{\zeta}_q + \tilde{N}(\varepsilon)\tilde{\zeta} + \tilde{g}(x,\tilde{\zeta},d(t);\varepsilon), \tag{9.3.38}$$

in the new state variables $\tilde{\zeta} = [\tilde{\zeta}_a', \tilde{\zeta}_s', \tilde{\zeta}_q']'$ where

$$\tilde{\zeta}_a = S_a(\varepsilon)Q_a^{-1}(\varepsilon)\bar{\zeta}_a, \tag{9.3.39}$$

$$\tilde{\zeta}_s = [\tilde{\zeta}'_{s1}, \tilde{\zeta}'_{s2}, \cdots, \tilde{\zeta}'_{sm}]', \; \tilde{\zeta}_{si} = [\tilde{\zeta}_{i1}, \tilde{\zeta}_{i2}, \cdots, \tilde{\zeta}_{iq_i-1}],$$

$$\tilde{\zeta}_{ij} = \varepsilon^{j-1}\zeta_{ij}, \tag{9.3.40}$$

$$\tilde{\zeta}_q = \zeta_q, \tag{9.3.41}$$

$$Q_a(\varepsilon) = \text{blkdiag}\{Q_{a1}(\varepsilon), Q_{a2}(\varepsilon), \cdots, Q_{al_a}(\varepsilon), I\}, \tag{9.3.42}$$

$$S_a(\varepsilon) = \text{blkdiag}\{S_{a1}(\varepsilon), S_{a2}(\varepsilon), \cdots, S_{al_a}(\varepsilon), I\}, \tag{9.3.43}$$

$$A^c_{aa}(\varepsilon) = S_a(\varepsilon)Q_a^{-1}(\varepsilon)[A_{aa} + L_{ad}F_a(\varepsilon)]Q_a(\varepsilon)S_a^{-1}(\varepsilon). \tag{9.3.44}$$

We recall that $Q_a(\varepsilon)$, $Q_a^{-1}(\varepsilon)$ and $S_a(\varepsilon)$ are all bounded for all $\varepsilon \in (0,1]$,

Also in the above representation, $A^c_s = \text{blkdiag}\{A^c_1, A^c_2, \cdots, A^c_m\}$ and A^c_i's are the companion matrices associated with the polynomials (9.3.28) and hence A_s is asymptotically stable, L_{sq} and L_{saa} are independent of ε, and $L_{sa}(\varepsilon)$ and $L_{ss}(\varepsilon)$ are ε dependent matrices of appropriate dimensions satisfying

$$|L_{sa}(\varepsilon)| \leq l_{sa}, \; |L_{ss}(\varepsilon)| \leq l_{ss}, \; \forall \varepsilon \in (0,1], \tag{9.3.45}$$

for some positive constants l_{sa} and l_{ss} independent of ε, $\tilde{N}(\varepsilon)$ is a matrix of appropriate dimensions defined in an obvious way, and finally, $\tilde{G}(x, \tilde{\zeta}, d(t); \varepsilon) = \bar{G}(x, \bar{\xi}, d(t))$ and $\tilde{g}(x, \tilde{\zeta}, d(t); \varepsilon) = \bar{g}(x, \bar{\xi}, d(t))$ are ε dependent $\mathcal{C}^1$ matrix function and $\mathcal{C}^1$ vector function respectively.

Let the compact sets $\mathcal{W}_x \subset \mathbb{R}^l$ and $\mathcal{W}_\xi \subset \mathbb{R}^n$ be such that $\mathcal{W} \subset \mathcal{W}_x \times \mathcal{W}_\xi$. Let the ε independent compact set $\mathcal{W}_{as} \in \mathbb{R}^{n-m}$ and the ε dependent compact set $\mathcal{W}_q(\varepsilon) \in \mathbb{R}^m$ be such that, for each $\varepsilon \in (0,1]$, $\xi \in \mathcal{W}_\xi$ implies that $(\tilde{\zeta}_a, \tilde{\zeta}_s) \in \mathcal{W}_{as}$ and $\tilde{\zeta}_q \in \mathcal{W}_q(\varepsilon)$. The existence of such sets $\mathcal{W}_{as}$ and $\mathcal{W}_q(\varepsilon)$ is guaranteed by the special form of the state transformation from ξ to $\tilde{\zeta}$. With these notations at hand, we next present a proposition.

Proposition 9.3.1. *There exists an $\varepsilon^* \in (0,1]$, such that for all $\varepsilon \in (0,1]$, the nonlinear system*

$$\dot{x} = f(x, \eta), \tag{9.3.46}$$

$$\dot{\tilde{\zeta}}_a = A^c_{aa}(\varepsilon)\tilde{\zeta}_a + S_a(\varepsilon)Q_a^{-1}(\varepsilon)L_{ad}\bar{\tilde{\eta}}, \; \bar{\tilde{\eta}}_i = \tilde{\zeta}_{i1}, \tag{9.3.47}$$

$$\dot{\tilde{\zeta}}_s = \frac{1}{\varepsilon}A^c_s\tilde{\zeta}_s + L_{sa}(\varepsilon)F_a(\varepsilon)Q_a(\varepsilon)S_a^{-1}(\varepsilon)\tilde{\zeta}_a$$

$$+ L_{saa}F_a(\varepsilon)A_{aa}Q_a(\varepsilon)S_a^{-1}(\varepsilon)\tilde{\zeta}_a + L_{ss}(\varepsilon)\tilde{\zeta}_s, \tag{9.3.48}$$

is locally asymptotically stable and $\mathcal{W}_x \times \mathcal{W}_{as}$ is contained in the basin of attraction $\tilde{\mathcal{W}}_0(\varepsilon)$ of its equilibrium $(x, \zeta_a, \zeta_s) = (0,0,0)$. $\boxed{\text{P}}$

Proof of Proposition 9.3.1. We prove this proposition by making use of Lemma 9.3.2. To start with, it follows from Section 3.3.2 (see (3.3.11)) that there exists a $P_a > 0$, independent of ε, such that

$$(A_{aa}^c(\varepsilon))'P_a + P_a A_{aa}^c(\varepsilon) \le -\frac{\varepsilon}{2}I, \quad \varepsilon \in (0, \varepsilon_a^*], \tag{9.3.49}$$

for some $\varepsilon_a^* \in (0, 1]$.

Let P_s be the positive definite solution to the Lyapunov equation

$$(A_s^c)'P_s + P_s A_s^c = -I. \tag{9.3.50}$$

The existence of such a P_s is due to the fact that A_s^c is asymptotically stable.

For the dynamics of $\tilde{\zeta}_a$ and $\tilde{\zeta}_s$, the linear system (9.3.47)-(9.3.48), we form a Lyapunov function candidate as,

$$V_{as}(\tilde{\zeta}_a, \tilde{\zeta}_s) = \tilde{\zeta}_a' P_a \tilde{\zeta}_a + \kappa \tilde{\zeta}_s' P_s \tilde{\zeta}_s, \tag{9.3.51}$$

where κ is a positive constant whose value is to be determined later.

We again recall that $Q_a(\varepsilon)$, $Q_a^{-1}(\varepsilon)$ and $S_a(\varepsilon)$ are all bounded for all $\varepsilon \in (0, 1]$. Also it follows from Lemma 2.2.4 that

$$|F_a(\varepsilon)Q_a(\varepsilon)S_a^{-1}(\varepsilon)| \le \alpha_a \varepsilon, \quad \varepsilon \in (0, 1], \tag{9.3.52}$$

$$|F_a(\varepsilon)A_{aa}Q_a(\varepsilon)S_a^{-1}(\varepsilon)| \le \beta_a \varepsilon, \quad \varepsilon \in (0, 1], \tag{9.3.53}$$

for some $\alpha_a, \beta_a > 0$ independent of ε. Hence, using (9.3.45), (9.3.49), and (9.3.50), we evaluate the derivative of V_{as} along the trajectories of (9.3.47)-(9.3.48) and yield that, for $\varepsilon \in (0, \varepsilon_a^*]$,

$$\dot{V}_{as} \le -[|\tilde{\zeta}_a|\ |\tilde{\zeta}_s|] \begin{bmatrix} \dfrac{\varepsilon}{2} & -\alpha_{12} - \alpha_{21}\kappa\varepsilon \\ -\alpha_{12} - \alpha_{21}\kappa\varepsilon & \kappa\dfrac{1}{\varepsilon} - \kappa\alpha_{22} \end{bmatrix} \begin{bmatrix} |\tilde{\zeta}_a| \\ |\tilde{\zeta}_s| \end{bmatrix}, \tag{9.3.54}$$

for some strictly positive numbers α_{12}, α_{21} and α_{22} independent of ε. Let $\kappa = 2a_{12}^2$. It is then straightforward to verify that there exists an $\varepsilon_b^* \in (0, \varepsilon_a^*]$ such that for all $\varepsilon \in (0, \varepsilon_b^*]$,

$$\dot{V}_{as} \le -\alpha_1 \varepsilon \left(|\tilde{\zeta}_a|^2 + |\tilde{\zeta}_s|^2 \right), \tag{9.3.55}$$

for some positive constant α_1 independent of ε. This shows that the linear system (9.3.47)-(9.3.48) is globally asymptotically stable and hence the asymptotic stability of the nonlinear system (9.3.46)-(9.3.48) follows trivially from, say, [6].

It remains to be shown that $\mathcal{W}_x \times \mathcal{W}_{as}$ is contained in the basin of attraction of the equilibrium $(x, \tilde{\zeta}_a, \tilde{\zeta}_s) = (0, 0, 0)$. It follows from the positive definiteness of the matrices P_a and P_s that

$$\dot{V}_{as} \leq -\alpha_2 \varepsilon V_{as}, \quad \forall \varepsilon \in (0, \varepsilon_b^*], \tag{9.3.56}$$

for some positive constant α_2, independent of ε. This in turn shows that

$$V_{as}(\tilde{\zeta}_a(t), \tilde{\zeta}_s(t)) \leq e^{-\alpha_2 \varepsilon t} V_{as}(0), \quad \forall \varepsilon \in (0, \varepsilon_b^*], \tag{9.3.57}$$

where we note that

$$V_{as}(0) \leq (|P_a| + \kappa |P_s|) R_1^2, \quad \forall (\tilde{\zeta}_a(0), \tilde{\zeta}_s(0)) \in \mathcal{W}_{as}, \forall \varepsilon \in (0, \varepsilon_b^*],$$

and $R_1 > 0$ is such that $(\tilde{\zeta}_a(0), \tilde{\zeta}_s(0)) \in \mathcal{W}_{as}$ implies that $|\tilde{\zeta}_a(0)| \leq R_1/2$ and $|\tilde{\zeta}_s(0)| \leq R_1/2$. It now follows form (9.3.51) that

$$\left| [\tilde{\zeta}_a'(t), \tilde{\zeta}_s'(t)]' \right| \leq \alpha_3 e^{-\alpha_4 \varepsilon t}, \quad \forall (\tilde{\zeta}_a(0), \tilde{\zeta}_s(0)) \in \mathcal{W}_{as}, \forall \varepsilon \in (0, \varepsilon_b^*], \tag{9.3.58}$$

where $\alpha_4 = \alpha_2/2$ and α_3 are some positive constants independent of ε.

Viewing $L_{sa}(\varepsilon) F_a(\varepsilon) Q_a(\varepsilon) S_a^{-1}(\varepsilon) \tilde{\zeta}_a + L_{saa} F_a(\varepsilon) A_{aa} Q_a(\varepsilon) S_a^{-1}(\varepsilon) \tilde{\zeta}_a + L_{ss}(\varepsilon) \tilde{\zeta}_s$ as an input to the dynamics of $\tilde{\zeta}_s$, (9.3.48), and using (9.3.45), (9.3.52) and (9.3.53), we can easily show that there exists an $\varepsilon_c^* \in (0, \varepsilon_b^*]$ such that,

$$|\tilde{\zeta}_s(t)| \leq \alpha_5 e^{-\alpha_6 t/\varepsilon} + \alpha_7 \varepsilon e^{-\alpha_4 \varepsilon t}, \quad \forall (\tilde{\zeta}_a(0), \tilde{\zeta}_s(0)) \in \mathcal{W}_{as}, \forall \varepsilon \in (0, \varepsilon_c^*], \tag{9.3.59}$$

for some positive numbers α_5, $\alpha_6 < 1$, and α_7 independent of ε.

It follows from (9.3.52) and (9.3.58) that

$$\begin{aligned} |F_a(\varepsilon) \zeta_a(t)| &= |F_a(\varepsilon) Q_a(\varepsilon) S_a^{-1}(\varepsilon) \tilde{\zeta}_a| \\ &\leq \alpha_8 \varepsilon e^{-\alpha_4 \varepsilon t}, \quad \forall (\tilde{\zeta}_a(0), \tilde{\zeta}_s(0)) \in \mathcal{W}_{as}, \forall \varepsilon \in (0, \varepsilon_c^*], \end{aligned} \tag{9.3.60}$$

where α_8 is some positive number independent of ε.

Finally, recalling that $\hat{\eta} = [\eta', \tilde{\eta}']'$ and $\hat{\eta} = \bar{\hat{\eta}} + F_a(\varepsilon) \zeta_a$, it follows readily from (9.3.59) and (9.3.60) that

$$|\eta(t)| \leq K[e^{-\alpha_6 t/\varepsilon} + \varepsilon e^{-\alpha_4 \varepsilon t}], \quad \forall (\tilde{\zeta}_a(0), \tilde{\zeta}_s(0)) \in \mathcal{W}_{as}, \forall \varepsilon \in (0, \varepsilon_c^*], \tag{9.3.61}$$

for some positive constant K independent of ε.

Taking $\beta(s) = K e^{-s}$, $\gamma(s) = s/\alpha_6$, $\alpha(s) = \alpha_4 s$, and letting $R > 0$ be such that $x(0) \in \mathcal{W}_x$ implies $|x(0)| \leq R$, we see from Lemma 9.3.2 that there exists an $\varepsilon^* \in (0, \varepsilon_c^*]$ such that for all $\varepsilon \in (0, \varepsilon^*]$, all the trajectories of x starting from $\mathcal{W}_x$ approach zero as t goes to ∞. This completes the proof of Proposition 9.3.1. $\boxtimes$

With Proposition 9.3.1 established, we return to the proof of Theorem 9.3.1. For the nonlinear system (9.3.46)-(9.3.48), the result of [36, Theorem 7] (see Remark 9.3.2) shows that, for each $\varepsilon \in (0, \varepsilon^*]$, there exists a smooth Lyapunov function $\tilde{V}_0(x, \tilde{\zeta}_a, \tilde{\zeta}_s; \varepsilon) : \tilde{\mathcal{W}}_0(\varepsilon) \to \mathbb{R}_+$, which is positive definite on $\tilde{\mathcal{W}}_0(\varepsilon) \setminus \{0\}$ and proper on $\tilde{\mathcal{W}}_0(\varepsilon)$ and its derivative along the trajectories of (9.3.46)-(9.3.48) satisfies

$$\dot{\tilde{V}}_0 \leq -\Psi_0(x, \tilde{\zeta}_a, \tilde{\zeta}_s; \varepsilon), \tag{9.3.62}$$

where $\Psi_0(x, \tilde{\zeta}_a, \tilde{\zeta}_s; \varepsilon)$ is continuous on $\tilde{\mathcal{W}}_0(\varepsilon)$ and positive definite on

$$\left\{ (x, \tilde{\zeta}_a, \tilde{\zeta}_s) : 0 < \tilde{V}_0(x, \tilde{\zeta}_a, \tilde{\zeta}_s; \varepsilon) \leq c_0 + 1 \right\}, \tag{9.3.63}$$

and where c_0 is an (arbitrarily large) positive number. Let $c_0(\varepsilon) \geq 1$ be such that $(x, \tilde{\zeta}_a, \tilde{\zeta}_s) \in \mathcal{W}_0 \times \mathcal{W}_{as}$ implies that $\tilde{V}_0(x, \tilde{\zeta}_a, \tilde{\zeta}_s; \varepsilon) \leq c_0(\varepsilon)$. Let $c_1(\varepsilon) \geq 1$ be such that $\tilde{\zeta}_q \in \mathcal{W}_q(\varepsilon)$ implies that $\tilde{\zeta}'_q \Gamma'_I P \Gamma_I \tilde{\zeta}_q / \lambda_{\min}(\Gamma'_I P \Gamma_I) \leq c_1(\varepsilon)$, where P is as defined in Assumption 9.2.4. With this P, it is straightforward to verify that

$$\Gamma'_I P \Gamma_I \tilde{G}(x, \tilde{\zeta}, d; \varepsilon) M \Gamma_I + \Gamma'_I M' \tilde{G}'(x, \tilde{\zeta}, d; \varepsilon) \Gamma'_I P \Gamma_I = -\Gamma'_I Q(x, \xi, d) \Gamma_I \leq -\Gamma'_I Q_0 \Gamma_I, \tag{9.3.64}$$

where again $Q(x, \xi, d)$ and Q_0 are also defined in Assumption 9.2.4 and Q_0 is positive definite.

Define

$$V_1(x, \tilde{\zeta}; \varepsilon) = \frac{c_0(\varepsilon) \tilde{V}_0(x, \tilde{\zeta}_a, \tilde{\zeta}_s; \varepsilon)}{c_0(\varepsilon) + 1 - \tilde{V}_0(x, \tilde{\zeta}_a, \tilde{\zeta}_s; \varepsilon)} + \frac{c_1(\varepsilon) \tilde{\zeta}'_q \Gamma'_I P \Gamma_I \tilde{\zeta}_q / \lambda_{\min}(\Gamma'_I P \Gamma_I)}{c_1(\varepsilon) + 1 - \tilde{\zeta}'_q \Gamma'_I P \Gamma_I \tilde{\zeta}_q / \lambda_{\min}(\Gamma'_I P \Gamma_I)}$$

and the set

$$\tilde{\mathcal{W}}_1(\varepsilon) = \left\{ (x, \tilde{\zeta}_a, \tilde{\zeta}_s) : \tilde{V}_0(x, \tilde{\zeta}_a, \tilde{\zeta}_s; \varepsilon) < c_0(\varepsilon) + 1 \right\}$$
$$\times \left\{ \tilde{\zeta}_q : \tilde{\zeta}'_q \Gamma'_I P \Gamma_I \tilde{\zeta}_q / \lambda_{\min}(\Gamma'_I P \Gamma_I) < c_1(\varepsilon) + 1 \right\}.$$

Then, $V_1 : \tilde{\mathcal{W}}_1(\varepsilon) \to \mathbb{R}_+$ is positive definite on $\tilde{\mathcal{W}}_1(\varepsilon) \setminus \{0\}$ and proper on $\tilde{\mathcal{W}}_1(\varepsilon)$. Furthermore, it follows from Lemma 9.3.3 that for any $\varepsilon \in (0, \varepsilon^*]$ and any $\rho \in (0, 1)$, there exists a $\mu_a^*(\varepsilon, \rho) > 0$ such that for each $\mu \in (0, \mu_a^*(\varepsilon, \rho)]$, the derivative of V_1 along the trajectories (9.3.35)-(9.3.38) satisfies

$$\dot{V}_1 \leq -\Psi_1(x, \tilde{\zeta}; \varepsilon, \mu),$$

where $\Psi_1(x, \tilde{\zeta}; \varepsilon, \mu)$ is positive definite on $\{(x, \tilde{\zeta}) : \rho \leq V_1(x, \tilde{\zeta}; \varepsilon) \leq c_0^2(\varepsilon) + c_1^2(\varepsilon) + 1\}$, which shows that the state $(x, \tilde{\zeta})$ will enter the set $\{(x, \tilde{\zeta}) : V_1(x, \tilde{\zeta}; \varepsilon) \leq \rho\}$ in a finite time and remain in it thereafter. Next, we note that for any

$\varepsilon \in (0, \varepsilon^*]$, there exists a $\rho^*(\varepsilon) \in (0, 1)$ such that $V_1(x, \tilde{\zeta}; \varepsilon) \leq \rho^*(\varepsilon)$ implies $(x, \xi) \in \mathcal{W}_0$. Now, for each $\varepsilon \in (0, \varepsilon^*]$, take $\mu^*(\varepsilon) = \mu_a^*(\varepsilon, \rho^*(\varepsilon))$ to complete the proof of Theorem 9.3.1. ⊠

Proof of Theorem 9.3.2. We will prove this theorem by showing that, under the extra Assumption 9.2.5, the family of state feedback laws (9.3.34) as constructed in the proof of Theorem 9.3.1, achieves semi-global asymptotic stabilization of the system (9.1.1). More specifically, we will show that for any *a priori* given (arbitrarily large) bounded set $\mathcal{W} \subset \mathbb{R}^{l+n}$, there exists an $\varepsilon^* > 0$ and for each $\varepsilon \in (0, \varepsilon^*]$ there exists a $\mu^*(\varepsilon) > 0$, such that, for all $\mu \in (0, \mu^*(\varepsilon)]$, $\varepsilon \in (0, \varepsilon^*]$, the equilibrium $(0, 0)$ of the closed-loop system is locally asymptotically stable and the set $\mathcal{W}$ is contained in its basin of attraction.

To this end, let us consider the closed-loop system (9.3.35)-(9.3.38) again. Let $\mathcal{U}_1 \subset \mathbb{R}^n$ be any compact set. Let the ε independent compact set $\mathcal{U}_{as} \subset \mathbb{R}^{n-m}$ and the ε dependent compact set $\mathcal{U}_q(\varepsilon) \in \mathbb{R}^m$ be such that, for each $\varepsilon \in (0, \varepsilon_a^*]$, $\xi \in \mathcal{U}_1$ implies that $(\tilde{\zeta}_a, \tilde{\zeta}_s) \in \mathcal{U}_{as}$ and $\tilde{\zeta}_q \in \mathcal{U}_q(\varepsilon)$. Now, the continuous differentiability of functions $\tilde{G}(x, \tilde{\zeta}, d(t); \varepsilon)$ and $\tilde{g}(x, \tilde{\zeta}, d(t); \varepsilon)$ and the compactness of the sets $\mathcal{U}_0$, $\mathcal{U}_{as}$, $\mathcal{U}_q(\varepsilon)$ and $\mathcal{D}$ show that there exist an ε independent positive number δ_1 and an ε dependent positive number $\delta_2(\varepsilon)$ such that for any $(x, \tilde{\zeta}, d) \in \mathcal{U}_0 \times \mathcal{U}_{as} \times \mathcal{U}_q(\varepsilon) \times \mathcal{D}$,

$$|f(x, \eta) - f(x, 0)| \leq \delta_1(|\tilde{\zeta}_a| + |\tilde{\zeta}_s|), \qquad (9.3.65)$$

$$|\tilde{g}(x, \tilde{\zeta}, d; \varepsilon) - \tilde{g}(x, 0, d; \varepsilon)| \leq \delta_2(\varepsilon)(|\tilde{\zeta}_a| + |\tilde{\zeta}_s| + |\tilde{\zeta}_q|). \qquad (9.3.66)$$

Noting that $\tilde{g}(x, 0, d; \varepsilon)$ is independent of ε, it follows from Assumption 9.2.3 that

$$|\tilde{g}(x, 0, d; \varepsilon)| \leq \gamma_{\tilde{g}} \psi(|x|), \ \forall x \in \mathcal{U}_0, \qquad (9.3.67)$$

for some positive number $\gamma_{\tilde{g}}$ independent of ε. We note here that $\mathcal{U}_0$ is as given in Assumption 9.2.5.

For the closed-loop system (9.3.35)-(9.3.38), we pick a Lyapunov function candidate as

$$V(x, \tilde{\zeta}; \varepsilon, \mu) = V_0(x) + \frac{1}{\sqrt{\mu}} V_{as}(\tilde{\zeta}_a, \tilde{\zeta}_s; \varepsilon) + V_q(\tilde{\zeta}_q), \qquad (9.3.68)$$

where V_0 is as given in Assumption 9.2.5, V_{as} is as given by (9.3.51), and

$$V_q(\tilde{\zeta}_q) = \tilde{\zeta}_q' \Gamma_1' P \Gamma_1 \tilde{\zeta}_q. \qquad (9.3.69)$$

Using (9.3.54), (9.3.65), (9.3.66), (9.3.67), and Assumption 9.2.4, we easily verify that, for $\varepsilon \in (0, \varepsilon_a^*]$ and $\mu \in (0, 1]$ the derivative of V along the trajectories inside $\mathcal{U}_0 \times \mathcal{U}_{as} \times \mathcal{U}_q(\varepsilon)$ can be evaluated as,

$$\dot{V} \leq -\psi^2(|x|) + \alpha_1 \psi(|x|)|\tilde{\zeta}| - [|\tilde{\zeta}_a| \, |\tilde{\zeta}_s| \, |\tilde{\zeta}_q|] \frac{\mathcal{R}(\varepsilon, \mu)}{\sqrt{\mu}} \begin{bmatrix} |\tilde{\zeta}_a| \\ |\tilde{\zeta}_s| \\ |\tilde{\zeta}_q| \end{bmatrix}, \qquad (9.3.70)$$

with

$$\mathcal{R}(\varepsilon, \mu) = \begin{bmatrix} \dfrac{\varepsilon}{2} & -\alpha_{12} - \alpha_{21}\kappa\varepsilon & -\alpha_{13}(\varepsilon) \\ -\alpha_{12} - \alpha_{21}\kappa\varepsilon & \dfrac{\kappa}{\varepsilon} - \kappa\alpha_{22} & -\alpha_{23}(\varepsilon) \\ -\alpha_{13}(\varepsilon) & -\alpha_{23}(\varepsilon) & \dfrac{\alpha_2}{\sqrt{\mu}} - \alpha_{33}(\varepsilon) \end{bmatrix},$$

where α_1, α_2, α_{12}, α_{21} and α_{22} are strictly positive constants independent of ε, κ is a positive constant whose value is to be determined, and all the other $\alpha_{ij}(\varepsilon)$'s are some ε dependent positive numbers. Let $\kappa = 2\alpha_{12}^2$. Clearly, there exists an $\varepsilon_a^* \in (0, 1]$ and for each fixed $\varepsilon \in (0, \varepsilon_a^*]$, there exists a $\mu_a^*(\varepsilon) \in (0, 1]$ such that the matrix $\mathcal{R}(\varepsilon, \mu)$ is positive definite for all $\mu \in (0, \mu_a^*(\varepsilon)]$, $\varepsilon \in (0, \varepsilon_b^*]$. Hence, there exists an ε dependent positive number $\alpha_3(\varepsilon)$ such that,

$$\dot{V} \leq -\psi^2(|x|) + \alpha_1 \psi(|x|)|\tilde{\zeta}| - \frac{\alpha_3(\varepsilon)}{\sqrt{\mu}}|\tilde{\zeta}|^2. \qquad (9.3.71)$$

Now simple completion of square arguments show that for each $\varepsilon \in (0, \varepsilon_b^*]$, there is a $\mu_b^*(\varepsilon) \in (0, \mu_a^*(\varepsilon)]$ such that, for all $\mu \in (0, \mu_b^*(\varepsilon)]$, $\varepsilon \in (0, \varepsilon_b^*]$,

$$-\frac{1}{2}\psi^2(|x|) + \alpha_1 \psi(|x|)|\tilde{\zeta}| - \frac{\alpha_3(\varepsilon)}{2\sqrt{\mu}}|\tilde{\zeta}|^2 \leq 0, \qquad (9.3.72)$$

and hence

$$\dot{V} \leq -\frac{1}{2}\left[\psi^2(|x|) + \frac{\alpha_3(\varepsilon)}{\sqrt{\mu}}|\tilde{\zeta}|^2\right], \quad \forall (x, \tilde{\zeta}) \in \mathcal{U}_0 \times \mathcal{U}_{as} \times \mathcal{U}_q(\varepsilon), \qquad (9.3.73)$$

which shows that the equilibrium of the closed-loop system (9.3.35)-(9.3.38) is locally asymptotically stable for all $\mu \in (0, \mu_b^*(\varepsilon)]$, $\varepsilon \in (0, \varepsilon_b^*]$.

We next proceed to show that, for sufficiently small ε and μ, the basin of attraction of the equilibrium of the closed-loop system (9.3.35)-(9.3.38) indeed contains $\mathcal{W}$. It follows from Theorem 9.3.1 that, under the family of the state feedback laws, for the given set $\mathcal{U}_0 \times \mathcal{U}_1$, there is an $\varepsilon_c^* \in (0, 1]$ and for each $\varepsilon \in (0, \varepsilon_c^*]$ there is a $\mu_c^*(\varepsilon) > 0$ such that, for all $\mu \in (0, \mu_c^*(\varepsilon)]$, $\varepsilon \in (0, \varepsilon_c^*]$, all the solutions of the closed-loop system with initial conditions $(x(0), \xi(0)) \in \mathcal{W}$ enter the set $\mathcal{U}_0 \times \mathcal{U}_1$ in a finite time and remain in it thereafter. Now choose

$\varepsilon^* = \min\{\varepsilon_b^*, \varepsilon_c^*\}$ and for each $\varepsilon \in (0, \varepsilon^*]$ choose $\mu^* = \min\{\mu_b^*(\varepsilon), \mu_c^*(\varepsilon)\}$. It is then trivial to see that, for all $\mu \in (0, \mu^*(\varepsilon)]$, $\varepsilon \in (0, \varepsilon^*]$, all the trajectories of the closed-loop system starting from the set $\mathcal{W}$ will enter the set $\mathcal{U}_0 \times \mathcal{U}_1$ in a finite time and remain in it thereafter, and due to (9.3.73) will approach the equilibrium $(x, \xi) = (0, 0)$ as t goes to ∞. This completes the proof of Theorem 9.3.2. $\boxtimes$

9.4. Output Feedback Results

The main results concerning semi-global practical stabilizability and semi-global asymptotic stabilizability via output feedback of the system (9.1.1) is given in the following two theorems.

Theorem 9.4.1. *If Assumptions 9.2.1-9.2.4 and 9.2.6 hold, then the point* $(0, 0)$ *is semi-globally practically stabilizable for the system (9.1.1) via dynamic output feedback of dynamical order n.* $\boxed{\text{T}}$

Theorem 9.4.2. *If Assumptions 9.2.1-9.2.6 hold, then the equilibrium* $(0, 0)$ *of the system (9.1.1) is semi-globally asymptotically stabilizable via dynamic output feedback of dynamical order n.* $\boxed{\text{T}}$

Remark 9.4.1. *In the proof of Theorems 9.4.1 and 9.4.2, a family of output feedback laws is explicitly constructed based solely on the knowledge of the matrices A, B, C and D, and no knowledge of the mappings f, g and G is required. In this sense, our output feedback design is robust.* $\boxed{\text{R}}$

Proof of Theorem 9.4.1. We will prove this theorem by first explicitly constructing a family of dynamic output feedback laws parameterized in ε, μ and v,

$$\begin{cases} \dot{z} = \varphi_{\mathcal{W}, \mathcal{W}_0}(z, y; \varepsilon, \mu, v), & z \in \mathbb{R}^n, \\ u = F_{\mathcal{W}, \mathcal{W}_0}(z, y; \varepsilon, \mu, v), \end{cases} \tag{9.4.1}$$

and then showing that this family of dynamic output feedback laws indeed achieves semi-global practical stabilization of the system (9.1.1). The family of output feedback laws we construct have observer based controller structure. More specifically, in the construction of the family of dynamic output feedback laws, we first design a family of static partial state feedback laws which would semi-globally practically stabilize the system (9.1.1) if the partial state ξ was available for feedback. These static partial state feedback laws are then implemented with the estimation of the state ξ using a high gain observer. The degree of the fastness of this observer crucially depends on the chosen partial state feedback laws. In other words, the design of observer as given here does not follow the traditional route of separation principle embedded in the conventional way of designing observer-based controllers. The effect of the peaking of the error dynamics is eliminated by saturating the implemented control signals.

The family of partial state feedback laws we choose is the same as the one constructed in the proof of Theorem 9.3.1, which is parameterized in ε and μ. With this family of partial state feedback laws, the closed-loop system is given by (9.3.35)-(9.3.38). We repeat this system below.

$$\dot{x} = f(x, \eta), \tag{9.4.2}$$

$$\dot{\tilde{\zeta}}_a = A_{aa}^c(\varepsilon)\tilde{\zeta}_a + S_a(\varepsilon)Q_a^{-1}(\varepsilon)L_{ad}\bar{\tilde{\eta}}, \ \ \bar{\tilde{\eta}}_i = \tilde{\zeta}_{i1}, \tag{9.4.3}$$

$$\dot{\tilde{\zeta}}_s = \frac{1}{\varepsilon}A_s^c\tilde{\zeta}_s + L_{sa}(\varepsilon)F_a(\varepsilon)Q_a(\varepsilon)S_a^{-1}(\varepsilon)\tilde{\zeta}_a$$

$$+ L_{saa}F_a(\varepsilon)A_{aa}Q_a(\varepsilon)S_a^{-1}(\varepsilon)\tilde{\zeta}_a + L_{ss}(\varepsilon)\tilde{\zeta}_s + L_{sq}\tilde{\zeta}_q, \tag{9.4.4}$$

$$\dot{\tilde{\zeta}}_q = \frac{1}{\mu}\tilde{G}(x, \tilde{\zeta}, d(t); \varepsilon)M\Gamma_I\tilde{\zeta}_q + \tilde{N}(\varepsilon)\tilde{\zeta} + \tilde{g}(x, \tilde{\zeta}, d(t); \varepsilon), \tag{9.4.5}$$

where $\varepsilon \in (0, 1]$, $\mu \in (0, 1]$, and all the notations follow from the proof of Theorem 9.3.1.

Let the compact sets $W_x \subset \mathbb{R}^l$, $W_\xi \subset \mathbb{R}^n$ and $W_{\hat{\xi}} \subset \mathbb{R}^n$ be such that $W \subset W_x \times W_\xi \times W_{\hat{\xi}}$. Let the ε independent compact set W_{as} and the ε dependent compact set $W_q(\varepsilon)$ be such that, for each $\varepsilon \in (0, \varepsilon_a^*]$, $\xi \in W_\xi$ implies that $(\tilde{\zeta}_a, \tilde{\zeta}_s) \in W_{as}$ and $\tilde{\zeta}_q \in W_q(\varepsilon)$. The existence of such sets W_{as} and $W_q(\varepsilon)$ is guaranteed by the special form of the state transformation from ξ to $\tilde{\zeta}$ as given in the proof of Theorem 9.3.1. We then have the following proposition, which was established in the proof of Theorem 9.3.1.

Proposition 9.4.1. *Consider the nonlinear system (9.4.2)-(9.4.5). Then, there exists an $\varepsilon_a^* \in (0, 1]$ such that for each $\varepsilon \in (0, \varepsilon_a^*]$, there exists a neighborhood of the origin of $\mathbb{R}^{n-m}$, $\tilde{W}_0(\varepsilon) \supset W_{as}$, and a smooth Lyapunov function $\tilde{V}_0(x, \tilde{\zeta}_a, \tilde{\zeta}_s; \varepsilon) : \tilde{W}_0(\varepsilon) \to \mathbb{R}_+$, which is positive definite on $\tilde{W}_0(\varepsilon) \setminus \{0\}$ and proper on $\tilde{W}_0(\varepsilon)$. Let $c_0(\varepsilon) \geq 1$ be such that $(x, \tilde{\zeta}_a, \tilde{\zeta}_s) \in W_0 \times W_{as}$ implies that $\tilde{V}_0(x, \tilde{\zeta}_a, \tilde{\zeta}_s; \varepsilon) \leq c_0(\varepsilon)$, and let $c_1(\varepsilon) \geq 1$ be such that $\tilde{\zeta}_q \in W_q(\varepsilon)$ implies that $\tilde{\zeta}_q' \Gamma_I' P \Gamma_I \tilde{\zeta}_q / \lambda_{\min}(\Gamma_I' P \Gamma_I) \leq c_1(\varepsilon)$, where P is as defined in Assumption 9.2.4 and Γ_I in the proof of Theorem 9.3.1. If we define*

$$V_1(x, \tilde{\zeta}; \varepsilon) = \frac{c_0(\varepsilon)\tilde{V}_0(x, \tilde{\zeta}_a, \tilde{\zeta}_s; \varepsilon)}{c_0(\varepsilon) + 1 - \tilde{V}_0(x, \tilde{\zeta}_a, \tilde{\zeta}_s; \varepsilon)} + \frac{c_1(\varepsilon)\tilde{\zeta}_q'\Gamma_I'P\Gamma_I\tilde{\zeta}_q / \lambda_{\min}(\Gamma_I'P\Gamma_I)}{c_1(\varepsilon) + 1 - \tilde{\zeta}_q'\Gamma_I'P\Gamma_I\tilde{\zeta}_q / \lambda_{\min}(\Gamma_I'P\Gamma_I)}$$

and the set

$$\tilde{W}_1(\varepsilon) = \left\{ (x, \tilde{\zeta}_a, \tilde{\zeta}_s) : \tilde{V}_0(x, \tilde{\zeta}_a, \tilde{\zeta}_s; \varepsilon) < c_0(\varepsilon) + 1 \right\}$$

$$\times \left\{ \tilde{\zeta}_q : \tilde{\zeta}_q' \Gamma_I' P \Gamma_I \tilde{\zeta}_q / \lambda_{\min}(\Gamma_I' P \Gamma_I) < c_1(\varepsilon) + 1 \right\},$$

then, $V_1 : \tilde{\mathcal{W}}_1(\varepsilon) \to \mathbb{R}_+$ is positive definite on $\tilde{\mathcal{W}}_1(\varepsilon) \setminus \{0\}$ and proper on $\tilde{\mathcal{W}}_1(\varepsilon)$. Furthermore, for any $\rho \in (0, 1)$, there exists a $\mu_a^*(\varepsilon, \rho) > 0$ such that for each $\mu \in (0, \mu_a^*(\varepsilon, \rho)]$, the derivative of V_1 along the trajectories (9.4.2)-(9.4.5) satisfies

$$\dot{V}_1 \leq -\Psi_1(x, \tilde{\zeta}; \varepsilon, \mu), \tag{9.4.6}$$

where $\Psi_1(x, \tilde{\zeta}; \varepsilon, \mu)$ is positive definite on the set

$$\left\{ (x, \tilde{\zeta}) : \rho \leq V_1(x, \tilde{\zeta}; \varepsilon) \leq c_0^2(\varepsilon) + c_1^2(\varepsilon) + 1 \right\}.$$

For future use, we define the set

$$\Gamma(\varepsilon) = \left\{ \tilde{\zeta}_q : c_1(\varepsilon) \frac{\tilde{\zeta}_q' \Gamma_I' P \Gamma_I \tilde{\zeta}_q / \lambda_{\min}(\Gamma_I' P \Gamma_I)}{c_1(\varepsilon) + 1 - \tilde{\zeta}_q' \Gamma_I' P \Gamma_I \tilde{\zeta}_q / \lambda_{\min}(\Gamma_I' P \Gamma_I)} \leq c_0^2(\varepsilon) + c_1^2(\varepsilon) + 1 \right\}.$$

$\boxed{\text{P}}$

We next proceed to build a high gain observer whose state is an estimate of the state $\bar{\xi}$ and is to be used for the implementation of the state feedback laws (9.3.34). To this end, we consider the system (9.3.20), the linearizable part of the system (9.1.1) on special coordinate basis (SCB),

$$\begin{cases} \dot{\bar{\xi}} = \bar{A}\bar{\xi} + \bar{B}[\bar{G}(x, \bar{\xi}, d(t))u + \bar{g}(x, \bar{\xi}, d(t))], \\ y = \bar{D}\bar{\xi}. \end{cases} \tag{9.4.7}$$

By Assumption 9.2.6, the triple $(\bar{A}, \bar{B}, \bar{D})$ is left invertible with no invariant zeros. We then perform a nonsingular state and output transformation on the system (9.4.7),

$$\bar{\xi} = \bar{\Gamma}_s \bar{\bar{\xi}}, \; \overset{\bullet}{y} = \bar{\Gamma}_o \bar{y}, \tag{9.4.8}$$

such that the system (9.4.7) can be rewritten as,

$$\begin{cases} \dot{\bar{\bar{\xi}}} = \bar{\bar{A}}\bar{\bar{\xi}} + \bar{\bar{B}}[\bar{G}(x, \bar{\Gamma}_s \bar{\bar{\xi}}, d(t))u + g(x, \bar{\Gamma}_s \bar{\bar{\xi}}, d(t))], \\ \bar{\bar{y}} = \bar{\bar{D}}\bar{\bar{\xi}}, \end{cases} \tag{9.4.9}$$

where $\bar{\bar{A}} = \bar{\Gamma}_s^{-1} \bar{A} \bar{\Gamma}_s$, $\bar{\bar{B}} = \bar{\Gamma}_s^{-1} \bar{B}$, $\bar{\bar{D}} = \bar{\Gamma}_o^{-1} \bar{D} \bar{\Gamma}_s$, and where $\bar{\Gamma}_s$ and $\bar{\Gamma}_o$ are such that $(\bar{\bar{A}}, \bar{\bar{B}}, \bar{\bar{D}})$ is in the SCB form of $(\bar{A}, \bar{B}, \bar{D})$. The existence of such transformation matrices $\bar{\Gamma}_s$ and $\bar{\Gamma}_o$ is established in Theorem A.1.1. Now in view of Theorem A.1.1, the system (9.4.9) has the following dynamic equations,

$$\bar{\bar{\xi}} = [\bar{\bar{\xi}}_b', \bar{\bar{\xi}}_d']',$$

$$\bar{\bar{\xi}}_b = [\bar{\bar{\xi}}_{b1}', \bar{\bar{\xi}}_{b2}', \cdots, \bar{\bar{\xi}}_{bp-m}']', \; \bar{\bar{\xi}}_{bi} = [\bar{\bar{\xi}}_{bi1}, \bar{\bar{\xi}}_{bi2}, \cdots, \bar{\bar{\xi}}_{bi\bar{r}_i}]',$$

$$\bar{\bar{\xi}}_d = [\bar{\bar{\xi}}_1', \bar{\bar{\xi}}_2', \cdots, \bar{\bar{\xi}}_m']', \; \bar{\bar{\xi}}_i = [\bar{\bar{\xi}}_{i1}, \bar{\bar{\xi}}_{i2}, \cdots, \bar{\bar{\xi}}_{i\bar{q}_i}]',$$

$$\bar{\bar{y}} = [\bar{\bar{y}}_b', \bar{\bar{y}}_d']', \; \bar{\bar{y}}_b = [\bar{\bar{y}}_{b1}, \bar{\bar{y}}_{b2}, \cdots, \bar{\bar{y}}_{bp-m}]', \; \bar{\bar{y}}_d = [\bar{\bar{y}}_1, \bar{\bar{y}}_2, \cdots, \bar{\bar{y}}_m]',$$

and for $i = 1$ to $p - m$,

$$\dot{\bar{\xi}}_{bi} = A_{\bar{r}_i} \bar{\xi}_{bi} + \bar{L}_{bib} \bar{y}_b + \bar{L}_{bid} \bar{y}_d, \tag{9.4.10}$$

$$\bar{y}_{bi} = C_{\bar{r}_i} \bar{\xi}_{bi} = \bar{\xi}_{bi1}, \tag{9.4.11}$$

for $i = 1$ to m,

$$\dot{\bar{\xi}}_i = A_{\bar{q}_i} \bar{\xi}_i + \bar{L}_{id} \bar{y}_d + B_{\bar{q}_i} [\omega_i(x, \bar{\Gamma}_s \bar{\xi}, u, d) + \bar{E}_{ib} \bar{\xi}_b + \bar{E}_{id} \bar{\xi}_d], \tag{9.4.12}$$

$$\bar{y}_i = C_{\bar{q}_i} \bar{\xi}_i = \bar{\xi}_{i1}, \tag{9.4.13}$$

where $\omega_i(x, \bar{\Gamma}_s \bar{\xi}, u, d)$ is the ith element of $\bar{G}(x, \bar{\Gamma}_s \bar{\xi}, d(t))u + \bar{g}(x, \bar{\Gamma}_s \bar{\xi}, d(t))$ as defined in the proof of Theorem 9.3.1.

We are now ready to construct the high gain observer. The construction is carried out in three steps.

Step 1. For $i = 1$ to $p - m$, choose $L_{bi} \in \mathbb{R}^{\bar{r}_i \times 1}$ such that

$$\lambda(A_{\bar{r}_i}^c) \in \mathbb{C}^-, \quad A_{\bar{r}_i}^c := A_{\bar{r}_i} + L_{bi} C_{\bar{r}_i}.$$

Note that the existence of such L_{bi} is guaranteed by the special structure of the matrix pair $(A_{\bar{r}_i}, C_{\bar{r}_i})$.

Similarly, for $i = 1$ to m, choose $L_{di} \in \mathbb{R}^{\bar{q}_i \times 1}$ such that

$$\lambda(A_{\bar{q}_i}^c) \in \mathbb{C}^-, \quad A_{\bar{q}_i}^c := A_{\bar{q}_i} + L_{di} C_{\bar{q}_i}.$$

Again, the existence of such L_{di} is guaranteed by the special structure of the matrix pair $(A_{\bar{q}_i}, C_{\bar{q}_i})$.

Step 2. For any $v \in (0, 1]$, define a matrix $\bar{L}(v) \in \mathbb{R}^{n \times p}$ as

$$\bar{L}(v) = \bar{\Gamma}_s \begin{bmatrix} -\bar{L}_{bb} + L_b(v) & -\bar{L}_{bd} \\ 0 & -\bar{L}_{dd} + L_d(v) \end{bmatrix} \bar{\Gamma}_o^{-1}, \tag{9.4.14}$$

where

$$\bar{L}_{bb} = \begin{bmatrix} \bar{L}_{b1b} \\ \bar{L}_{b2b} \\ \vdots \\ \bar{L}_{bp-mb} \end{bmatrix}, \quad \bar{L}_{bd} = \begin{bmatrix} \bar{L}_{b1d} \\ \bar{L}_{b2d} \\ \vdots \\ \bar{L}_{bmd} \end{bmatrix}, \quad \bar{L}_{dd} = \begin{bmatrix} \bar{L}_{1d} \\ \bar{L}_{2d} \\ \vdots \\ \bar{L}_{md} \end{bmatrix},$$

$$L_b(v) = \begin{bmatrix} \dfrac{1}{v^{\bar{r}_1}} S_{\bar{r}_1}(v) L_{b1} & 0 & \cdots & 0 \\ 0 & \dfrac{1}{v^{\bar{r}_2}} S_{\bar{r}_2}(v) L_{b2} & \cdots & 0 \\ \vdots & \vdots & \ddots & \vdots \\ 0 & 0 & \cdots & \dfrac{1}{v^{\bar{r}_{p-m}}} S_{\bar{r}_{p-m}}(v) L_{bp-m} \end{bmatrix},$$

$$L_d(v) = \begin{bmatrix} \frac{1}{v^{\bar{q}_1}}S_{\bar{q}_1}(v)L_{d1} & 0 & \cdots & 0 \\ 0 & \frac{1}{v^{\bar{q}_2}}S_{\bar{q}_2}(v)L_{d2} & \cdots & 0 \\ \vdots & \vdots & \ddots & \vdots \\ 0 & 0 & \cdots & \frac{1}{\mu^{\bar{q}_m}}S_{\bar{q}_m}(v)L_{dm} \end{bmatrix},$$

and where for any integer $r \geq 1$,

$$S_r(v) = \begin{bmatrix} v^{r-1} & 0 & \cdots & 0 \\ 0 & v^{r-2} & \cdots & 0 \\ \vdots & \vdots & \ddots & \vdots \\ 0 & 0 & \cdots & 1 \end{bmatrix}.$$

Step 3. At this step, we are ready to construct the high gain observer we need. This high gain observer takes the following form,

$$\dot{\hat{\xi}} = \bar{A}\hat{\xi} - \bar{L}(v)(y - \bar{D}\hat{\xi}), \quad \hat{\xi}(0) \in \mathcal{W}_{\hat{\xi}}, \tag{9.4.15}$$

where the matrix $\bar{L}(v)$ is as given in (9.4.14). $\quad\boxed{\text{A}}$

We are now to implement the state feedback laws (9.3.34) with the state of the high gain observer (9.4.15), $\hat{\xi}$. To this end, we partition the state vector $\hat{\xi}$ according to that of $\bar{\xi}$ as,

$$\hat{\xi} = [\hat{\xi}_a', \hat{\xi}_d']', \quad \hat{\xi}_d = [\hat{\xi}_1', \hat{\xi}_2', \cdots, \hat{\xi}_m']', \quad \hat{\xi}_i = [\hat{\xi}_{i1}, \hat{\xi}_{i2}, \cdots, \hat{\xi}_{iq_i}]',$$

and define a positive number h as,

$$h = \sqrt{c_1(\varepsilon) + 1}.$$

Clearly, for $\tilde{\zeta}_q \in \Gamma(\varepsilon)$,

$$|\tilde{\zeta}_q| \leq \sqrt{(c_1(\varepsilon) + 1)\frac{c_0^2(\varepsilon) + c_1^2(\varepsilon) + 1}{c_0^2(\varepsilon) + c_1^2(\varepsilon) + c_1(\varepsilon) + 1}} \leq \sqrt{c_1(\varepsilon) + 1} = h.$$

We then implement the state feedback laws (9.3.34) as,

$$u = \frac{1}{\mu}M\Gamma_1\sigma_h(\hat{\zeta}_q), \tag{9.4.16}$$

where

$$\hat{\zeta}_q = [\hat{\zeta}_{1q_1}, \hat{\zeta}_{2q_2}, \cdots, \hat{\zeta}_{mq_m}]',$$

and and $\hat{\zeta}_{iq_i}$'s are the $\bar{\zeta}_{iq_i}$'s defined in (9.3.27) with variables ξ_a and ξ_{ij}'s replaced by the corresponding $\hat{\xi}_a$ and $\hat{\xi}_{ij}$'s. Also, in (9.4.16), $\sigma_h(s)$ is a vector saturation function defined as,

$$\sigma_h(s) = [\text{sat}(s_1), \text{sat}(s_2), \cdots, \text{sat}(s_m)]',$$

with sat being a standard saturation function, i.e., $\text{sat}(s) = \text{sign}(s)\min\{h, |s|\}$, and $s = [s_1, s_2, \cdots, s_m]'$. It will become clear as we proceed that the saturation function has been included in our implementation of the state feedback laws (9.3.34) to eliminate the peaking effect of the error dynamics. This idea was originally proposed in [30] for semi-global stabilization of fully linearizable systems and used in [115] for more general single input nonlinear systems with zero dynamics.

So far, we have constructed our family of dynamic output feedback laws (9.4.1) as given by (9.4.15) and (9.4.16). It remains to show that this family of dynamic output feedback laws indeed achieves semi-global practical stabilization for the system (9.1.1). More specifically, we will show that for the given sets $\mathcal{W}$ and $\mathcal{W}_0$, there exists an $\varepsilon^* > 0$, for each $\varepsilon \in (0, \varepsilon^*]$ there exists a $\mu^*(\varepsilon) > 0$, and for each pair $\mu \in (0, \mu^*(\varepsilon)]$, $\varepsilon \in (0, \varepsilon^*]$, there exists a $\upsilon^*(\varepsilon, \mu)$ such that, for all $\upsilon \in (0, \upsilon^*(\varepsilon, \mu)]$, $\mu \in (0, \mu^*(\varepsilon)]$, $\varepsilon \in (0, \varepsilon^*]$, all the solutions of the closed-loop system with initial conditions in $\mathcal{W}$ enter the given set $\mathcal{W}_0$ in a finite time and remain in it thereafter. To this end, let $e = \bar{\xi} - \hat{\xi}$. The error dynamics can be written as,

$$\dot{e} = (\bar{A} + \bar{L}(\upsilon)\bar{D})e + \bar{B}[\bar{G}(x, \bar{\xi}, d(t))u + \bar{g}(x, \bar{\xi}, d(t))], \quad e(0) \in \mathcal{W}_e, \quad (9.4.17)$$

where $\mathcal{W}_e = \{e = \bar{\xi} - \hat{\xi} : \xi \in \mathcal{W}_\xi, \hat{\xi} \in \mathcal{W}_{\hat{\xi}}\}$ is independent of the controller parameters ε, μ and υ.

Letting

$$e = \bar{\Gamma}_s \bar{e}, \quad \bar{e} = [e_b', e_d']',$$

$$e_b = [e_{b1}', e_{b2}', \cdots, e_{bp-m}']', \quad \bar{e}_d = [e_1', e_2' \cdots, e_m']',$$

and recalling (9.4.2)-(9.4.5), we write the closed-loop system as follows,

$$\dot{x} = f(x, \eta), \tag{9.4.18}$$

$$\dot{\tilde{\zeta}}_a = A_{aa}^c(\varepsilon)\tilde{\zeta}_a + S_a(\varepsilon)Q_a^{-1}(\varepsilon)L_{ad}\bar{\tilde{\eta}}, \quad \bar{\tilde{\eta}}_i = \tilde{\zeta}_{i1}, \tag{9.4.19}$$

$$\dot{\tilde{\zeta}}_s = \frac{1}{\varepsilon}A_s^c\tilde{\zeta}_s + L_{sa}(\varepsilon)F_a(\varepsilon)Q_a(\varepsilon)S_a^{-1}(\varepsilon)\tilde{\zeta}_a$$
$$+ L_{saa}F_a(\varepsilon)A_{aa}Q_a(\varepsilon)S_a^{-1}(\varepsilon)\tilde{\zeta}_a + L_{ss}(\varepsilon)\tilde{\zeta}_s + L_{sq}\tilde{\zeta}_q, \tag{9.4.20}$$

$$\dot{\tilde{\zeta}}_q = \frac{1}{\mu}\tilde{G}(x, \tilde{\zeta}, d(t); \varepsilon)M\Gamma_1\sigma_h(\tilde{\zeta}_q + \tilde{E}(\varepsilon)\bar{e}) + \tilde{N}(\varepsilon)\tilde{\zeta} + \tilde{g}(x, \tilde{\zeta}, d(t); \varepsilon), \tag{9.4.21}$$

$$\dot{e}_{bi} = A_{\bar{r}_i}e_{bi} + \frac{1}{\upsilon^{\bar{r}_i}}\dot{S}_{\bar{r}_i}(\upsilon)L_{bi}C_{\bar{r}_i}e_{bi}, \quad i = 1, 2, \cdots, p-m, \tag{9.4.22}$$

$$\dot{e}_i = A_{\bar{q}_i}e_i + \frac{1}{\upsilon^{\bar{q}_i}}S_{\bar{q}_i}(\upsilon)L_iC_{\bar{q}_i}e_i + B_{\bar{q}_i}\left[\frac{1}{\mu}\bar{\rho}_i(x, \tilde{\zeta}, d, \bar{e}; \varepsilon) + \bar{E}_{ib}e_b + \bar{E}_{id}e_d\right],$$

$$i = 1, 2, \cdots, m, \tag{9.4.23}$$

where $\tilde{E}(\varepsilon)$ is a matrix defined in an obvious way, and $\bar{\rho}_i(x, \tilde{\zeta}, d, \bar{e}; \varepsilon)$ is the ith element of $\tilde{G}(x, \tilde{\zeta}; \varepsilon) M \Gamma_1 \sigma_h(\tilde{\zeta}_q + \tilde{E}(\varepsilon)\bar{e}) + \tilde{g}(x, \tilde{\zeta}, d(t); \varepsilon)$.

Consider now the following scalings of the variables e_{bi} and e_i,

$$\tilde{e}_{bi} = S_{\bar{r}_i}^{-1}(v)e_{bi}, \quad i = 1, 2, \cdots, p - m, \tag{9.4.24}$$

$$\tilde{e}_i = S_{\bar{q}_i}^{-1}(v)e_i, \quad i = 1, 2, \cdots, m, \tag{9.4.25}$$

and define

$$S(v) = \text{blkdiag}\{S_b(v), S_f(v)\},$$

$$S_b(v) = \text{blkdiag}\{S_{\bar{r}_1}(v), S_{\bar{r}_2}(v), \cdots, S_{\bar{r}_{p-m}}(v)\},$$

$$S_d(v) = \text{blkdiag}\{S_{\bar{q}_1}(v), S_{\bar{q}_2}(v), \cdots, S_{\bar{q}_m}(v)\}.$$

Then, the closed-loop system (9.4.18)-(9.4.23) can be rewritten as,

$$\dot{x} = f(x, \eta), \tag{9.4.26}$$

$$\dot{\tilde{\zeta}}_a = A_{aa}^c(\varepsilon)\tilde{\zeta}_a + S_a(\varepsilon)Q_a^{-1}(\varepsilon)L_{ad}\bar{\tilde{\eta}}, \quad \bar{\tilde{\eta}}_i = \tilde{\zeta}_{i1}, \tag{9.4.27}$$

$$\dot{\tilde{\zeta}}_s = \frac{1}{\varepsilon}A_s^c\tilde{\zeta}_s + L_{sa}(\varepsilon)F_a(\varepsilon)Q_a(\varepsilon)S_a^{-1}(\varepsilon)\tilde{\zeta}_a$$

$$\qquad + L_{saa}F_a(\varepsilon)A_{aa}Q_a(\varepsilon)S_a^{-1}(\varepsilon)\tilde{\zeta}_a + L_{ss}(\varepsilon)\tilde{\zeta}_s + L_{sq}\tilde{\zeta}_q, \tag{9.4.28}$$

$$\dot{\tilde{\zeta}}_q = \frac{1}{\mu}\tilde{G}(x, \tilde{\zeta}, d(t); \varepsilon)M\Gamma_1\sigma_h(\tilde{\zeta}_q + \tilde{E}(\varepsilon)S(v)\tilde{e}) + \tilde{N}(\varepsilon)\tilde{\zeta} + \tilde{g}(x, \tilde{\zeta}, d(t); \varepsilon),$$

$$\tag{9.4.29}$$

$$\dot{\tilde{e}}_{bi} = \frac{1}{v}A_{\bar{r}_i}^c\tilde{e}_{bi}, \quad i = 1, 2, \cdots, p - m, \tag{9.4.30}$$

$$\dot{\tilde{e}}_i = \frac{1}{v}A_{\bar{q}_i}^c\tilde{e}_i + B_{\bar{q}_i}\left[\frac{1}{\mu}\tilde{\rho}_i(x, \tilde{\zeta}, d, \tilde{e}; \varepsilon) + \bar{E}_{ib}S_b(v)\tilde{e}_b + \bar{E}_{id}S_d(v)\tilde{e}_d\right],$$

$$i = 1, 2, \cdots, m, \tag{9.4.31}$$

where $\tilde{\rho}_i(x, \tilde{\zeta}, d, \tilde{e}; \varepsilon)$ is the ith element of $\tilde{G}(x, \tilde{\zeta}, d(t); \varepsilon)M\Gamma_1\sigma_h(\tilde{\zeta}_q + \tilde{E}(\varepsilon)S(v)\tilde{e})$ $+\tilde{g}(x, \tilde{\zeta}, d(t); \varepsilon)$. Letting

$$\tilde{e} = [\tilde{e}_b', \ \tilde{e}_d']', \quad \tilde{e}_b = [\tilde{e}_{b1}', \tilde{e}_{b2}', \cdots, \tilde{e}_{bp-m}']', \quad \tilde{e}_d = [\tilde{e}_1', \tilde{e}_2', \cdots, \tilde{e}_m']',$$

$$A^c = \text{blkdiag}\{A_{\bar{r}_1}^c, A_{\bar{r}_2}^c, \cdots, A_{\bar{r}_{p-m}}^c, A_{\bar{q}_1}^c, A_{\bar{q}_2}^c, \cdots, A_{\bar{q}_m}^c\},$$

$$B^c = [0, \text{blkdiag}\{B_{\bar{q}_1}, B_{\bar{q}_2}, \cdots, B_{\bar{q}_m}\}']',$$

we write the closed-loop system in a compact form as,

$$\dot{x} = f(x, \eta), \tag{9.4.32}$$

$$\dot{\tilde{\zeta}}_a = A_{aa}^c(\varepsilon)\tilde{\zeta}_a + L_{ad}\bar{\tilde{\eta}}, \quad \bar{\tilde{\eta}}_i = \tilde{\zeta}_i, \tag{9.4.33}$$

$$\dot{\tilde{\zeta}}_s = \frac{1}{\varepsilon} A_s^c \tilde{\zeta}_s + L_{sa}(\varepsilon) F_a(\varepsilon) Q_a(\varepsilon) S_a^{-1}(\varepsilon) \tilde{\zeta}_a$$

$$+ L_{saa} F_a(\varepsilon) A_{aa} Q_a(\varepsilon) S_a^{-1}(\varepsilon) \tilde{\zeta}_a + L_{ss}(\varepsilon) \tilde{\zeta}_s + L_{sq} \tilde{\zeta}_q, \tag{9.4.34}$$

$$\dot{\tilde{\zeta}}_q = \frac{1}{\mu} \tilde{G}(x, \tilde{\zeta}, d(t); \varepsilon) M \Gamma_1 \sigma_h(\tilde{\zeta}_q + \tilde{E}(\varepsilon) S(\upsilon) \tilde{e}) + \tilde{N}(\varepsilon) \tilde{\zeta} + \tilde{g}(x, \tilde{\zeta}, d(t); \varepsilon), \tag{9.4.35}$$

$$\dot{\tilde{e}} = \frac{1}{\upsilon} A^c \tilde{e} + \frac{1}{\mu} B^c \tilde{G}(x, \tilde{\zeta}, d(t); \varepsilon) M \Gamma_1 \sigma_h(\tilde{\zeta}_q + E(\varepsilon) S(\upsilon) \tilde{e})$$

$$+ B^c [\bar{E}_b S_b(\upsilon) \tilde{e}_b + \bar{E}_d S_d(\upsilon) \tilde{e}_d + \tilde{g}(x, \tilde{\zeta}, d(t); \varepsilon)], \tag{9.4.36}$$

where

$$\bar{E}_b = \begin{bmatrix} \bar{E}_{1b} \\ \bar{E}_{2b} \\ \vdots \\ \bar{E}_{mb} \end{bmatrix}, \ \bar{E}_d = \begin{bmatrix} \bar{E}_{1d} \\ \bar{E}_{2d} \\ \vdots \\ \bar{E}_{md} \end{bmatrix}.$$

Clearly, A^c is an asymptotically stable matrix since all the submatrices $A_{\bar{r}_i}^c$ and $A_{\bar{q}_i}^c$ are asymptotically stable. Denoting $\tilde{x} = [x', \tilde{\zeta}']'$, the closed-loop system (9.4.32)-(9.4.36) can be written in the form of

$$\dot{\tilde{x}} = f_\sigma(\tilde{x}, \tilde{e}, d(t)), \tag{9.4.37}$$

$$\dot{\tilde{e}} = \frac{1}{\upsilon} A^c \tilde{e} + g_\sigma(\tilde{x}, \tilde{e}, d(t)), \tag{9.4.38}$$

where the functions f_σ and g_σ are defined in an obvious way and are continuous. We are in a position to apply Lemma A.2.1. Note that the system

$$\dot{\tilde{x}} = f_\sigma(\tilde{x}, 0, d(t)) \tag{9.4.39}$$

coincides with (9.4.2)-(9.4.5) if $\tilde{\zeta}_q$ is restricted in the set $\Gamma(\varepsilon)$. Hence, by Proposition 9.4.1, for each $\mu \in (0, \mu_a^*(\varepsilon, \rho)]$, $\varepsilon \in (0, \varepsilon_a^*]$, $\rho \in (0, 1)$, the C^1 function $V_1 : \tilde{\mathcal{W}}_1(\varepsilon) \to \mathbb{R}_+$ is positive definite on $\tilde{\mathcal{W}}_1(\varepsilon) \setminus \{0\}$ and proper on $\tilde{\mathcal{W}}_1(\varepsilon)$ and satisfies

$$\frac{\partial V_1}{\partial \tilde{x}} f_\sigma(\tilde{x}, 0, d(t)) \leq -\Psi_1(\tilde{x}; \varepsilon, \mu),$$

where $\Psi_1(\tilde{x}; \varepsilon, \mu)$ is continuous on $\tilde{\mathcal{W}}_1(\varepsilon)$ and positive definite on $\{\rho < V_1(\tilde{x}; \varepsilon, \mu) \leq c_0^2(\varepsilon) + c_1^2(\varepsilon) + 1\}$.

We also have that

$$|f_\sigma(\tilde{x}, e, d) - f_\sigma(\tilde{x}, 0, d)| = \frac{1}{\mu} |\tilde{G}(x, \tilde{\zeta}, d; \varepsilon) M \Gamma_1 [\sigma_h(\tilde{\zeta}_q + \tilde{E}(\varepsilon) S(\upsilon) \tilde{e})) - \sigma_h(\tilde{\zeta}_q)]|. \tag{9.4.40}$$

By Lipschitz continuity of σ_h, continuity of $\tilde{G}$ and the compactness of the sets $L_{V_1}(c_0^2(\varepsilon) + c_1^2(\varepsilon) + 1)$ and $\mathcal{D}$, there are positive numbers $\delta_1(\varepsilon, \mu) > 0$ and $\delta(\varepsilon, \mu) > 0$ such that

$$|f_\sigma(\tilde{x}, e, d) - f_\sigma(\tilde{x}, 0, d)| = \delta_1(\varepsilon, \mu) \min\{\delta_2(\varepsilon, \mu)|\tilde{e}|, 2h\} \leq \gamma_\sigma(|\tilde{e}|; \varepsilon, \mu),$$

$$\forall(\tilde{x}, \tilde{e}, d) \in L_{V_1}(c_0^2(\varepsilon) + c_1^2(\varepsilon) + 1) \times \mathbf{R}^n \times \mathcal{D}, \qquad (9.4.41)$$

for some bounded function $\gamma_\sigma(\cdot\,; \varepsilon, \mu) : \mathbf{R}_+ \to \mathbf{R}_+$ satisfying $\gamma_\sigma(0; \varepsilon, \mu) = 0$. Also by the compactness of the sets $L_{V_1}(c_0^2(\varepsilon) + c_1^2(\varepsilon) + 1)$ and $\mathcal{D}$ and the boundedness of σ_h, there exist positive numbers $\alpha_\sigma(\varepsilon, \mu)$ and $\beta_\sigma(\varepsilon, \mu)$, both independent of v, such that

$$|g_\sigma(\tilde{x}, \tilde{e}, d)| \le \alpha_\sigma(\varepsilon, \mu)|\tilde{e}| + \beta_\sigma(\varepsilon, \mu), \quad \forall(\tilde{x}, \tilde{e}, d) \in L_{V_1}(c_0^2(\varepsilon) + c_1^2(\varepsilon) + 1) \times \mathbf{R}^n \times \mathcal{D}.$$
$$(9.4.42)$$

Hence the condition (A.2.3) of Lemma A.2.1 is satisfied.

Now let P_e satisfy the Lyapunov equation $(A^c)'P_e + P_eA^c = -I$. We choose $c_2(1/v) = \ln(1 + \lambda_{\max}(P_e)R^2/v^{2r})$, where R is such that $e \in \mathcal{W}_{\hat{\xi}} - \mathcal{W}_\xi$ implies $|\tilde{e}| \le R$ and $r = \max\{\bar{r}_1, \bar{r}_2, \cdots, \bar{r}_{p-m}, \bar{q}_1, \bar{q}_2, \cdots, \bar{q}_m\}$. Then we have $\ln(1 + \tilde{e}(0)P_e\tilde{e}(0)) \le c_2(1/v)$ and

$$\lim_{v \to 0} v c_2^4(1/v) = 0. \qquad (9.4.43)$$

We then define the Lyapunov function

$$V_2(\tilde{x}, \tilde{e}; \varepsilon, v) = \frac{(c_0^2(\varepsilon) + c_1^2(\varepsilon))V_1(\tilde{x}; \varepsilon)}{c_0^2(\varepsilon) + c_1^2(\varepsilon) + 1 - V_1(\tilde{x}; \varepsilon)} + \frac{c_2(1/v)\ln(1 + \tilde{e}'P_e\tilde{e})}{c_2(1/v) + 1 - \ln(1 + \tilde{e}'P_e\tilde{e})}$$
$$(9.4.44)$$

and the set

$$\tilde{\mathcal{W}}_2(\varepsilon, v) = \{\tilde{x} : V_1(\tilde{x}; \varepsilon) < c_0^2(\varepsilon) + c_1^2(\varepsilon) + 1\}$$
$$\times \{\tilde{e} : \ln(1 + \tilde{e}'P_e\tilde{e}) < c_2(1/v) + 1\}.$$

It then follows from Lemma A.2.1 that, for any fixed $\mu \in (0, \mu_a^*(\varepsilon, \rho)], \varepsilon \in (0, \varepsilon_a^*]$, $\rho \in (0, 1)$, there exists a $v_a^*(\varepsilon, \mu, \rho) > 0$, such that, for all $v \in (0, v_a^*]$,

$$\dot{V}_2 \le -\Psi_2(\tilde{x}, \tilde{e}; \varepsilon, \mu, v), \qquad (9.4.45)$$

where $\Psi_2(\tilde{x}, \tilde{e}; \varepsilon, \mu, v)$ is positive definite on the set $\{(\tilde{x}, \tilde{e}) : 2\rho \le V_2(\tilde{x}, \tilde{e}; \varepsilon, \mu, v) \le (c_0^2(\varepsilon) + c_1^2(\varepsilon))^2 + c_2^2(1/v) + 1\}$, which shows that the state $(\tilde{x}, \tilde{e})$ will remain in this set. Moreover, the state $(\tilde{x}, \tilde{e})$ will enter the set $\{(\tilde{x}, \tilde{e}) : V_2(\tilde{x}, \tilde{e}; \varepsilon, \mu, v) \le 2\rho\} \subset \{\tilde{x} : V_1(\tilde{x}; \varepsilon) \le 4\rho\} \times \{\tilde{e} : \ln(1 + \tilde{e}'P_e\tilde{e}) \le 4\rho\}$ in a finite time and remain in it thereafter. Let the v independent $\rho^*(\varepsilon) > 0$ be such that $V_1(\tilde{x}; \varepsilon) \le 4\rho^*(\varepsilon)$ and $\ln(1 + \tilde{e}'P_e\tilde{e}) \le 4\rho^*(\varepsilon)$ imply that $(x, \xi, \hat{\xi}) \in \mathcal{W}_0$. The existence of such a v independent $\rho^*(\varepsilon)$ is due to the special form of state transformation from $\bar{e}$ to $\tilde{e}$. Then, take $\varepsilon^* = \varepsilon_a^*$, for each $\varepsilon \in (0, \varepsilon^*]$ take $\mu^*(\varepsilon) = \mu_a^*(\varepsilon, \rho^*(\varepsilon))$ and for each pair $\mu \in (0, \mu^*(\varepsilon)], \varepsilon \in (0, \varepsilon^*]$ take $v^*(\varepsilon, \mu) = v_a^*(\varepsilon, \mu, \rho^*(\varepsilon))$. This completes the proof of Theorem 9.4.1. ⊠

Proof of Theorem 9.4.2. We prove this theorem by showing that under the extra Assumption 9.2.5 the family of dynamic output feedback laws constructed in the proof of Theorem 9.4.1 also achieves semi-global asymptotic stabilization of the system (9.1.1). To this end, let us consider the closed-loop system (9.4.32)-(9.4.36) as repeated below.

$$\dot{x} = f(x, \eta), \tag{9.4.46}$$

$$\dot{\tilde{\zeta}}_a = A_{aa}^c(\varepsilon)\tilde{\zeta}_a + S_a(\varepsilon)Q_a^{-1}(\varepsilon)L_{af}\bar{\tilde{\eta}}, \ \bar{\tilde{\eta}}_i = \tilde{\zeta}_i, \tag{9.4.47}$$

$$\dot{\tilde{\zeta}}_s = \frac{1}{\varepsilon}A_s^c\tilde{\zeta}_s + L_{sa}(\varepsilon)F_a(\varepsilon)Q_a(\varepsilon)S_a^{-1}(\varepsilon)\tilde{\zeta}_a$$
$$+L_{saa}F_a(\varepsilon)A_{aa}Q_a(\varepsilon)S_a^{-1}(\varepsilon)\tilde{\zeta}_a + L_{ss}(\varepsilon)\tilde{\zeta}_s + L_{sq}\tilde{\zeta}_q, \tag{9.4.48}$$

$$\dot{\tilde{\zeta}}_q = \frac{1}{\mu}\tilde{G}(x, \tilde{\zeta}, d(t); \varepsilon)M\Gamma_1\sigma_h(\tilde{\zeta}_q + \tilde{E}(\varepsilon)S(v)\tilde{e})$$
$$+\tilde{N}(\varepsilon)\tilde{\zeta} + \tilde{g}(x, \tilde{\zeta}, d(t); \varepsilon), \tag{9.4.49}$$

$$\dot{e} = \frac{1}{v}A^c\tilde{e} + \frac{1}{\mu}B^c\tilde{G}(x, \tilde{\zeta}, d(t); \varepsilon)M\Gamma_1\sigma_h(\tilde{\zeta}_q + E(\varepsilon)S(v)\tilde{e})$$
$$+B^c[\bar{E}_b S_b(v)\tilde{e}_b + \bar{E}_d S_d(v)\tilde{e}_d + \tilde{g}(x, \tilde{\zeta}, d(t); \varepsilon)], \tag{9.4.50}$$

where $\varepsilon, \mu, v \in (0, 1]$ and all notations follow from the proof of Theorem 9.4.1.

Let $\mathcal{U}_1 \subset \mathbb{R}^{2n}$ be any compact set such that $\mathcal{U}_0 \times \mathcal{U}_1 \subset \mathcal{W}$, where the set $\mathcal{U}_0$ is as defined in Assumption 9.2.4. Let $c_1(\varepsilon)$ be as defined in Proposition 9.4.1. It follows trivially from $\mathcal{U}_0 \times \mathcal{U}_1 \subset \mathcal{W}$ that, for any $\varepsilon \in (0, \varepsilon_a^*]$, $(x, \xi, \hat{\xi}) \in \mathcal{U}_0 \times \mathcal{U}_1$ implies $\tilde{\zeta}_q \in \Gamma(\varepsilon)$. We recall here that $\Gamma(\varepsilon)$ is also defined in Proposition 9.4.1. It follows from the continuity of functions $\tilde{G}$ and $\tilde{g}$, the Lipschitz continuity of the function σ_h and the compactness of the set $\mathcal{U}_0 \times \mathcal{U}_1$ that there exist an ε independent positive number δ_2 and some ε dependent positive numbers $\delta_1(\varepsilon)$, $\delta_3(\varepsilon)$ and $\delta_4(\varepsilon)$ such that for all $(x, \xi, \hat{\xi}, d) \in \mathcal{U}_0 \times \mathcal{U}_1 \times \mathcal{D}$,

$$|\tilde{G}(x, \tilde{\zeta}, d; \varepsilon)| \leq \delta_1(\varepsilon), \tag{9.4.51}$$

$$|f(x, \eta) - f(x, 0)| \leq \delta_2(|\tilde{\zeta}_a| + |\tilde{\zeta}_s|), \tag{9.4.52}$$

$$|\tilde{g}(x, \tilde{\zeta}, d; \varepsilon) - \tilde{g}(x, 0, d; \varepsilon)| \leq \delta_3(\varepsilon)(|\tilde{\zeta}_a| + |\tilde{\zeta}_s| + |\tilde{\zeta}_q|), \tag{9.4.53}$$

$$|\sigma_h(\tilde{\zeta}_q + \tilde{E}(\varepsilon)S(v)\tilde{e}) - \sigma_h(\tilde{\zeta}_q)| \leq \delta_4(\varepsilon)|\tilde{e}|. \tag{9.4.54}$$

Noting that $\tilde{g}(x, 0, d; \varepsilon)$ is independent of ε, it follows from Assumption 9.2.5 that

$$|\tilde{g}(x, 0, d; \varepsilon)| \leq \tilde{\gamma}_{\tilde{g}}\psi(|x|), \tag{9.4.55}$$

for some positive number $\gamma_{\tilde{g}}$ independent of ε.

For the nonlinear system (9.4.46)-(9.4.50), we form the following Lyapunov function candidate

$$V(x, \tilde{\zeta}; \varepsilon, \mu) = V_0(x) + \frac{1}{\sqrt{\mu}} V_{as}(\tilde{\zeta}_a, \tilde{\zeta}_s; \varepsilon) + V_q(\tilde{\zeta}_q) + V_e(\tilde{e}), \qquad (9.4.56)$$

where $V_0(x)$ is defined in Assumption 9.2.5, $V_a(\tilde{\zeta}_a, \tilde{\zeta}_s; \varepsilon)$ and $V_q(\tilde{\zeta}_q)$ are given by (9.3.51) and (9.3.69) respectively, and

$$V_e(\tilde{e}) = \tilde{e}' P_e \tilde{e}, \qquad (9.4.57)$$

with P_e given in the proof of Theorem 9.4.1.

Using (9.3.70) and (9.4.51)-(9.4.55), it follows readily from Assumption 9.2.5 that for $\varepsilon \in (0, \varepsilon_a^*]$ and $\mu, \upsilon \in (0, 1]$, the derivative of V along the trajectories of (9.4.46)-(9.4.50) such that $(x, \xi, \hat{\xi}) \in \mathcal{U}_0 \times \mathcal{U}_1$ can be evaluated as,

$$\dot{V} \leq -\psi^2(|x|) + \alpha_1 \psi(|x|)(|\tilde{\zeta}| + |\tilde{e}|) - \left[|\tilde{\zeta}_a| \, |\tilde{\zeta}_s| \, |\tilde{\zeta}_q| \right] \frac{\mathcal{R}(\varepsilon, \mu)}{\sqrt{\mu}} \begin{bmatrix} |\tilde{\zeta}_a| \\ |\tilde{\zeta}_s| \\ |\tilde{\zeta}_q| \end{bmatrix}$$

$$+ \frac{\alpha_3(\varepsilon)}{\mu} |\tilde{\zeta}| |\tilde{e}| - \left(\frac{1}{\upsilon} - \alpha_2(\varepsilon, \mu) \right) |\tilde{e}|^2, \qquad (9.4.58)$$

with

$$\mathcal{R}(\varepsilon, \mu) = \begin{bmatrix} \dfrac{\varepsilon}{2} & -\alpha_{12} - \alpha_{21} \kappa \varepsilon & -\alpha_{13}(\varepsilon) \\ -\alpha_{12} - \alpha_{21} \kappa \varepsilon & \dfrac{1}{\varepsilon} - \kappa \alpha_{22} & -\alpha_{23}(\varepsilon) \\ -\alpha_{13}(\varepsilon) & -\alpha_{23}(\varepsilon) & \dfrac{\alpha_4}{\sqrt{\mu}} - \alpha_{33}(\varepsilon) \end{bmatrix},$$

where α_1, α_4, α_{12}, α_{21} and α_{22} are positive constants independent of ε, μ and υ, $\alpha_2(\varepsilon, \mu)$ is a positive number dependent on ε and μ, and all the other $\alpha_{ij}(\varepsilon)$'s and $\alpha_3(\varepsilon)$ are some ε dependent positive numbers. Let $\kappa = 2\alpha_{12}^2$. Clearly, there exists an $\varepsilon_b^* \in (0, \varepsilon_a^*]$ and for each fixed $\varepsilon \in (0, \varepsilon_b^*]$, there exists a $\mu_a^*(\varepsilon) \in (0, 1]$ such that the matrix $\mathcal{R}(\varepsilon, \mu)$ is positive definite for all $\mu \in (0, \mu_a^*(\varepsilon)]$, $\varepsilon \in (0, \varepsilon_b^*]$. Hence, there exists an ε dependent positive number $\alpha_5(\varepsilon)$ such that

$$\dot{V} \leq -\psi^2(|x|) + \alpha_1 \psi(|x|) \left(|\tilde{\zeta}| + |\tilde{e}| \right) - \frac{\alpha_5(\varepsilon)}{\sqrt{\mu}} |\tilde{\zeta}|^2 + \frac{\alpha_3(\varepsilon)}{\mu} |\tilde{\zeta}| |\tilde{e}|$$

$$- \left(\frac{1}{\upsilon} - \alpha_2(\varepsilon, \mu) \right) |\tilde{e}|^2. \qquad (9.4.59)$$

Simple completion of square arguments show that for each $\mu \in (0, \mu_a^*(\varepsilon)]$, $\varepsilon \in (0, \varepsilon_b^*]$, there exists a $\upsilon_a^*(\varepsilon, \mu) \in (0, 1]$, such that for all $\upsilon \in (0, \upsilon_a^*(\varepsilon, \mu)]$, $\mu \in (0, \mu_a^*(\varepsilon)]$, $\varepsilon \in (0, \varepsilon_b^*]$,

$$- \frac{\alpha_5(\varepsilon)}{2\sqrt{\mu}} |\tilde{\zeta}|^2 + \frac{\alpha_3(\varepsilon)}{\mu} |\tilde{\zeta}| |\tilde{e}| - \left(\frac{1}{2\upsilon} - \alpha_{22}(\varepsilon, \mu) \right) |\tilde{e}|^2 \leq 0, \qquad (9.4.60)$$

and hence

$$\dot{V} \leq -\psi^2(|x|) + \alpha_1\psi(|x|)\left(|\tilde{\zeta}| + |\tilde{e}|\right) - \frac{\alpha_5(\varepsilon)}{2\sqrt{\mu}}|\tilde{\zeta}|^2 - \frac{1}{2\upsilon}|\tilde{e}|^2. \qquad (9.4.61)$$

Again by simple completion of square arguments, for any $\varepsilon \in (0, \varepsilon_b^*]$, there exists a $\mu_b^*(\varepsilon) \in (0, \mu_a^*(\varepsilon)]$ such that for all $\mu \in (0, \mu_b^*(\varepsilon)]$, $\varepsilon \in (0, \varepsilon_b^*]$,

$$-\frac{1}{4}\psi^2(|x|) + \alpha_1\psi(|x|)|\tilde{\zeta}| - \frac{\alpha_5(\varepsilon)}{4\sqrt{\mu}}|\tilde{\zeta}|^2 \leq 0, \qquad (9.4.62)$$

and there exists a $\upsilon_b^*(\varepsilon, \mu) \in (0, \upsilon_a^*(\varepsilon, \mu)]$ such that for all $\upsilon \in (0, u_b^*(\varepsilon, \mu)]$, $\mu \in (0, \mu_b^*(\varepsilon)]$, $\varepsilon \in (0, \varepsilon_b^*]$,

$$-\frac{1}{4}\psi^2(|x|) + \alpha_1\psi(x)|\tilde{e}| - \frac{1}{4\upsilon}|\tilde{e}|^2 \leq 0. \qquad (9.4.63)$$

We finally have that for all $\upsilon \in (0, u_b^*(\varepsilon, \mu)]$, $\mu \in (0, \mu_b^*(\varepsilon)]$, $\varepsilon \in (0, \varepsilon_b^*]$,

$$\dot{V} \leq -\frac{1}{2}\psi^2(|x|) - \frac{\alpha_5(\varepsilon)}{4\sqrt{\mu}}|\tilde{\zeta}|^2 - \frac{1}{4\upsilon}|\tilde{e}|^2, \qquad (9.4.64)$$

which shows that the equilibrium of the closed-loop system (9.4.46)-(9.4.50) is locally asymptotically stable for all $\upsilon \in (0, \upsilon_b^*(\varepsilon, \mu)]$, $\mu \in (0, \mu_b^*(\varepsilon)]$, $\varepsilon \in (0, \varepsilon_b^*]$.

It remains to show that the basin of attraction of the equilibrium $(0, 0, 0)$ of the closed-loop system indeed contains the given set $\mathcal{W}$ for sufficiently small ε, μ and υ. It follows from Theorem 9.4.1 that, under the family of dynamic output feedback laws, for the given set $\mathcal{U}_0 \times \mathcal{U}_1$, there is an $\varepsilon_c^* > 0$, for each $\varepsilon \in (0, \varepsilon_c^*]$ there is a $\mu_c^*(\varepsilon) > 0$, and for each pair $\mu \in (0, \mu_c^*(\varepsilon)]$, $\varepsilon \in (0, \varepsilon_c^*]$ there is a $\upsilon_c^*(\varepsilon, \mu)$ such that for all $\upsilon \in (0, \upsilon_c^*]$, $\mu \in (0, \mu_c^*(\varepsilon)]$, $\varepsilon \in (0, \varepsilon_c^*]$, all the solutions of the closed-loop system with initial conditions $(x(0), \xi(0), \hat{\xi}(0)) \in \mathcal{W}$ will enter the set $\mathcal{U}_0 \times \mathcal{U}_1$ in a finite time and remain in it thereafter. Now choose $\varepsilon^* = \min\{\varepsilon_b^*, \varepsilon_c^*\}$, for each $\varepsilon \in (0, \varepsilon^*]$ choose $\mu^*(\varepsilon) = \min\{\mu_b^*(\varepsilon), \mu_c^*(\varepsilon)\}$, and for each pair $\mu \in (0, \mu^*(\varepsilon)]$, $\varepsilon \in (0, \varepsilon^*]$ choose $\upsilon^*(\varepsilon, \mu) = \min\{\upsilon_b^*(\varepsilon, \mu), \upsilon_c^*(\varepsilon, \mu)\}$. Then, for all $\upsilon \in (0, \upsilon^*(\varepsilon, \mu)]$, $\mu \in (0, \mu^*(\varepsilon)]$, $\varepsilon \in (0, \varepsilon^*]$, all the trajectories of the closed-loop system starting from $(x(0), \xi(0), \hat{\xi}(0)) \in \mathcal{W}$ will enter the set $\mathcal{U}_0 \times \mathcal{U}_1$ in a finite time and remain in it thereafter, and due to (9.4.64) will approach the equilibrium $(x, \xi, \hat{\xi}) = (0, 0, 0)$ as t goes to ∞. This completes the proof of Theorem 9.4.2. ☒

9.5. Concluding Remarks

In this chapter we have utilized the low gain feedback design technique, along with other design techniques, to establish semi-global asymptotic stabilizability and/or practical stabilizability of multi-input multi-output minimum-phase input output linearizable systems under some fairly weak conditions, most of which are necessary. Other design techniques utilized include asymptotic time-scale and eigenstructure assignment (ATEA) [91], high-gain observer theory [93], and saturating control during the peaking of the error dynamics [30].

Chapter 10

Perfect Regulation and H_2-Suboptimal Control

10.1. Introduction

Optimization of a cost criterion is a classical design tool. Design philosophy based on optimization has a rich and long history. For example, linear quadratic control problems (LQCP) belong to such a design philosophy. In an LQCP, the cost criterion is an infinite horizon integral whose integrand consists of a quadratic function in state and control vectors. The quadratic aspect of such a cost function is rooted in the well known least squares method for infimization of measurement error. A non-negative definite LQCP is the LQCP in which the quadratic integrand is non-negative definite. For a non-negative definite LQCP, the cost criterion can be reformulated or thought of as the L_2-norm of a controlled output vector.

In the LQCP literature, a *perfect regulation* problem refers to the case when a state feedback controller is utilized and when the resulting infimum of the cost criterion (the L_2-norm of the controlled output) is zero. The problem of perfect regulation has a prominent position in the H_2-optimal control literature due to the fact that any H_2-suboptimal control problem can be cast into and solved as a perfect regulation problem [94]. If solvable, in general, the problem of perfect regulation requires a sequence or a family of state feedback control laws so that one can select a control law from the family such that the resulting L_2-norm of the controlled output is arbitrarily small. There exist two methods for constructing this family of feedback laws: an Algebraic Riccati Equation (ARE) based method [17,37], and a direct eigenstructure assignment methodology [32].

Francis [17], using continuity arguments, approaches the design problem, via a "regularization technique." He provides an ARE based method of solving the problem of perfect regulation when there is no feedthrough from the input to the controlled output. To be explicit, in this method, a sequence or a family of state feedback controllers is produced by solving an ARE parameterized in a parameter, say ε. For each specified $\varepsilon \neq 0$, one has to solve the ARE in order to obtain a member of the family of controllers. As ε tends to zero, the corresponding value of the cost criterion tends to zero. Thus, the method requires repetitive solutions of AREs. Such an ARE based method has a major numerical problem in that the solution of the concerned ARE, especially for small ε, is in general numerically cumbersome and becomes 'stiff' owing to the low and high gain nature of the state feedback gain. Besides 'stiffness', there are other issues such as the lack of freedom in assigning the resulting asymptotically infinite eigenstructure. These issues were discussed carefully in [86].

Kimura [32] approaches the design for perfect regulation from the perspective of direct closed-loop pole and eigenvector assignment. He does this for a number of reasons, e.g., (1) to alleviate the inherent computational difficulties of an ARE based method, (2) to gain freedom or flexibility in assigning the asymptotically infinite eigenstructure, and (3) to gain insights into the problem. However, one limitation of Kimura's method is that it excludes *a priori* the presence of invariant zeros of the given system on the imaginary axis. Kimura [32] does this by cleverly restricting himself to a class of parameterized state feedback gains having the property that as the gain tends to infinity the limits of all the resulting closed-loop root loci remain in the open left-half complex plane. Kimura acknowledges the latent difficulties associated with the presence of invariant zeros on the imaginary axis, especially those having nontrivial Jordan blocks.

In this chapter, we propose another direct eigenstructure assignment procedure to achieve perfect regulation whenever it is achievable. Unlike the existing literature, our design procedure is applicable to any general linear multivariable system. That is, it allows the the presence of invariant zeros on the imaginary axis (or on the unit circle in the discrete-time context). Our ability to handle invariant zeros on the imaginary axis (unit circle) is due to the low gain design techniques of Chapter 2.

The chapter is organized as follows. Sections 10.2 and 10.3 deal with continuous-time systems and discrete-time systems respectively. In each of the these two sections, we first state the problem, then construct a family of feedback laws and show that it solves the problem of perfect regulation, and finally illustrate the solution with a numerical example. Section 10.4 contains some concluding remarks.

10.2. Continuous-Time Systems

10.2.1. Problem Statement

Consider the continuous-time linear system

$$\Sigma : \begin{cases} \dot{x} = Ax + Bu, & x(0) = x_0, \\ z = Cx + Du, \end{cases} \tag{10.2.1}$$

where $x \in \mathbb{R}^n$ is the state, $u \in \mathbb{R}^m$ is the input, and $z \in \mathbb{R}^p$ is the controlled output. Let us also consider an associated cost criterion,

$$J(x_0, u) = \int_0^\infty |z(t)|^2 dt. \tag{10.2.2}$$

We make a standing assumption that the pair (A, B) is stabilizable. Moreover, without loss of generality, we assume that the matrices $[C, \ D]$ and $\begin{bmatrix} B \\ D \end{bmatrix}$ are of full rank.

The problem of perfect regulation is a classical one and is defined as follows.

Problem 10.2.1. *Consider the system Σ as given by (10.2.1) along with the associated cost $J(x_0, u)$ as given by (10.2.2). The problem of perfect regulation via state feedback is to find a family of parameterized linear state feedback laws $u = F(\varepsilon)x$ having the following properties:*

1. *There exists an $\varepsilon^* > 0$ such that for all $\varepsilon \in (0, \varepsilon^*]$, the closed-loop system comprising Σ and $u = F(\varepsilon)x$ is internally stable;*

2. *For each $x_0 \in \mathbb{R}^n$, one has*

$$J(x_0, u) \to 0 \quad \text{as} \ \varepsilon \to 0. \tag{10.2.3}$$

P

We recall from [63] the necessary and sufficient conditions under which the problem of perfect regulation as formulated above is solvable.

Theorem 10.2.1. *Consider the system Σ as given in (10.2.1) along with the associated cost $J(x_0, u)$ as given in (10.2.2). The problem of perfect regulation via state feedback $u = F(\varepsilon)x$ is solvable if and only if the given system Σ is right invertible and has all its invariant zeros located in the closed left-half plane.* T

Our objective here is to develop a low gain feedback based direct eigenstructure assignment design procedure for constructing a family of state feedback laws that, under the solvability condition of Theorem 10.2.1, solves the problem of perfect regulation.

10.2.2. Solution of the Problem of Perfect Regulation

We present an eigenstructure assignment based design algorithm and show that the resulting state feedback laws indeed solve the problem of perfect regulation. The algorithm we propose is based on the eigenstructure assignment based low gain feedback design technique of Chapter 2 and consists of the following three steps.

Step 1: Construction of the SCB of Σ. Perform a nonsingular state, input and controlled output transformation on the system Σ according to Theorem A.1.1,

$$x = \Gamma_s \bar{x}, \ z = \Gamma_o \bar{z}, \ u = \Gamma_i \bar{u},$$

such that the system Σ can be written in the following SCB form,

$$\bar{x} = [x_a', x_c', x_d']', \ \bar{x}_d = [x_1', x_2', \cdots, x_{m_d}']', \ x_i = [x_{i1}, x_{i2}, \cdots, x_{iq_i}]',$$
$$\bar{z} = [z_0', z_d'], \ \bar{z}_d = [z_1, z_2, \cdots, z_{m_d}]', \ z_i = x_{i1},$$
$$\bar{u} = [u_0', u_d', u_c']', \ \bar{u}_d = [u_1, u_2, \cdots, u_{m_d}]',$$

and

$$\dot{x}_a = A_{aa} x_a + B_{0a} z_0 + L_{ad} z_d, \qquad\qquad (10.2.4)$$

$$\dot{x}_i = A_{q_i} x_i + B_{0id} z_0 + L_{id} z_d + B_{q_i} \left[E_{ia} x_a + E_{ic} x_c + \sum_{j=1}^{m_d} E_{ij} x_j + u_i \right],$$

$$i = 1, 2, \cdots, m_d, \qquad\qquad (10.2.5)$$

$$z_0 = C_{0a} x_a + C_{0c} x_c + \sum_{j=1}^{m_d} C_{0jd} x_j + u_0, \qquad\qquad (10.2.6)$$

where C_{ojd} is defined as,

$$C_{0d} = [\, C_{01d} \quad C_{02d} \quad \cdots \quad C_{0m_d d} \,].$$

We note that the output z_b, and hence the state x_b, is not present as Σ is right invertible.

Step 2: Construction of a parameterized low gain matrix $F_a(\varepsilon)$. By Property A.1.1 of the SCB, the pair $(A_{aa}, [B_{0a}, L_{ad}])$ is stabilizable. Moreover, by Property A.1.2 of the SCB, the eigenvalues of A_{aa} are the invariant zeros of the system Σ and hence, are all located in the closed left-half plane. Following the eigenstructure assignment based low gain design technique of Section 2.2.1, we design a feedback gain $F_a(\varepsilon)$ for the pair $(A_{aa}, B_a) = (A_{aa}, [B_{0a}, L_{ad}])$ as follows.

Step 2.1. Find the nonsingular transformation matrices Γ_{sa} and Γ_{Ia} such that (A_{aa}, B_a) can be transformed into the block diagonal control canonical form,

$$
\Gamma_{sa}^{-1} A_{aa} \Gamma_{sa} = \begin{bmatrix} A_1 & 0 & \cdots & 0 & 0 \\ 0 & A_2 & \cdots & 0 & 0 \\ \vdots & \vdots & \ddots & \vdots & \vdots \\ 0 & 0 & \cdots & A_l & 0 \\ 0 & 0 & \cdots & 0 & A_0 \end{bmatrix},
$$

$$
\Gamma_{sa}^{-1} B_a \Gamma_{Ia}^0 = \begin{bmatrix} B_1 & B_{12} & \cdots & B_{1l} & \star \\ 0 & B_2 & \cdots & B_{2l} & \star \\ \vdots & \vdots & \ddots & \vdots & \vdots \\ 0 & 0 & \cdots & B_l & \star \\ B_{01} & B_{02} & \cdots & B_{0l} & \star \end{bmatrix},
$$

where A_0 contains all the open left-half plane eigenvalues of A_{aa}, l is an integer and for $i = 1, 2, \cdots, l$, all eigenvalues of A_i are on the $j\omega$ axis and hence (A_i, B_i) is controllable as given by,

$$
A_i = \begin{bmatrix} 0 & 1 & 0 & \cdots & 0 \\ 0 & 0 & 1 & \cdots & 0 \\ \vdots & \vdots & \vdots & \ddots & \vdots \\ 0 & 0 & 0 & \cdots & 1 \\ -a_{n_i}^i & -a_{n_i-1}^i & -a_{n_i-2}^i & \cdots & -a_1^i \end{bmatrix}, \quad B_i = \begin{bmatrix} 0 \\ 0 \\ \vdots \\ 0 \\ 1 \end{bmatrix}.
$$

Here the $\star$'s represent sub-matrices of less interest. We also note that the existence of the above canonical form was shown in Wonham [126] while its software realization can be found in Chen [9].

Step 2.2. For each (A_i, B_i), let $F_i(\varepsilon) \in \mathbb{R}^{1 \times n_i}$ be the state feedback gain such that

$$
\lambda(A_i + B_i F_i(\varepsilon)) = -\varepsilon + \lambda(A_i) \in \mathbb{C}^-. \tag{10.2.7}
$$

Note that $F_i(\varepsilon)$ is unique.

Step 2.3. Compose $F_a(\varepsilon)$ as follows,

$$F_a(\varepsilon) = \Gamma_{Ia} \begin{bmatrix} F_1(\varepsilon) & 0 & \cdots & 0 & 0 \\ 0 & F_2(\varepsilon) & \cdots & 0 & 0 \\ \vdots & \vdots & \ddots & \vdots & \vdots \\ 0 & 0 & \cdots & F_{l-1}(\varepsilon) & 0 \\ 0 & 0 & \cdots & 0 & F_l(\varepsilon) \\ 0 & 0 & \cdots & 0 & 0 \end{bmatrix} \Gamma_{sa}^{-1}, \quad (10.2.8)$$

where $\varepsilon \in (0,1]$ is a design parameter whose value is to be specified later.

For later use, partition the matrix $F_a(\varepsilon)$ as,

$$F_a(\varepsilon) = [F'_{a0}(\varepsilon), F'_{ad}(\varepsilon)]' = [F_{a0}(\varepsilon)', F'_{a1}(\varepsilon), F'_{a2}(\varepsilon), \cdots, F'_{am_d}(\varepsilon)]',$$

where $F_{a0}(\varepsilon) \in \mathbb{R}^{m_o \times n_a}$ and for each $i = 1, 2, \cdots, m_d$, $F_{ai}(\varepsilon) \in \mathbb{R}^{1 \times n_a}$.

Step 3: Construction of a parameterized gain matrix $F(\varepsilon)$. By Theorem A.1.1, the pair (A_{cc}, B_c) is controllable, hence one can choose a feedback gain matrix F_c such that $A_{cc} + B_c F_c$ is asymptotically stable and has a chosen set of eigenvalues. Also, for $i = 1$ to m_d, choose F_i such that $A_{q_i} + B_{q_i} F_i$ is asymptotically stable. The existence of such gain matrices F_i's is guaranteed by the special form of (A_{q_i}, B_{q_i}). For further use, let the first element of F_i be F_{i1}.

Finally, a composite state feedback gain is formed for the system Σ. This state feedback gain takes the form of

$$F(\varepsilon) = \Gamma_I \begin{bmatrix} F_{u_0}(\varepsilon) \\ F_{u_d}(\varepsilon) \\ F_{u_c} \end{bmatrix} \Gamma_s^{-1}, \quad (10.2.9)$$

where

$$F_{u_0} = -[C_{0a} - F_{a0}(\varepsilon) \quad C_{0c} \quad C_{01d} \quad C_{02d} \quad \cdots \quad C_{0m_d d}],$$

$$F_{u_d} = \begin{bmatrix} F_{u_1}(\varepsilon) \\ F_{u_2}(\varepsilon) \\ \vdots \\ F_{u_{m_d}}(\varepsilon) \end{bmatrix},$$

$$F_{u_c} = [-E_{ca} \quad F_c \quad 0 \quad 0 \quad \cdots \quad 0],$$

and for $i = 1$ to m_d,

$$F_{u_i} = -\left[E_{ia} + \frac{F_{i1}}{\varepsilon^{q_i}} F_{ai} \quad E_{ic} \quad E_{i1} \quad E_{i2} \quad \cdots \quad E_{ii} - \frac{F_i}{\varepsilon^{q_i}} S_{q_i}(\varepsilon) \quad \cdots \quad E_{im_d} \right],$$

$$S_{q_i}(\varepsilon) = \text{Diag}\left\{1, \varepsilon, \varepsilon^2, \cdots, \varepsilon^{q_i-1}\right\}.$$

A

We choose a family of state feedback laws, parameterized in ε, as,

$$u = F(\varepsilon)x, \qquad (10.2.10)$$

where $F(\varepsilon)$ is as given by (10.2.9).

The following theorem establishes that the family of state feedback laws as given by (10.2.10) indeed achieves perfect regulation for the system Σ.

Theorem 10.2.2. *Consider the system Σ as given in (10.2.1) along with the associated cost $J(x_0, u)$ as given in (10.2.2). Also, assume that Σ is right invertible and has all its invariant zeros in the close left-half plane. Then, the family of state feedback laws as given by (10.2.10) achieves perfect regulation for the system Σ.* T

Proof of Theorem 10.2.2. With the state feedback laws (10.2.10), the closed-loop system in the special coordinate basis can be written as,

$$\dot{x}_a = A_{aa}x_a + B_{0a}z_0 + L_{ad}z_d \qquad (10.2.11)$$

$$\dot{x}_c = (A_{cc} + B_cF_c)x_c + B_{0c}z_0 + L_{cd}z_d \qquad (10.2.12)$$

$$\dot{x}_i = A_{q_i}x_i + B_{0id}\bar{z}_0 + L_{id}z_d + B_{q_i}\left[\frac{-F_{i1}}{\varepsilon^{q_i}}F_{ai}x_a + \frac{F_i}{\varepsilon^{q_i}}S_{q_i}(\varepsilon)x_i\right],$$

$$i = 1, 2, \cdots, m_d, \qquad (10.2.13)$$

$$z_0 = F_{a0}(\varepsilon)x_a. \qquad (10.2.14)$$

We next consider the following scaling and redefinition of variables,

$$\tilde{x} = [\tilde{x}_a', \tilde{x}_c', \tilde{x}_d']', \qquad (10.2.15)$$

$$\tilde{x}_a = [\tilde{x}_{a1}', \tilde{x}_{a2}', \cdots \tilde{x}_{al}', \tilde{x}_{a0}']' = S_a(\varepsilon)Q_a^{-1}(\varepsilon)\Gamma_{sa}^{-1}x_a, \qquad (10.2.16)$$

$$S_a(\varepsilon) = \text{blkdiag}\{S_{a1}(\varepsilon), S_{a2}(\varepsilon), \cdots, S_{al}(\varepsilon), I\},$$

$$Q_a(\varepsilon) = \text{blkdiag}\{Q_{a1}(\varepsilon), Q_{a2}(\varepsilon), \cdots, Q_{al}(\varepsilon), I\},$$

$$\tilde{x}_c = x_c, \qquad (10.2.17)$$

$$\tilde{x}_d = [\tilde{x}_1', \tilde{x}_2', \cdots, \tilde{x}_{m_d}']', \ \tilde{x}_i = [\tilde{x}_{i1}, \tilde{x}_{di2}, \cdots, \tilde{x}_{iq_i}]', \qquad (10.2.18)$$

$$\tilde{x}_{i1} = x_{i1} - F_{ai}\Gamma_{sa}Q_a(\varepsilon)S_a^{-1}(\varepsilon)\tilde{x}_a, \ i = 1, 2, \cdots, m_d, \qquad (10.2.19)$$

$$\tilde{x}_{ij} = \varepsilon^{j-1}x_{ij}, \ j = 2, 3, \cdots, q_i, \ i = 1, 2, \cdots, m_d, \qquad (10.2.20)$$

where $Q_{ai}(\varepsilon)$ and $S_{ai}(\varepsilon)$ are the $Q(\varepsilon)$ and $S(\varepsilon)$ of Lemmas 2.2.2 and 2.2.3 for the triple $(A_i, B_i, F_i(\varepsilon))$. Hence, Lemmas 2.2.2-2.2.5 all apply. In these new state variables, the closed-loop system becomes,

$$\dot{\tilde{x}}_a = \tilde{J}_a(\varepsilon)\tilde{x}_a + \tilde{B}_a(\varepsilon)\tilde{x}_a + \tilde{L}_{ad}(\varepsilon)\tilde{z}_d,$$

$$z_d = \tilde{z}_d + F_{ad}(\varepsilon)\Gamma_{sa}Q_a(\varepsilon)S_a^{-1}(\varepsilon)\tilde{x}_a, \tag{10.2.21}$$

$$\dot{\tilde{x}}_c = (A_{cc} + B_c F_c)\tilde{x}_c + L_{cd}\tilde{z}_d + B_{0c}F_{a0}(\varepsilon)\Gamma_{sa}Q_a(\varepsilon)S_a^{-1}(\varepsilon)\tilde{x}_a$$

$$+ L_{cd}F_{ad}(\varepsilon)\Gamma_{sa}Q_a(\varepsilon)S_a^{-1}(\varepsilon)\tilde{x}_a, \tag{10.2.22}$$

$$\dot{\tilde{x}}_d = \frac{1}{\varepsilon}A_d^c\tilde{x}_d + D_{da}(\varepsilon)F_a(\varepsilon)\Gamma_{sa}Q_a(\varepsilon)S_a^{-1}(\varepsilon)\tilde{x}_a$$

$$+ D_{daa}F_a(\varepsilon)A_{aa}\Gamma_{sa}Q_a(\varepsilon)S_a^{-1}(\varepsilon)\tilde{x}_a + D_{dd}(\varepsilon)\tilde{x}_d, \tag{10.2.23}$$

$$z_0 = F_{a0}(\varepsilon)\Gamma_{sa}Q_a(\varepsilon)S_a^{-1}(\varepsilon)\tilde{x}_a, \tag{10.2.24}$$

where

$$A_d^c = \text{blkdiag}\{A_{q_1} + B_{q_1}F_1, \ A_{q_2} + B_{q_2}F_2, \cdots, A_{q_{m_d}} + B_{q_{m_d}}F_{m_d}\}$$

is asymptotically stable,

$$\tilde{J}_a(\varepsilon) = \text{blkdiag}\{\varepsilon\tilde{J}_{a1}(\varepsilon), \varepsilon\tilde{J}_{a2}(\varepsilon), \cdots, \varepsilon\tilde{J}_{al}(\varepsilon), A_0\},$$

$$\tilde{B}_a(\varepsilon) = \begin{bmatrix} 0 & \tilde{B}_{12}(\varepsilon) & \tilde{B}_{13}(\varepsilon) & \cdots & \tilde{B}_{1l}(\varepsilon) \\ 0 & 0 & \tilde{B}_{23}(\varepsilon) & \cdots & \tilde{B}_{2l}(\varepsilon) \\ \vdots & \vdots & \vdots & \ddots & \vdots \\ 0 & 0 & 0 & \cdots & 0 \\ \tilde{B}_{01}(\varepsilon) & \tilde{B}_{02}(\varepsilon) & \tilde{B}_{03}(\varepsilon) & \cdots & \tilde{B}_{0l}(\varepsilon) \end{bmatrix},$$

$$\tilde{B}_{0j} = B_{0j}F_j(\varepsilon)Q_{aj}(\varepsilon)S_{aj}^{-1}(\varepsilon),$$

$$\tilde{B}_{ij}(\varepsilon) = S_{ai}(\varepsilon)Q_{ai}^{-1}(\varepsilon)B_{ij}F_j(\varepsilon)Q_{aj}(\varepsilon)S_{aj}^{-1}(\varepsilon),$$

$$i = 1, 2, \cdots, l, j = i+1, i+2, \cdots, l,$$

$$\tilde{L}_{ad}(\varepsilon) = S_a(\varepsilon)Q_a^{-1}(\varepsilon)\Gamma_{sa}^{-1}L_{ad},$$

$D_{da}(\varepsilon)$, D_{daa}, $D_{dd}(\varepsilon)$ and $\tilde{L}_{ad}(\varepsilon)$ are some matrices of appropriate dimensions satisfying

$$|D_{da}(\varepsilon)| \leq d_{da}, \ |D_{daa}| \leq d_{daa},$$

$$|D_{dd}(\varepsilon)| \leq d_{dd}, \ |\tilde{L}_{ad}(\varepsilon)| \leq \tilde{l}_{ad}, \ \forall \varepsilon \in (0, \varepsilon_a^*], \tag{10.2.25}$$

and for $i = 1$ to l, $j = i + 1$ to l,

$$|\tilde{B}_{ij}(\varepsilon)| \leq \tilde{b}_{ij}\varepsilon, \tag{10.2.26}$$

for some positive constants d_{da}, d_{daa}, d_{dd}, $\tilde{l}_{ad}$ and $\tilde{b}_{ij}$ independent of ε.

It follows from Lemma 2.2.4 and the special form of the above state transformation that

$$|\tilde{x}(0)| \leq \alpha_0 |x(0)|, \quad \forall \varepsilon \in (0,1], \tag{10.2.27}$$

for some positive constant α_0 independent of ε.

To prove the theorem, we need to establish properties 1 and 2 of Problem 10.2.1. To this end, let us construct a Lyapunov function for the closed-loop system (10.2.21)-(10.2.24). We do this by composing Lyapunov functions for the subsystems. For the subsystem of $\tilde{x}_a = [\tilde{x}'_{a1}, \tilde{x}'_{a2}, \cdots, \tilde{x}'_{al}, \tilde{x}'_{a0}]'$, we choose a Lyapunov function,

$$V_a(\tilde{x}_a) = \sum_{i=0}^{l} \kappa_a^i \tilde{x}'_{ai} P_{ai} \tilde{x}_{ai}, \tag{10.2.28}$$

where κ_a is a positive scalar, whose value is to be determined later, $P_{a0} > 0$ is such that $A'_0 P_{a0} + P_{a0} A_0 = -I$, and each $i = 1$ to l, P_{ai} is the unique solution to the Lyapunov equation

$$\tilde{J}_{ai}(\varepsilon)' P_{ai} + P_{ai} \tilde{J}_{ai}(\varepsilon) = -I, \tag{10.2.29}$$

which, by Lemma 2.2.3, is independent of ε. Similarly, for the subsystem $\tilde{x}_c$, choose a Lyapunov function

$$V_c(\tilde{x}_c) = \tilde{x}'_c P_c \tilde{x}_c, \tag{10.2.30}$$

where $P_c > 0$ is the unique solution to the Lyapunov equation

$$(A_{cc} + B_c F_c)' P_c + P_c (A_{cc} + B_c F_c) = -I. \tag{10.2.31}$$

The existence of such a P_c is again guaranteed by the fact that $A_{cc} + B_c F_c$ is asymptotically stable. Finally, for the subsystem of $\tilde{x}_d$, choose a Lyapunov function

$$V_d(\tilde{x}_d) = \tilde{x}'_d P_d \tilde{x}_d, \tag{10.2.32}$$

where each P_d is the unique solution to the Lyapunov equation

$$(A_d^c)' P_d + P_d A_d^c = -I. \tag{10.2.33}$$

Once again, the existence of such P_d is due to the fact that A_d^c is asymptotically stable.

We now construct a Lyapunov function for the closed-loop system (10.2.21)-(10.2.24) as follows,

$$V(\tilde{x}_a, \tilde{x}_c, \tilde{x}_d) = V_a(\tilde{x}_a) + V_c(\tilde{x}_c) + \kappa_d V_d(\tilde{x}_d), \tag{10.2.34}$$

where the value of κ_d is to be determined.

Let us first consider the derivative of $V_a(\tilde{x}_a)$ along the trajectories of the subsystem $\tilde{x}_a$ and obtain that,

$$
\dot{V}_a(\tilde{x}_a) = \sum_{i=1}^{l} \left[-\kappa_a^i \varepsilon \tilde{x}_{ai}' \tilde{x}_{ai} + 2 \sum_{j=i+1}^{l} \kappa_a^i \tilde{x}_{ai}' P_{ai} \tilde{B}_{ij}(\varepsilon) \tilde{x}_{aj} \right]
$$
$$
-\tilde{x}_{a0}' \tilde{x}_{a0} + 2 \sum_{j=1}^{l} \tilde{x}_{a0}' P_{a0} \tilde{B}_{0j}(\varepsilon) \tilde{x}_{aj}
$$
$$
+2 \sum_{i=0}^{l} \kappa_a^i \tilde{x}_{ai}' P_{ai} \tilde{L}_{adi}(\varepsilon) \tilde{z}_d. \tag{10.2.35}
$$

where

$$
\tilde{L}_{ad}(\varepsilon) = \begin{bmatrix} \tilde{L}_{ad1}(\varepsilon) \\ \tilde{L}_{ad2}(\varepsilon) \\ \vdots \\ \tilde{L}_{adl}(\varepsilon) \\ \tilde{L}_{ad0}(\varepsilon) \end{bmatrix}.
$$

Using (10.2.26) it is straightforward to show that, there exists a $\kappa > 0$ such that

$$
\dot{V}_a(\tilde{x}_a) \le -\frac{3}{4}\varepsilon |\tilde{x}_a|^2 + \alpha_1(\kappa_a)|\tilde{x}_a|\,|\tilde{z}_d|, \tag{10.2.36}
$$

for some nonnegative constant $\alpha_1(\kappa_a)$ independent of ε.

In view of (10.2.36), the derivative of V along the trajectory of the closed-loop system (10.2.21)-(10.2.24) can be evaluated as follows,

$$
\dot{V} \le -\frac{3}{4}\varepsilon |\tilde{x}_a|^2 + \alpha_1(\kappa_a)|\tilde{x}_a|\,|\tilde{z}_d|
$$
$$
-\tilde{x}_c' \tilde{x}_c + 2\tilde{x}_c' P_c L_{cd} \tilde{z}_d + 2\tilde{x}_c' P_c B_{0c} F_a(\varepsilon) \Gamma_{sa} Q_a(\varepsilon) S_a^{-1}(\varepsilon) \tilde{x}_a
$$
$$
+2\tilde{x}_c' P_c L_{cd} F_{ad}(\varepsilon) \Gamma_{sa} Q_a(\varepsilon) S_a^{-1}(\varepsilon) \tilde{x}_a
$$
$$
-\frac{\kappa_d}{\varepsilon} \tilde{x}_d' \tilde{x}_d + 2\kappa_d \tilde{x}_d' P_d D_{da}(\varepsilon) F_a(\varepsilon) \Gamma_{sa} Q_a(\varepsilon) S_a^{-1}(\varepsilon) \tilde{x}_a
$$
$$
+2\kappa_d \tilde{x}_d' P_d D_{daa} F_a(\varepsilon) A_{aa} \Gamma_{sa} Q_a(\varepsilon) S_a^{-1}(\varepsilon) \tilde{x}_a
$$
$$
+2\kappa_d \tilde{x}_d' P_d D_{dd}(\varepsilon) \tilde{x}_d. \tag{10.2.37}
$$

Using the majorizations (10.2.25) and Lemma 2.2.4, we can easily verify that, there exist a $\kappa_d > 0$ and an $\varepsilon^* \in (0,1]$ such that, for all $\varepsilon \in (0, \varepsilon^*]$,

$$
\dot{V} \le -\frac{1}{2}\varepsilon |\tilde{x}_a|^2 - \frac{1}{2}|\tilde{x}_c|^2 - \frac{1}{2\varepsilon}|\tilde{x}_d|^2, \tag{10.2.38}
$$

which implies that the closed-loop system is asymptotically stable for all $\varepsilon \in (0, \varepsilon^*]$. This establishes property 1 of Problem 10.2.1.

It remains to establish property 2 of Problem 10.2.1. It follows from (10.2.38) and (10.2.27) that, for all $\varepsilon \in (0, \varepsilon^*]$,

$$\dot{V} \leq -\alpha_3 \varepsilon V, \ |V(0)| \leq \alpha_4 |x(0)|^2, \tag{10.2.39}$$

for some positive constants α_3 and α_4 independent of ε.

Standard comparison theorems then show that

$$V(t) \leq \alpha_4 e^{-\alpha_3 \varepsilon t} |x(0)|^2, \ t \geq 0, \tag{10.2.40}$$

which implies that

$$|\tilde{x}_a(t)| \leq \alpha_5 e^{-\alpha_3 \varepsilon t} |x(0)|, \ |\tilde{x}_d(t)| \leq \alpha_6 e^{-\alpha_3 \varepsilon t} |x(0)|, \ t \geq 0. \tag{10.2.41}$$

for some positive constants α_5 and α_6 independent of ε.

Viewing the second, third and fourth terms of the right hand side of (10.2.23) as the inputs to the x_d dynamics, and using Lemma 2.2.4, we can easily verify that

$$|\tilde{x}_d(t)| \leq \alpha_7 e^{-\frac{\alpha_8}{\varepsilon} t} |x(0)| + \alpha_9 e^{-\alpha_3 \varepsilon t} |x(0)|, \tag{10.2.42}$$

for some positive constants α_7, α_8 and α_9, all independent of ε.

Finally, recalling that $z_d = \tilde{z}_d + F_{ad}(\varepsilon)\Gamma_{sa}Q_a(\varepsilon)S_a^{-1}(\varepsilon)\tilde{x}_a$, $z_0 = F_{a0}(\varepsilon)\Gamma_{sa}Q_a(\varepsilon)S_a^{-1}(\varepsilon)\tilde{x}_a$, and using once again Lemma 2.2.4, we can verify that

$$|\tilde{z}| \leq \alpha_{10} \left[e^{-\frac{\alpha_8}{\varepsilon} t} + \varepsilon e^{-\alpha_3 \varepsilon t} \right] |x(0)|, \tag{10.2.43}$$

for some positive constant α_{10} independent of ε, and hence

$$J(x_0, u) = \int_0^\infty |z(t)|^2 dt = \int_0^\infty |\Gamma_o \tilde{z}(t)|^2 dt \leq |\Gamma_o|^2 \int_0^\infty |\tilde{z}(t)|^2 dt \to 0, \ \text{as} \ \varepsilon \to 0. \tag{10.2.44}$$

☒

10.2.3. An Example

Consider the system

$$\begin{cases} \dot{x} = \begin{bmatrix} 0 & 1 \\ 1 & 0 \end{bmatrix} x + \begin{bmatrix} 0 \\ 1 \end{bmatrix} u, \\ z = \begin{bmatrix} 0 & 1 \end{bmatrix} x. \end{cases} \tag{10.2.45}$$

It can be easily seen that this system is already in the SCB form with $x_a = x_1$ and $x_d = x_2$. The system is square invertible with only one invariant zero at $s = 0$. By Theorem 10.2.1 the problem of perfect regulation is solvable for this

system. Following the design procedure of Section 10.2.2, we readily obtain the
following family of feedback laws,

$$u = -2x_1 - \left(\varepsilon + \frac{1}{\varepsilon}\right)x_2, \tag{10.2.46}$$

which is explicitly parameterized in terms of ε.

We next show that the above family of state feedback laws actually solves
the problem of perfect regulation for the given system. In fact, under this family
of feedback laws, the closed-loop system is given by

$$\begin{cases} \dot{x} = \begin{bmatrix} 0 & 1 \\ -1 & -\varepsilon - \frac{1}{\varepsilon} \end{bmatrix} x, \\ z = [\, 0 \quad 1\,]\, x. \end{cases} \tag{10.2.47}$$

Let $T_{zw}(s,\varepsilon)$ be the transfer function from the disturbance w to the controlled
output z of the following auxiliary system,

$$\begin{cases} \dot{x} = \begin{bmatrix} 0 & 1 \\ -1 & -\varepsilon - \frac{1}{\varepsilon} \end{bmatrix} x + Iw, \\ z = [\, 0 \quad 1\,]\, x. \end{cases} \tag{10.2.48}$$

Then verifying that the family of feedback laws (10.2.46) solves the problem of
perfect regulation for the system (10.2.45) is tantamount to verifying that the
H_2-norm of $T_{zw}(s,\varepsilon)$ approaches zero as ε goes to zero. Fig. 10.2.1 shows that
this is indeed the case.   ▣

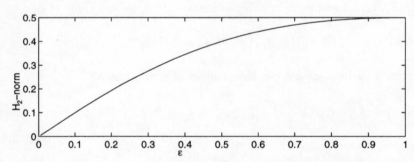

Figure 10.2.1: Plot of the H_2-norm of $T_{zw}(s,\varepsilon)$ with respect to ε.

10.3. Discrete-Time Systems

10.3.1. Problem Statement

Consider the discrete-time linear system

$$\Sigma : \begin{cases} x^+ = Ax + Bu, \quad x(0) = x_0, \\ z = Cx + Du, \end{cases} \tag{10.3.1}$$

where $x \in \mathbb{R}^n$ is the state, $u \in \mathbb{R}^m$ is the input, and $z \in \mathbb{R}^p$ is the controlled output. Let us also consider an associated cost criterion,

$$J(x_0, u) = \sum_{k=0}^{\infty} |z(k)|^2. \qquad (10.3.2)$$

We make a standing assumption that the pair (A, B) is stabilizable. Moreover, without loss of generality, we assume that the matrices $[C, \; D]$ and $\begin{bmatrix} B \\ D \end{bmatrix}$ are of full rank.

The problem of perfect regulation for the discrete-time linear system (10.3.1) can then be formulated as follows.

Problem 10.3.1. *Consider the system Σ as given by (10.3.1) along with the associated cost $J(x_0, u)$ as given by (10.3.2). Then the problem of perfect regulation via state feedback is to find a family of parameterized linear state feedback laws $u = F(\varepsilon)x$ having the following properties:*

1. *There exists an $\varepsilon^* > 0$ such that for all $\varepsilon \in (0, \varepsilon^*]$, the closed-loop system comprising of Σ and $u = F(\varepsilon)x$ is internally stable;*

2. *For each $x_0 \in \mathbb{R}^n$,*

$$J(x_0, u) \to 0 \quad \text{as} \quad \varepsilon \to 0. \qquad (10.3.3)$$

P

We recall from [64] the necessary and sufficient conditions under which the problem of perfect regulation as formulated above is solvable.

Theorem 10.3.1. *Consider the system Σ given in (10.3.1) along with the associated cost $J(x_0, u)$ given in (10.3.2). Then the problem of perfect regulation via state feedback $u = F(\varepsilon)x$ is solvable if and only if the given system Σ is right invertible, has all its invariant zeros located inside or on the unit circle, and has no infinite zero of order greater than or equal to one.* T

Our objective in this section is to develop a low gain feedback based direct eigenstructure assignment design procedure for constructing a family of state feedback laws that, under the solvability condition of Theorem 10.3.1, solves the problem of perfect regulation.

10.3.2. Solution of the Problem of Perfect Regulation

We present an eigenstructure assignment based design algorithm and show that the resulting state feedback laws indeed solve the problem of perfect regulation. The algorithm we propose is based on the eigenstructure assignment based low gain feedback design technique of Chapter 2 and consists of the following three steps.

Step 1: Construction of the SCB of Σ. Perform a nonsingular state, input and controlled output transformation on the system Σ. That is, let

$$x = \Gamma_s \bar{x}, \; z = \Gamma_o \bar{z}, \; u = \Gamma_\iota \bar{u}$$

be such that the system Σ can be written in the following SCB form,

$$\bar{x} = [x_a', x_c']', \; \bar{z} = z_0', \; \bar{u} = [u_0', u_c']',$$

and

$$x_a^+ = A_{aa}\bar{x}_a + B_{0a}z_0, \tag{10.3.4}$$

$$x_c^+ = A_{cc}x_c + B_{0c}z_0 + B_c[E_{ca}x_a + u_c], \tag{10.3.5}$$

$$z_0 = C_{0a}x_a + C_{0c}x_c + u_0. \tag{10.3.6}$$

We note that the state x_b is not present as Σ is right invertible, and that x_d is absent also since Σ has no infinite zero of order greater than or equal to one.

Step 2: Construction of a parameterized low gain matrix $F_a(\varepsilon)$. By Property A.1.1 of the SCB, the pair (A_{aa}, B_{0a}) is stabilizable. Moreover, by Property A.1.2 of the SCB, the eigenvalues of A_{aa} are the invariant zeros of the system Σ and hence, are all located inside or on the unit circle. Following the eigenstructure assignment based low gain feedback design procedure of Section 2.3.1, we can design a feedback gain $F_a(\varepsilon)$ for the pair (A_{aa}, B_{0a}) as follows.

Step 2.1. Find the nonsingular transformation matrices Γ_{sa} and $\Gamma_{\iota a}$ such that (A_{aa}, B_{0a}) can be transformed into the block diagonal control canonical form,

$$\Gamma_{sa}^{-1} A_{aa} \Gamma_{sa} = \begin{bmatrix} A_1 & 0 & \cdots & 0 & 0 \\ 0 & A_2 & \cdots & 0 & 0 \\ \vdots & \vdots & \ddots & \vdots & \vdots \\ 0 & 0 & \cdots & A_l & 0 \\ 0 & 0 & \cdots & 0 & A_0 \end{bmatrix},$$

$$\Gamma_{sa}^{-1} B_a \Gamma_{Ia}^0 = \begin{bmatrix} B_1 & B_{12} & \cdots & B_{1l} & \star \\ 0 & B_2 & \cdots & B_{2l} & \star \\ \vdots & \vdots & \ddots & \vdots & \vdots \\ 0 & 0 & \cdots & B_l & \star \\ B_{01} & B_{02} & \cdots & B_{0l} & \star \end{bmatrix},$$

where A_0 contains all the eigenvalues of A_{aa} that are strictly inside the unit circle, l is an integer and for $i = 1, 2, \cdots, l$, all eigenvalues of A_i are on the unit circle and hence (A_i, B_i) is controllable as given by,

$$A_i = \begin{bmatrix} 0 & 1 & 0 & \cdots & 0 \\ 0 & 0 & 1 & \cdots & 0 \\ \vdots & \vdots & \vdots & \ddots & \vdots \\ 0 & 0 & 0 & \cdots & 1 \\ -a_{n_i}^i & -a_{n_i-1}^i & -a_{n_i-2}^i & \cdots & -a_1^i \end{bmatrix}, \quad B_i = \begin{bmatrix} 0 \\ 0 \\ \vdots \\ 0 \\ 1 \end{bmatrix}.$$

Here the $\star$'s represent sub-matrices of less interest. We also note that the existence of the above canonical form was shown in Wonham [126] while its software realization can be found in Chen [9].

Step 2.2. For each (A_i, B_i), let $F_i(\varepsilon) \in \mathbb{R}^{1 \times n_i}$ be the state feedback gain such that

$$\lambda(A_i + B_i F_i(\varepsilon)) = (1 - \varepsilon)\lambda(A_i) \in \mathbb{C}^{\odot}. \tag{10.3.7}$$

Note that $F_i(\varepsilon)$ is unique.

Step 2.3. Compose $F_a(\varepsilon)$ as follows,

$$F_a(\varepsilon) = \Gamma_{Ia} \begin{bmatrix} F_1(\varepsilon) & 0 & \cdots & 0 & 0 \\ 0 & F_2(\varepsilon) & \cdots & 0 & 0 \\ \vdots & \vdots & \ddots & \vdots & \vdots \\ 0 & 0 & \cdots & F_{l-1}(\varepsilon) & 0 \\ 0 & 0 & \cdots & 0 & F_l(\varepsilon) \\ 0 & 0 & \cdots & 0 & 0 \end{bmatrix} \Gamma_{sa}^{-1}, \tag{10.3.8}$$

where $\varepsilon \in (0, 1]$ is a design parameter whose value is to be specified later.

Step 3: Construction of a parameterized gain matrix $F(\varepsilon)$: By Theorem A.1.1, the pair (A_{cc}, B_c) is controllable, hence one can choose a feedback gain matrix F_c such that $A_{cc} + B_c F_c$ is asymptotically stable and has any chosen set of eigenvalues.

Next a composite static state feedback gain is formed for the system Σ. This state feedback gain takes the form of

$$F(\varepsilon) = \Gamma_{\mathrm{I}} \begin{bmatrix} F_{u_0}(\varepsilon) \\ F_{u_c} \end{bmatrix} \Gamma_{\mathrm{s}}^{-1}, \qquad (10.3.9)$$

where $F_{u_0} = [\, -C_{0a} + F_a(\varepsilon) \quad -C_{0c} \,]$ and $F_{u_c} = [\, -E_{ca} \quad F_c \,]$.

This concludes the description of a low gain based state feedback design method that leads to a parameterized gain $F(\varepsilon)$. ▣

Next we choose a family of state feedback laws, parameterized in ε, as

$$u = F(\varepsilon)x, \qquad (10.3.10)$$

where $F(\varepsilon)$ is as given by (10.3.9).

The following theorem then establishes that the family of state feedback laws as given by (10.3.10) indeed solves the problem of perfect regulation for the system Σ as given by (10.3.1).

Theorem 10.3.2. *Consider the system Σ as given in (10.3.1) along with the associated cost $J(x_0, u)$ as given in (10.3.2). Also, assume that Σ is right invertible, has all its invariant zeros located inside or on the unit circle, and has no infinite zero of order greater than or equal to one. Then, the family of state feedback laws as given by (10.3.10) solves the problem of perfect regulation for the system Σ.* ▣

Proof of Theorem 10.3.2. With the state feedback laws (10.3.10), the closed-loop system in the special coordinate basis can be written as,

$$x_a^+ = (A_{aa} + B_{0a}F_a(\varepsilon))x_a, \qquad (10.3.11)$$

$$x_c^+ = (A_{cc} + B_c F_c)x_c + B_{0c}F_a(\varepsilon)x_a, \qquad (10.3.12)$$

$$z_0 = F_a(\varepsilon)x_a. \qquad (10.3.13)$$

The stability of this closed-loop system follows immediately from the fact that both $A_{aa} + B_{0a}F_a(\varepsilon)$ and $A_{cc} + B_c F_c$ are asymptotically stable.

We next establish the second property of Problem 10.3.1, i.e.,

$$\lim_{\varepsilon \to 0} J(x_0, F(\varepsilon)x) = 0. \qquad (10.3.14)$$

To do so, we need only to consider the dynamics of x_a. Let us perform a state transformation on x_a as follows,

$$\tilde{x}_a = [\tilde{x}'_{a1}, \tilde{x}'_{a2}, \cdots, \tilde{x}'_{al}, \tilde{x}'_{a0}]' = S_a(\varepsilon)Q_a^{-1}(\varepsilon)(\Gamma_{sa}^0)^{-1}x_a,$$

$$S_a(\varepsilon) = \mathrm{blkdiag}\{S_{a1}(\varepsilon), S_{a2}(\varepsilon), \cdots, S_{al}(\varepsilon), I\},$$

$$Q_a(\varepsilon) = \mathrm{blkdiag}\{Q_{a1}(\varepsilon), Q_{a2}(\varepsilon), \cdots, Q_{al}(\varepsilon), I\}, \qquad (10.3.15)$$

where $Q_{ai}(\varepsilon)$ and $S_{ai}(\varepsilon)$ are the $Q(\varepsilon)$ and $S(\varepsilon)$ of Lemmas 2.3.2 and 2.3.3 for the triple $(A_i, B_i, F_i(\varepsilon))$. Hence, Lemmas 2.3.2-2.3.5 all apply. In these new state variables, the dynamics of x_a becomes,

$$\tilde{x}_a^+ = \tilde{J}_a(\varepsilon)\tilde{x}_a + \tilde{B}_a(\varepsilon)\tilde{x}_a, \tag{10.3.16}$$

where

$$\tilde{J}_a(\varepsilon) = \text{blkdiag}\{\tilde{J}_{a1}(\varepsilon), \tilde{J}_{a2}(\varepsilon), \cdots, \tilde{J}_{al}(\varepsilon), A_0\}, \tag{10.3.17}$$

$$\tilde{B}_a(\varepsilon) = \begin{bmatrix} 0 & \tilde{B}_{12}(\varepsilon) & \tilde{B}_{13}(\varepsilon) & \cdots & \tilde{B}_{1l}(\varepsilon) \\ 0 & 0 & \tilde{B}_{23}(\varepsilon) & \cdots & \tilde{B}_{2l}(\varepsilon) \\ \vdots & \vdots & \vdots & \ddots & \vdots \\ 0 & 0 & 0 & \cdots & 0 \\ \tilde{B}_{01}(\varepsilon) & \tilde{B}_{02}(\varepsilon) & \tilde{B}_{03}(\varepsilon) & \cdots & \tilde{B}_{0l}(\varepsilon) \end{bmatrix}, \tag{10.3.18}$$

$$\tilde{B}_{0j}(\varepsilon) = B_{0j}F_j(\varepsilon)Q_{aj}(\varepsilon)S_{aj}^{-1}(\varepsilon),$$

$$\tilde{B}_{ij}(\varepsilon) = S_{ai}(\varepsilon)Q_{ai}^{-1}(\varepsilon)B_{ij}F_j(\varepsilon)Q_{aj}(\varepsilon)S_{aj}^{-1}(\varepsilon),$$

$$i = 1, 2, \cdots, l, j = i+1, i+2, \cdots, l, \tag{10.3.19}$$

and where, for $i = 1$ to l, $\tilde{J}_{ai}(\varepsilon)$ is as defined in Lemma 2.3.2.

By Lemma 2.2.4, we have that, for all $\varepsilon \in (0,1]$,

$$|F_a(\varepsilon)\Gamma_{sa}Q_a(\varepsilon)S_a^{-1}(\varepsilon)| \le f_a\varepsilon, \tag{10.3.20}$$

and for $i = 1$ to l, $j = i+1$ to l,

$$|\tilde{B}_{ij}(\varepsilon)| \le \tilde{b}_{ij}\varepsilon, \tag{10.3.21}$$

where f_a and $\tilde{b}_{ij}$ are some positive constants, independent of ε.

It follows from Lemma 2.3.4 and the special form of the above state transformation that

$$|\tilde{x}_a(0)| \le \alpha_0|x(0)|, \quad \forall\varepsilon \in (0,1], \tag{10.3.22}$$

for some positive constant α_0 independent of ε.

We then construct a Lyapunov function

$$V_a(\tilde{x}_a) = \sum_{i=0}^{l} \kappa_a^{i-1}\tilde{x}_{ai}'P_{ai}(\varepsilon)\tilde{x}_{ai}, \tag{10.3.23}$$

where κ_a is a positive scalar, whose value is to be determined later, $P_{a0} > 0$ is such that $A_0'P_{a0}A_0 - P_{a0} = -I$, and each $i = 1$ to l, $P_{ai}(\varepsilon)$ is the unique solution to the Lyapunov equation

$$\tilde{J}_{ai}(\varepsilon)'P_{ai}\tilde{J}_{ai}(\varepsilon) - P_{ai} = -\varepsilon I, \tag{10.3.24}$$

which, by Lemma 2.3.3, satisfies

$$P_{ai1} \leq P_{ai}(\varepsilon) \leq P_{ai2}, \ \forall \varepsilon \in (0, \varepsilon^*], \tag{10.3.25}$$

for some $\varepsilon^* \in (0, 1]$ and some positive definite matrices P_{ai1} and P_{ai2} independent of ε.

The evaluation of the difference of $V_a(\tilde{x}_a)$ along the trajectories of the subsystem $\tilde{x}_a$ yields that,

$$\Delta V_a = \sum_{i=0}^{l} \left[-\kappa_a^{i-1} \tilde{x}'_{ai} \tilde{x}_{ai} + 2 \sum_{j=i+1}^{l} \kappa_a^{i-1} \tilde{x}'_{ai} \tilde{J}'_{ai}(\varepsilon) P_{ai}(\varepsilon) \tilde{B}_{ij}(\varepsilon) \tilde{x}_{aj} \right.$$
$$\left. + \kappa_a^{i-1} \left(\sum_{j=i+1}^{l} \tilde{B}_{ij}(\varepsilon) \tilde{x}_{aj}(\varepsilon) \right)' P_{ai}(\varepsilon) \left(\sum_{j=i+1}^{l} \tilde{B}_{ij}(\varepsilon) \tilde{x}_{aj}(\varepsilon) \right) \right]. \tag{10.3.26}$$

Using (10.3.20), (10.3.21) and (10.3.25), it is straightforward to show that, there exists an $\kappa_a > 0$ such that,

$$\Delta V_a \leq -\frac{3}{4} \varepsilon \tilde{x}'_a \tilde{x}_a. \tag{10.3.27}$$

It follows from (10.3.27) and (10.3.22) that, for all $\varepsilon \in (0, \varepsilon^*]$,

$$\Delta V_a \leq -\alpha_1 \varepsilon V_a, \ |V(0)| \leq \alpha_2 |x(0)|^2, \tag{10.3.28}$$

for some positive constants α_1 and α_2 independent of ε such that $\alpha_1 \varepsilon \leq 1/2$. Standard comparison theorems then show that

$$V_a(k) \leq \alpha_2 (1 - \alpha_1 \varepsilon)^k |x(0)|^2, \ k \geq 0, \tag{10.3.29}$$

which, by (10.3.25) implies that

$$|\tilde{x}_a(k)| \leq \alpha_3 (1 - \alpha_1 \varepsilon)^k |x(0)|, \ k \geq 0, \tag{10.3.30}$$

for some positive constant α_3 independent of ε.

Finally, recalling that $z_0 = F_a(\varepsilon) \Gamma_{sa} Q_a(\varepsilon) S_a^{-1}(\varepsilon) \tilde{x}_a$, and using once again (10.3.20), we can verify that

$$|\bar{z}(k)| = |z_0(k)| \leq \alpha_4 \varepsilon (1 - \alpha_1 \varepsilon)^k |x(0)|, \ k \geq 0, \tag{10.3.31}$$

for some positive constant α_4 independent of ε, and hence

$$J(x_0, u) = \sum_{k=0}^{\infty} |z(k)|^2 dt = \sum_{k=0}^{\infty} |\Gamma_o \bar{z}(k)|^2 dt \leq |\Gamma_o|^2 \sum_{k=0}^{\infty} |\bar{z}(k)|^2 dt \to 0, \ \text{as } \varepsilon \to 0. \tag{10.3.32}$$

$$\boxtimes$$

10.3.3. An Example

Consider the system

$$
\begin{cases}
x(k+1) = \begin{bmatrix} 1 & 1 & 0 \\ 1 & 0 & 1 \\ 0 & -1 & 3 \end{bmatrix} x(k) + \begin{bmatrix} 0 & 0 \\ 1 & 0 \\ 1 & 1 \end{bmatrix} u(k), \\
z(k) = \begin{bmatrix} 1 & -1 & 1 \end{bmatrix} x(k) + \begin{bmatrix} 1 & 0 \end{bmatrix} u(k).
\end{cases}
\tag{10.3.33}
$$

It can be easily seen that this system is already in the SCB form with

$$
A_{aa} = \begin{bmatrix} 1 & 1 \\ 0 & 1 \end{bmatrix}, \quad B_{0a} = \begin{bmatrix} 0 \\ 1 \end{bmatrix},
$$

$$
A_c = 2, \ B_{0c} = 1, \ B_c = 1, \ E_{ca} = \begin{bmatrix} -1 & 0 \end{bmatrix}, \ C_{0a} = \begin{bmatrix} 1 & -1 \end{bmatrix}, \ C_{0c} = 1.
$$

Hence, by the properties of SCB, this system has two repeated invariant zeros at $z = 1$ and has no infinite zeros of order greater than or equal to one. Following the proposed design procedure, we readily obtain the desired family of feedback laws,

$$
u = \begin{bmatrix} -1 - \varepsilon^2 & 1 - 2\varepsilon & -1 \\ 1 & 0 & -2 \end{bmatrix} x.
\tag{10.3.34}
$$

This feedback gain is parameterized explicitly in terms of ε. We next show that the family of state feedback control laws $u = F(\varepsilon)x$ actually solves the problem of perfect regulation for the system (10.3.33). In fact, under the control law $u = F(\varepsilon)x$, the closed-loop system is given by

$$
\begin{cases}
x^+ = \begin{bmatrix} 1 & 1 & 0 \\ -\varepsilon^2 & 1 - 2\varepsilon & 0 \\ -\varepsilon^2 & -2\varepsilon & 0 \end{bmatrix} x, \\
z = \begin{bmatrix} -\varepsilon^2 & -2\varepsilon & 0 \end{bmatrix} x.
\end{cases}
\tag{10.3.35}
$$

Let $T_{zw}(z, \varepsilon)$ be the transfer function from the disturbance w to the controlled variable z of the auxiliary system,

$$
\begin{cases}
x^+ = \begin{bmatrix} 1 & 1 & 0 \\ -\varepsilon^2 & 1 - 2\varepsilon & 0 \\ -\varepsilon^2 & -2\varepsilon & 0 \end{bmatrix} x + Iw \\
z = \begin{bmatrix} -\varepsilon^2 & -2\varepsilon & 0 \end{bmatrix} x.
\end{cases}
\tag{10.3.36}
$$

Then, verifying that the family of feedback laws (10.3.34) solve the problem of perfect regulation for the system (10.3.33) is tantamount to verifying that the H_2-norm of $T_{zw}(z, \varepsilon)$ approaches zero as ε goes to zero. Fig. 10.3.1 shows that this is indeed the case.   ▣

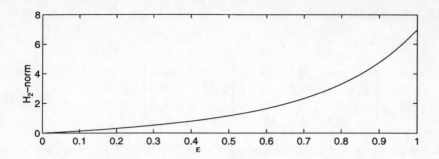

Figure 10.3.1: Plot of the H_2-norm of $T_{zw}(z, \varepsilon)$ with respect to ε.

10.4. Concluding Remarks

For general linear systems, we developed a direct eigenstructure assignment design procedure for constructing the state feedback laws that achieve perfect regulation. The heart of our design procedure is an appropriate parameterization of state feedback gain $F(\varepsilon)$ with a tuning parameter ε. The procedure used to construct the components of $F(\varepsilon)$ utilizes the eigenstructure assignment based low gain feedback design techniques of Chapter 2 and does not require explicit values of the parameter ε. In this sense, ε truly acts as a tuning parameter, and can be adjusted either off-line or on-line to achieve performance as close as required to the ideal design objective.

Chapter 11

Almost Disturbance Decoupling with Internal Stability for Linear and Nonlinear Systems

11.1. Introduction

Earlier in this monograph, we considered the problem of almost disturbance decoupling with internal stability for linear systems with saturating actuators. The problem was solved in a semi-global framework in Chapter 6 and in a global framework in Chapter 7.

In this chapter, we revisit the problem of H_∞ almost disturbance decoupling with internal stability for general linear systems, both continuous-time and discrete-time, and for a class of nonlinear systems. As discussed in Chapter 6, the problem of almost disturbance decoupling has a vast history behind it, occupying a central part of classical as well as modern control theory. Several important problems, such as robust control, decentralized control, non-interactive control, model reference or tracking control, H_2- and H_∞-optimal control problems can all be recast into an almost disturbance decoupling problem. Roughly speaking, the basic almost disturbance decoupling problem is to find an output feedback control law such that in the closed-loop system the disturbances are quenched, say in an L_p sense, up to any pre-specified degree of accuracy while maintaining internal stability. Such a problem was originally formulated by Willems [124,125] and labeled as ADDPMS (the almost disturbance decoupling problem with measurement feedback and internal stability). In the case that, instead of a measurement feedback, a state feedback is used, the above problem is termed as ADDPS (the almost disturbance decoupling problem with internal stability). The prefix H_∞ in the acronyms H_∞-ADDPMS and H_∞-ADDPS is used to specify that the degree of accuracy in disturbance quenching is measured in L_2-sense.

For continuous-time linear systems, there is extensive literature on the almost disturbance decoupling problem (see, for example, the recent works [82,83,122] and the references therein). In [122], several variations of the disturbance decoupling problems and their solvability conditions are summarized, and the necessary and sufficient conditions are given, under which the H_∞-ADDPMS and H_∞-ADDPS for continuous-time linear systems are solvable. These conditions are given in terms of geometry subspaces and for strictly proper systems (i.e., without direct feedthrough terms from the control input to the to be controlled output and from the disturbance input to the measurement output). Under these conditions, [82] constructs feedback laws, parameterized explicitly in a single parameter ε, that solve the H_∞-ADDPMS and the H_∞-ADDPS. These results were later extended to proper systems (i.e., with direct feedthrough terms) in [83]. We emphasize that in all the results mentioned above, the internal stability was always with respect a closed set in the complex plane. Such a closeness restriction, while facilitating the development of the above results, excludes systems with disturbance affected purely imaginary invariant zero dynamics from consideration. Only recently was this "final" restriction on the internal stability restriction removed by Scherer [97], thus allowing purely imaginary invariant zero dynamics to be affected by the disturbance. More specifically, Scherer [97] gave a set of necessary and sufficient conditions under which the H_∞-ADDPMS and the H_∞-ADDPS, with internal stability being with respect to the open left-half plane, is solvable for general proper linear systems. When the stability is with respect to the open left-half plane, the H_∞-ADDPMS and the H_∞-ADDPS will be referred to as the general H_∞-ADDPMS and the general H_∞-ADDPS, respectively. The explicit construction algorithm for feedback control laws that solve these general H_∞-ADDPMS and H_∞-ADDPS under Scherer's necessary and sufficient conditions still does not exist. One of the objectives of this chapter is to show how low gain feedback design techniques of Chapter 2 can be used to explicitly construct feedback laws that solve the general H_∞-ADDPMS.

For discrete-time linear systems, the problem of almost disturbance decoupling with internal stability is much less studied. Only recently has the necessary and sufficient conditions under which the H_∞-ADDPMS for general discrete-time systems been derived [11]. Although the problem considered in

[11] is general in that the system is allowed to have invariant zeros on the unit circle, the problem of constructing feedback laws that solve the H_∞-ADDPMS for discrete-time linear systems was unattempted. The second objective of this chapter is to demonstrate how discrete-time low gain feedback design technique of Chapter 2 can be utilized to explicitly construct feedback laws that solve the the H_∞-ADDPMS for general discrete-time systems, in which its subsystems are allowed to have invariant zeros on the unit circle.

Following the formulation for linear systems, the problem of almost disturbance decoupling with internal stability has also been formulated and solved for various classes of nonlinear systems, including linear systems subject to actuator saturation as addressed in Chapters 6 and 7 and nonlinear systems in certain normal forms. (see, for example, [25,26,74,75,90] and the references therein.) The third objective in this chapter is to utilize low gain feedback design technique of Chapter 2 to solve the H_∞-ADDPS for a new class of nonlinear systems. This result compliments some of the recent breakthroughs in the solution of the H_∞-ADDPS for nonlinear systems.

In the remainder of this chapter, Sections 11.2, 11.3 and 11.4 deal with continuous-time linear systems, discrete-time linear systems, and nonlinear systems, respectively. Finally, Section 11.5 contains some concluding remarks.

11.2. Continuous-Time Linear Systems

11.2.1. Problem Statement

Consider the following general continuous-time linear system,

$$\Sigma : \begin{cases} \dot{x} = A\,x + B\,u + E\,w, \\ y = C_1\,x \qquad\quad + D_1\,w, \\ z = C_2\,x + D_2\,u + D_{22}\,w, \end{cases} \tag{11.2.1}$$

where $x \in \mathbb{R}^n$ is the state, $u \in \mathbb{R}^m$ is the control input, $y \in \mathbb{R}^\ell$ is the measurement, $w \in \mathbb{R}^q$ is the disturbance and $z \in \mathbb{R}^p$ is the output to be controlled, and A, B, E, C_1, C_2, D_1, D_2, and D_{22} are constant matrices of appropriate dimensions. For convenient references in the future development, throughout this section, we define Σ_P to be the subsystem characterized by the matrix quadruple (A, B, C_2, D_2) and Σ_Q to be the subsystem characterized by the matrix quadruple (A, E, C_1, D_1). The following dynamic feedback control laws are investigated,

$$\Sigma_c : \begin{cases} \dot{x}_c = A_c\,x_c + B_c\,y, \\ u = C_c\,x_c + D_c\,y. \end{cases} \tag{11.2.2}$$

The controller Σ_c of (11.2.2) is said to be internally stabilizing when applied to the system Σ, if the following matrix is asymptotically stable,

$$A_{\mathrm{cl}} = \begin{bmatrix} A + BD_cC_1 & BC_c \\ B_cC_1 & A_c \end{bmatrix}, \tag{11.2.3}$$

i.e., all its eigenvalues lie in the open left-half complex plane. Denote by T_{zw} the corresponding closed-loop transfer matrix from the disturbance w to the to be controlled output z, i.e.,

$$T_{zw} = [\, C_2 + D_2D_cC_1 \quad D_2C_c \,] \left(sI - \begin{bmatrix} A + BD_cC_1 & BC_c \\ B_cC_1 & A_c \end{bmatrix} \right)^{-1}$$
$$\times \begin{bmatrix} E + BD_cD_1 \\ B_cD_1 \end{bmatrix} + D_2D_cD_1 + D_{22}. \tag{11.2.4}$$

The H_∞-norm of the transfer matrix T_{zw} is given by

$$\|T_{zw}\|_\infty = \sup_{\omega \in [0,\infty)} \sigma_{\max}[T_{zw}(j\omega)], \tag{11.2.5}$$

where $\sigma_{\max}[\cdot]$ denotes the largest singular value. Then the general H_∞-ADDPMS and the general H_∞-ADDPS can be formally defined as follows.

Problem 11.2.1. *The general H_∞ almost disturbance decoupling problem with measurement feedback and internal stability (the general H_∞-ADDPMS) for (11.2.1) is defined as follows. For any given positive scalar $\gamma > 0$, find a controller of the form (11.2.2) such that,*

1. *in the absence of disturbance, the closed-loop system comprising the system (11.2.1) and the controller (11.2.2) is asymptotically stable, i.e., the matrix A_{cl} as given by (11.2.3) is asymptotically stable;*

2. *the closed-loop system has an L_2-gain, from the disturbance w to the controlled output z, that is less than or equal to γ, i.e.,*

$$\|z\|_{L_2} \le \gamma \|w\|_{L_2}, \ \ \forall w \in L_2 \text{ and for } (x(0), x_c(0)) = (0,0). \tag{11.2.6}$$

Equivalently, the H_∞-norm of the closed-loop transfer matrix from w to z, T_{zw}, is less than or equal to γ, i.e., $\|T_{zw}\|_\infty \le \gamma$.

In the case that $C_1 = I$ and $D_1 = 0$, the general H_∞-ADDPMS as defined above becomes the general H_∞-ADDPS, where only a static state feedback, $u = Fx$, instead the dynamic output feedback (11.2.2) is necessary. $\boxed{\mathrm{P}}$

A set of necessary and sufficient conditions under which the above problems are solvable was recently given in [97]. To state these conditions, we need to define the following geometric subspaces.

Definition 11.2.1. (Geometric Subspaces) *Consider a linear time-invariant system Σ_* characterized by a matrix quadruple (A, B, C, D). The weakly unobservable subspaces of Σ_*, $\mathcal{V}^\times$, and the strongly controllable subspaces of Σ_*, $\mathcal{S}^\times$, are defined as follows:*

1. *$\mathcal{V}^\times(\Sigma_*)$ is the maximal subspace of $\mathbb{R}^n$ which is $(A + BF)$-invariant and contained in $\mathrm{Ker}\,(C + DF)$ such that the eigenvalues of $(A + BF)|\mathcal{V}^\times$ are contained in $\mathbb{C}^\times \subseteq \mathbb{C}$ for some constant matrix F.*

2. *$\mathcal{S}^\times(\Sigma_*)$ is the minimal $(A + LC)$-invariant subspace of $\mathbb{R}^n$ containing $\mathrm{Im}\,(B + LD)$ such that the eigenvalues of the map which is induced by $(A + LC)$ on the factor space $\mathbb{R}^n/\mathcal{S}^\times$ are contained in $\mathbb{C}^\times \subseteq \mathbb{C}$ for some constant matrix L.*

Furthermore, we denote $\mathcal{V}^- = \mathcal{V}^\times$ and $\mathcal{S}^- = \mathcal{S}^\times$, if $\mathbb{C}^\times = \mathbb{C}^- \cup \mathbb{C}^0$; $\mathcal{V}^+ = \mathcal{V}^\times$ and $\mathcal{S}^+ = \mathcal{S}^\times$, if $\mathbb{C}^\times = \mathbb{C}^+$; and finally $\mathcal{V}^* = \mathcal{V}^\times$ and $\mathcal{S}^* = \mathcal{S}^\times$, if $\mathbb{C}^\times = \mathbb{C}$. ▣

Definition 11.2.2. *Consider a linear system Σ_* characterized by a matrix quadruple (A, B, C, D). For any $\lambda \in \mathbb{C}$, we define*

$$S_\lambda(\Sigma_*) = \left\{ x \in \mathbb{C}^n \,\middle|\, \exists u \in \mathbb{C}^{n+m} : \begin{pmatrix} x \\ 0 \end{pmatrix} = \begin{bmatrix} A - \lambda I & B \\ C & D \end{bmatrix} u \right\} \qquad (11.2.7)$$

and

$$V_\lambda(\Sigma_*) = \left\{ x \in \mathbb{C}^n \,\middle|\, \exists u \in \mathbb{C}^m : 0 = \begin{bmatrix} A - \lambda I & B \\ C & D \end{bmatrix} \begin{pmatrix} x \\ u \end{pmatrix} \right\}. \qquad (11.2.8)$$

$V_\lambda(\Sigma_*)$ and $S_\lambda(\Sigma_*)$ are associated with the so-called state zero directions of Σ_* if λ is an invariant zero of Σ_*. ▣

The following results concerning the solvability of the general H_∞-ADDPMS and the general H_∞-ADDPS are recalled from [97].

Theorem 11.2.1. *Consider the general measurement feedback system (11.2.1) with $D_{22} = 0$. Then the general H_∞ almost disturbance decoupling problem for (11.2.1) with internal stability (the general H_∞-ADDPMS) is solvable, if and only if the following conditions are satisfied:*

1. *(A, B) is stabilizable;*

2. (A, C_1) is detectable;

3. $\text{Im}(E) \subset \mathcal{S}^+(\Sigma_P) \cap \{\cap_{\lambda \in \mathbb{C}^0} \mathcal{S}_\lambda(\Sigma_P)\}$;

4. $\text{Ker}(C_2) \supset \mathcal{V}^+(\Sigma_Q) \cup \{\cup_{\lambda \in \mathbb{C}^0} \mathcal{V}_\lambda(\Sigma_Q)\}$;

5. $\mathcal{V}^+(\Sigma_Q) \subset \mathcal{S}^+(\Sigma_P)$. ▦

It is simple to verify that for the case that all states of the system (11.2.1) are fully measurable, i.e., $C_1 = I$ and $D_1 = 0$, then the solvability conditions for the general H_∞-ADDPS reduce to the following: 1) (A, B) is stabilizable; 2) $D_{22} = 0$; and 3) $\text{Im}(E) \subset \mathcal{S}^+(\Sigma_P) \cap \{\cap_{\lambda \in \mathbb{C}^0} \mathcal{S}_\lambda(\Sigma_P)\}$. Moreover, in this case, a static state feedback control law, i.e., $u = Fx$, where F is a constant matrix and might be parameterized by certain tuning parameters, exists that solves the general H_∞-ADDPS.

The objective of this section is to demonstrate how the eigenstructure assignment based low gain feedback design technique of Section 2.2.1 can be employed to construct families of feedback control laws of the form (11.2.2), parameterized in a single parameter, say ε, that, under the necessary and sufficient conditions of Theorem 11.2.1, solve the above defined general H_∞-ADDPMS and H_∞-ADDPS for general systems whose subsystems Σ_P and Σ_Q may have invariant zeros on the imaginary axis.

11.2.2. Solutions to the General H_∞-ADDPS

The general H_∞-ADDPMS is solved by explicit construction of feedback laws. The feedback laws we are to construct are observer-based. A family of static state feedback control laws parameterized in a single parameter is first constructed to solve the general H_∞-ADDPS. A class of observers parameterized in the same parameter ε is then constructed to implement the state feedback control laws and thus obtain a family of dynamic measurement feedback control laws parameterized in a single parameter ε that solve the general H_∞-ADDPMS. The observer gains for this family of observers are constructed by applying the algorithm for constructing state feedback gains to the dual system. Besides the low gain feedback design technique of Chapter 2, another basic tool we use in the construction of such families of feedback control laws is the special coordinate basis (SCB) [92,95], in which a linear system is decomposed into several subsystems corresponding to its finite and infinite zero structures as well as its invertibility structures. A summary of SCB and its properties is given in Appendix A.

Since our objective here is to demonstrate how low gain feedback design technique can be used in the construction of observer based feedback laws that solve the general H_∞-ADDPMS and since both the state feedback gain matrix and the observer gain matrix can be constructed using the same algorithm, we will only present the algorithm for explicitly constructing a family of state feedback laws that solves the general H_∞-ADDPS. The solution to the general H_∞-ADDPMS can be found in [12].

More specifically, we present a design procedure that constructs a family of parameterized static state feedback control laws,

$$u = F(\varepsilon)x, \tag{11.2.9}$$

that solves the general H_∞-ADDPS for the following system,

$$\begin{cases} \dot{x} = A\,x + B\,u + E\,w, \\ y = x, \\ z = C_2\,x + D_2\,u + D_{22}\,w. \end{cases} \tag{11.2.10}$$

That is, under this family of state feedback control laws, the resulting closed-loop system is asymptotically stable for sufficiently small ε and the H_∞-norm of the closed-loop transfer matrix from w to z, $T_{zw}(s,\varepsilon)$, tends to zero as ε tends to zero, where

$$T_{zw}(s,\varepsilon) = [C_2 + D_2 F(\varepsilon)][sI - A - BF(\varepsilon)]^{-1}E + D_{22}. \tag{11.2.11}$$

Clearly, $D_{22} = 0$ is a necessary condition for the solvability of the general H_∞-ADDPS.

The algorithm for constructing a family of state feedback laws that solves the H_∞-ADDPS consists of the following six steps.

Step 1: Decomposition of Σ_P. Transform the subsystem Σ_P, i.e., the quadruple (A, B, C_2, D_2), into the special coordinate basis (SCB) by Theorem A.1.1 of the Appendix. Denote the state, output and input transformation matrices as Γ_{SP}, Γ_{OP} and Γ_{IP}, respectively.

Step 2: Gain matrix for the subsystem associated with $\mathcal{X}_c$. Let F_c be any arbitrary $m_c \times n_c$ matrix subject to the constraint that

$$A_{cc}^c = A_{cc} + B_c F_c, \tag{11.2.12}$$

is an asymptotically stable matrix. Note that the existence of such an F_c is guaranteed by the property of SCB, i.e., (A_{cc}, B_c) is controllable.

Step 3: Gain matrix for the subsystems associated with $\mathcal{X}_a^+$ and $\mathcal{X}_b$. Let

$$
F_{ab}^+ = \begin{bmatrix} F_{a0}^+ & F_{b0} \\ F_{ad}^+ & F_{bd} \end{bmatrix}, \tag{11.2.13}
$$

be any arbitrary $(m_0 + m_d) \times (n_a^+ + n_b)$ matrix subject to the constraint that

$$
A_{ab}^{+c} = \begin{bmatrix} A_{aa}^+ & L_{ab}^+ C_b \\ 0 & A_{bb} \end{bmatrix} + \begin{bmatrix} B_{0a}^+ & L_{ad}^+ \\ B_{0b} & L_{bd} \end{bmatrix} F_{ab}^+ \tag{11.2.14}
$$

is an asymptotically stable matrix. Again, note that the existence of such an F_{ab}^+ is guaranteed by the stabilizability of (A, B) and Property A.1.1 of the special coordinate basis. For future use, let us partition $[\, F_{ad}^+ \;\; F_{bd}\,]$ as,

$$
[\, F_{ad}^+ \;\; F_{bd}\,] = \begin{bmatrix} F_{ad1}^+ & F_{bd1} \\ F_{ad2}^+ & F_{bd2} \\ \vdots & \vdots \\ F_{adm_d}^+ & F_{bdm_d} \end{bmatrix}, \tag{11.2.15}
$$

where F_{adi}^+ and F_{bdi} are of dimensions $1 \times n_a^+$ and $1 \times n_b$, respectively.

Step 4: Gain matrix for the subsystem associated with $\mathcal{X}_a^0$. The construction of this gain matrix is carried out in the following sub-steps.

Step 4.1: Preliminary coordinate transformation. Recalling the definition of $(A_{\mathrm{con}}, B_{\mathrm{con}})$, i.e., (A.1.28), we have

$$
A_{\mathrm{con}} + B_{\mathrm{con}} F_{ab}^+ = \begin{bmatrix} A_{aa}^- & 0 & A_{aab}^- \\ 0 & A_{aa}^0 & A_{aab}^0 \\ 0 & 0 & A_{ab}^{+c} \end{bmatrix}, \quad B_{\mathrm{con}} = \begin{bmatrix} B_{0a}^- & L_{ad}^- \\ B_{0a}^0 & L_{ad}^0 \\ B_{0ab}^+ & L_{abd}^+ \end{bmatrix},
$$

where

$$
B_{0ab}^+ = \begin{bmatrix} B_{0a}^+ \\ B_{0b} \end{bmatrix}, \quad L_{abd}^+ = \begin{bmatrix} L_{ad}^+ \\ L_{bd} \end{bmatrix},
$$
$$
A_{aab}^0 = [\, 0 \;\; L_{ab}^0 C_b\,] + [\, B_{0a}^0 \;\; L_{ad}^0\,] F_{ab}^+,
$$

and

$$
A_{aab}^- = [\, 0 \;\; L_{ab}^- C_b\,] + [\, B_{0a}^- \;\; L_{ad}^-\,] F_{ab}^+.
$$

Clearly $(A_{\mathrm{con}} + B_{\mathrm{con}} F_{ab}^+, B_{\mathrm{con}})$ remains stabilizable. Construct the following nonsingular transformation matrix,

$$
\Gamma_{ab} = \begin{bmatrix} I_{n_a^-} & 0 & 0 \\ 0 & 0 & I_{n_a^+ + n_b} \\ 0 & I_{n_a^0} & T_a^0 \end{bmatrix}^{-1}, \tag{11.2.16}
$$

where T_a^0 is the unique solution to the following Lyapunov equation,

$$A_{aa}^0 T_a^0 - T_a^0 A_{ab}^{+c} = A_{aab}^0. \qquad (11.2.17)$$

We note here that such a unique solution to the above Lyapunov equation always exists since all the eigenvalues of A_{aa}^0 are on the imaginary axis and all the eigenvalues of A_{ab}^{+c} are in the open left-half plane. It is now easy to verify that

$$\Gamma_{ab}^{-1}(A_{\text{con}} + B_{\text{con}} F_{ab}^+) \Gamma_{ab} = \begin{bmatrix} A_{aa}^- & A_{aab}^- & 0 \\ 0 & A_{ab}^{+c} & 0 \\ 0 & 0 & A_{aa}^0 \end{bmatrix}, \qquad (11.2.18)$$

$$\Gamma_{ab}^{-1} B_{\text{con}} = \begin{bmatrix} B_{0a}^- & L_{ad}^- \\ B_{0ab}^+ & L_{abd}^+ \\ B_{0a}^0 + T_a^0 B_{0ab}^+ & L_{ad}^0 + T_a^0 L_{abd}^+ \end{bmatrix}. \qquad (11.2.19)$$

Hence, the matrix pair (A_{aa}^0, B_a^0) is controllable, where

$$B_a^0 = [\, B_{0a}^0 + T_a^0 B_{0ab}^+ \quad L_{ad}^0 + T_a^0 L_{abd}^+ \,].$$

Step 4.2: Further coordinate transformation. Find the nonsingular transformation matrices Γ_{sa}^0 and Γ_{Ia}^0 such that (A_{aa}^0, B_a^0) can be transformed into the block diagonal control canonical form,

$$(\Gamma_{sa}^0)^{-1} A_{aa}^0 \Gamma_{sa}^0 = \begin{bmatrix} A_1 & 0 & \cdots & 0 \\ 0 & A_2 & \cdots & 0 \\ \vdots & \vdots & \ddots & \vdots \\ 0 & 0 & \cdots & A_l \end{bmatrix},$$

$$(\Gamma_{sa}^0)^{-1} B_a^0 \Gamma_{Ia}^0 = \begin{bmatrix} B_1 & B_{12} & \cdots & B_{1l} & \star \\ 0 & B_2 & \cdots & B_{2l} & \star \\ \vdots & \vdots & \ddots & \vdots & \vdots \\ 0 & 0 & \cdots & B_l & \star \end{bmatrix},$$

where l is an integer and for $i = 1, 2, \cdots, l$,

$$A_i = \begin{bmatrix} 0 & 1 & 0 & \cdots & 0 \\ 0 & 0 & 1 & \cdots & 0 \\ \vdots & \vdots & \vdots & \ddots & \vdots \\ 0 & 0 & 0 & \cdots & 1 \\ -a_{n_i}^i & -a_{n_i-1}^i & -a_{n_i-2}^i & \cdots & -a_1^i \end{bmatrix}, \quad B_i = \begin{bmatrix} 0 \\ 0 \\ \vdots \\ 0 \\ 1 \end{bmatrix}.$$

We note that all the eigenvalues of A_i are on the imaginary axis. Here the $\star$'s represent sub-matrices of less interest. We also note that the existence of the above canonical form was shown in Wonham [126] while its software realization can be found in Chen [9].

Step 4.3: Subsystem design. For each (A_i, B_i), let $F_i(\varepsilon) \in \mathbb{R}^{1 \times n_i}$ be the state feedback gain such that

$$\lambda(A_i + B_i F_i(\varepsilon)) = -\varepsilon + \lambda(A_i) \in \mathbb{C}^-. \tag{11.2.20}$$

Note that $F_i(\varepsilon)$ is unique.

Step 4.4: Composition of gain matrix for subsystem associated with $\mathcal{X}_a^0$. Let

$$F_a^0(\varepsilon) = \Gamma_{Ia}^0 \begin{bmatrix} F_1(\varepsilon) & 0 & \cdots & 0 & 0 \\ 0 & F_2(\varepsilon) & \cdots & 0 & 0 \\ \vdots & \vdots & \ddots & \vdots & \vdots \\ 0 & 0 & \cdots & F_{l-1}(\varepsilon) & 0 \\ 0 & 0 & \cdots & 0 & F_l(\varepsilon) \end{bmatrix} (\Gamma_{sa}^0)^{-1}, \tag{11.2.21}$$

where $\varepsilon \in (0, 1]$ is a design parameter whose value is to be specified later.

We recall that the above Steps 4.2-4.4 are the application to the pair (A_a^0, B_a^0) of the eigenstructure assignment based low gain design procedure of Section 2.2.1. Clearly, we have

$$|F_a^0(\varepsilon)| \le f_a^0 \varepsilon, \ \ \varepsilon \in (0, 1], \tag{11.2.22}$$

for some positive constant f_a^0, independent of ε. For future use, we define and partition $F_{ab}(\varepsilon) \in \mathbb{R}^{(m_0 + m_d) \times (n_a + n_b)}$ as

$$F_{ab}(\varepsilon) = \begin{bmatrix} F_{ab0}(\varepsilon) \\ F_{abd}(\varepsilon) \end{bmatrix} = \begin{bmatrix} 0_{m_0 \times n_a^-} & 0_{m_0 \times (n_a^+ + n_b)} & F_{a0}^0(\varepsilon) \\ 0_{m_d \times n_a^-} & 0_{m_d \times (n_a^+ + n_b)} & F_{ad}^0(\varepsilon) \end{bmatrix} \Gamma_{ab}^{-1}, \tag{11.2.23}$$

and

$$F_{abd}(\varepsilon) = \begin{bmatrix} F_{abd1}(\varepsilon) \\ F_{abd2}(\varepsilon) \\ \vdots \\ F_{abdm_d}(\varepsilon) \end{bmatrix}, \tag{11.2.24}$$

where $F_{a0}^0(\varepsilon)$ and $F_{ad}^0(\varepsilon)$ are defined as

$$F_a^0(\varepsilon) = \begin{bmatrix} F_{a0}^0(\varepsilon) \\ F_{ad}^0(\varepsilon) \end{bmatrix}. \tag{11.2.25}$$

We also partition $F_{ad}^0(\varepsilon)$ as,

$$F_{ad}^0(\varepsilon) = \begin{bmatrix} F_{ad1}^0(\varepsilon) \\ F_{ad2}^0(\varepsilon) \\ \vdots \\ F_{adm_d}^0(\varepsilon) \end{bmatrix}. \tag{11.2.26}$$

Step 5: Gain matrix for the subsystem associated with $\mathcal{X}_d$. This step makes use of subsystems, $i = 1$ to m_d, represented by (A.1.16) of the Appendix. Let $\Lambda_i = \{\lambda_{i1}, \lambda_{i2}, \cdots, \lambda_{iq_i}\}$, $i = 1$ to m_d, be the sets of q_i elements all in $\mathbf{C}^-$, which are closed under complex conjugation, where q_i and m_d are as defined in Theorem A.1.1 but associated with the special coordinate basis of Σ_P. Let $\Lambda_d = \Lambda_1 \cup \Lambda_2 \cup \cdots \cup \Lambda_{m_d}$. For $i = 1$ to m_d, we define

$$p_i(s) = \prod_{j=1}^{q_i} (s - \lambda_{ij}) = s^{q_i} + F_{i1} s^{q_i-1} + \cdots + F_{iq_i-1} s + F_{iq_i}, \quad (11.2.27)$$

and

$$\tilde{F}_i(\varepsilon) = \frac{1}{\varepsilon^{q_i}} F_i S_i(\varepsilon), \quad (11.2.28)$$

where

$$F_i = [\, F_{iq_i} \quad F_{iq_i-1} \quad \cdots \quad F_{i1}\,], \quad S_i(\varepsilon) = \operatorname{diag}\left\{1, \varepsilon, \varepsilon^2, \cdots, \varepsilon^{q_i-1}\right\}, \quad (11.2.29)$$

Step 6: Composition of parameterized gain matrix $F(\varepsilon)$. In this step, various gains calculated in Steps 2 to 5 are put together to form a composite state feedback gain matrix $F(\varepsilon)$. Let

$$\tilde{F}_{abd}(\varepsilon) = \begin{bmatrix} F_{abd1}(\varepsilon) F_{1q_1} / \varepsilon^{q_1} \\ F_{abd2}(\varepsilon) F_{2q_2} / \varepsilon^{q_2} \\ \vdots \\ F_{abdm_d}(\varepsilon) F_{m_d q_{m_d}} / \varepsilon^{q_{m_d}} \end{bmatrix}, \quad (11.2.30)$$

$$\tilde{F}_{ad}^+(\varepsilon) = \begin{bmatrix} F_{ad1}^+ F_{1q_1} / \varepsilon^{q_1} \\ F_{ad2}^+ F_{2q_2} / \varepsilon^{q_2} \\ \vdots \\ F_{adm_d}^+ F_{m_d q_{m_d}} / \varepsilon^{q_{m_d}} \end{bmatrix}, \quad (11.2.31)$$

and

$$\tilde{F}_{bd}(\varepsilon) = \begin{bmatrix} F_{bd1} F_{1q_1} / \varepsilon^{q_1} \\ F_{bd2} F_{2q_2} / \varepsilon^{q_2} \\ \vdots \\ F_{bdm_d} F_{m_d q_{m_d}} / \varepsilon^{q_{m_d}} \end{bmatrix}. \quad (11.2.32)$$

Then define the state feedback gain $F(\varepsilon)$ as

$$F(\varepsilon) = -\Gamma_{iP} \left(\tilde{F}_{abcd}^\star(\varepsilon) + \tilde{F}_{abcd}(\varepsilon) \right) \Gamma_{sP}^{-1}, \quad (11.2.33)$$

where

$$\tilde{F}^{\star}_{abcd}(\varepsilon)=$$
$$\begin{bmatrix} C^-_{0a} & C^0_{0a} & C^+_{0a} - F^+_{a0} & C_{0b} - F_{b0} & C_{0c} & C_{0d} \\ E^-_{da} & E^0_{da} & E^+_{da} - \tilde{F}^+_{ad}(\varepsilon) & E_{db} - \tilde{F}_{bd}(\varepsilon) & E_{dc} & E_d - \tilde{F}_d(\varepsilon) \\ E^-_{ca} & E^0_{ca} & E^+_{ca} & 0 & -F_c & 0 \end{bmatrix},$$

$$(11.2.34)$$

$$\tilde{F}_{abcd}(\varepsilon) = \begin{bmatrix} F_{ab0}(\varepsilon) & 0 & 0 \\ \tilde{F}_{abd}(\varepsilon) & 0 & 0 \\ 0 & 0 & 0 \end{bmatrix}, \qquad (11.2.35)$$

and where

$$E_d = \begin{bmatrix} E_{11} & \cdots & E_{1m_d} \\ \vdots & \ddots & \vdots \\ E_{m_d 1} & \cdots & E_{m_d m_d} \end{bmatrix}, \qquad (11.2.36)$$

$$\tilde{F}_d(\varepsilon) = \mathrm{diag}\Big\{ \tilde{F}_1(\varepsilon), \ \tilde{F}_2(\varepsilon), \ \cdots, \ \tilde{F}_{m_d}(\varepsilon) \Big\}. \qquad (11.2.37)$$

Ⓐ

The following theorem then shows that the above algorithm indeed yields a family of state feedback laws that solves the general H_∞-ADDPS.

Theorem 11.2.2. *Consider the given system (11.2.10) satisfying the following conditions: 1) (A, B) is stabilizable; 2) $D_{22} = 0$; and 3) $\mathrm{Im}\,(E) \subset \mathcal{S}^+(\Sigma_{\mathrm{P}}) \cap \{\cap_{\lambda \in \mathbf{C}^0} \mathcal{S}_\lambda(\Sigma_{\mathrm{P}})\}$. Then the closed-loop system comprising (11.2.10) and the static state feedback control law $u = F(\varepsilon)x$, with $F(\varepsilon)$ given by (11.2.33), has the following properties: For any given $\gamma > 0$, there exists a positive scalar $\varepsilon^* > 0$ such that for all $\varepsilon \in (0, \varepsilon^*]$,*

1. *the closed-loop system is asymptotically stable, i.e., $\lambda(A + BF(\varepsilon)) \subset \mathbf{C}^-$; and*

2. *the H_∞-norm of the closed-loop transfer matrix from the disturbance w to the controlled output z is less than or equal to γ, i.e., $\|T_{zw}(s, \varepsilon)\|_\infty \leq \gamma$.*

Hence, the family of control laws $u = F(\varepsilon)x$ solves the general H_∞-ADDPMS for (11.2.10). Ⓣ

Proof of Theorem 11.2.2. Under the feedback control law $u = F(\varepsilon)x$, the closed-loop system on the special coordinate basis can be written as follows,

$$\dot{x}^-_a = A^-_{aa}x^-_a + B^-_{0a}z_0 + L^-_{ad}z_d + L^-_{ab}z_b + E^-_a w, \qquad (11.2.38)$$

$$\dot{x}_a^0 = A_{aa}^0 x_a^0 + B_{0a}^0 z_0 + L_{ad}^0 z_d + L_{ab}^0 z_b + E_a^0 w, \tag{11.2.39}$$

$$\dot{x}_{ab}^+ = A_{ab}^{+c} x_{ab}^+ + B_{0ab}^+ F_{a0}^0(\varepsilon)[x_a^0 + T_a^0 x_{ab}^+] + L_{abd}^+[F_{ad}^+, F_{bd}] x_{ab}^+ + L_{abd}^+ z_d$$
$$+ E_{ab}^+ w, \quad z_b = [0_{m_b \times n_a^+}, \ C_b] x_{ab}^+, \tag{11.2.40}$$

$$\dot{x}_c = A_{cc}^c + B_{0c} z_0 + L_{cb} z_b + L_{cd} z_d + E_c w, \tag{11.2.41}$$

$$z_0 = [F_{a0}^+, \ F_{b0}] x_{ab}^+ + F_{a0}^0(\varepsilon)(x_a^0 + T_a^0 x_{ab}^+), \tag{11.2.42}$$

$$\dot{x}_i = A_{q_i} x_i + B_{0id} z_0 + L_{id} z_d + \frac{1}{\varepsilon^{q_i}} B_{q_i} \Big[F_{adi}^+ F_{iq_i} x_a^+ + F_{bdi} F_{iq_i} x_b$$
$$+ F_{adi}^0(\varepsilon) F_{iq_i} [x_a^0 + T_a^0 x_{ab}^+] + F_i S_i(\varepsilon) x_i \Big] + E_i w, \tag{11.2.43}$$

$$z_i = C_{q_i} x_i, \quad i = 1, 2, \cdots, m_d, \tag{11.2.44}$$

where $x_{ab}^+ = [(x_a^+)', x_b']'$ and B_{0ab}^+ and L_{abd}^+ are as defined in Step 4.1 of the state feedback design algorithm. We have also used Condition 2 of the theorem, i.e., $D_{22} = 0$, and E_a^-, E_a^0, E_{ab}^+, E_b, E_c and E_i, $i = 1, 2, \cdots, m_d$, are defined as follows,

$$\Gamma_{\mathrm{SP}}^{-1} E = [(E_a^-)' \quad (E_a^0)' \quad (E_{ab}^+)' \quad E_c' \quad E_1' \quad E_2' \quad \cdots \quad E_{m_d}']'. \tag{11.2.45}$$

Condition 3 of the theorem then implies that

$$E_{ab}^+ = 0, \tag{11.2.46}$$

and

$$\mathrm{Im}(E_a^0) \subset \mathcal{S}(A_{aa}^0) = \cap_{\omega \in \lambda(A_{aa}^0)} \mathrm{Im}\{\omega I - A_{aa}^0\}. \tag{11.2.47}$$

To complete the proof, we will make two state transformations on the closed-loop system (11.2.38)-(11.2.44). The first state transformation is given as follows,

$$\bar{x}_{ab} = \Gamma_{ab}^{-1} x_{ab}, \quad \bar{x}_c = x_c, \tag{11.2.48}$$

$$\bar{x}_{i1} = x_{i1} - F_{adi}^+ x_a^+ - F_{bdi} x_b - F_{adi}^0(\varepsilon)[x_a^0 + T_a^0 x_{ab}^+],$$
$$i = 1, 2, \cdots, m_d, \tag{11.2.49}$$

$$\bar{x}_{ij} = x_{ij}, \ j = 2, 3, \cdots, q_i, \ i = 1, 2, \cdots, m_d, \tag{11.2.50}$$

where $x_{ab} = [(x_a^-)', (x_a^0)', (x_{ab}^+)']'$ and $\bar{x}_{ab} = [(\bar{x}_a^-)', (\bar{x}_{ab}^+)', (\bar{x}_a^0)']'$. In the new state variables (11.2.48)-(11.2.50), the closed-loop system becomes,

$$\dot{\bar{x}}_a^- = A_{aa}^- \bar{x}_a^- + A_{aab}^- \bar{x}_{ab}^+ + [B_{0a}^-, \ L_{ad}^-] F_a^0(\varepsilon) \bar{x}_a^0 + L_{ad}^- \bar{z}_d + E_a^- w, \tag{11.2.51}$$

$$\dot{\bar{x}}_{ab}^+ = A_{ab}^{+c} \bar{x}_{ab}^+ + [B_{0ab}^+, \ L_{abd}^+] F_a^0(\varepsilon) \bar{x}_a^0 + L_{abd}^+ \bar{z}_d,$$
$$z_b = [0_{m_b \times n_a^+}, \ C_b] \bar{x}_{ab}^+, \tag{11.2.52}$$

$$\dot{\bar{x}}_a^0 = (A_{aa}^0 + B_a^0 F_a^0(\varepsilon)) \bar{x}_a^0 + (L_{ad}^0 + T_a^0 L_{abd}^+) \bar{z}_d + E_a^0 w, \tag{11.2.53}$$

$$\dot{\bar{x}}_c = A_{cc}^c \bar{x}_c + \left(L_{cb}[0, C_b] + [B_{0c}, L_{cd}]F_{ab}^+ \right)\bar{x}_{ab}^+ + [B_{0c}, L_{cd}]F_a^0(\varepsilon)\bar{x}_a^0$$
$$+ L_{cd}\bar{z}_d + E_c w, \tag{11.2.54}$$

$$z_0 = [F_{a0}^+, F_{b0}]\bar{x}_{ab}^+ + F_{a0}^0(\varepsilon)\bar{x}_a^0, \tag{11.2.55}$$

$$\dot{\bar{x}}_i = A_{q_i}\bar{x}_i + \frac{1}{\varepsilon^{q_i}}B_{q_i}F_i S_i(\varepsilon)\bar{x}_i + L_{iab}^+(\varepsilon)\bar{x}_{ab}^+ + L_{ia}^{01}(\varepsilon)F_a^0(\varepsilon)\bar{x}_a^0$$
$$+ L_{ia}^{02}(\varepsilon)F_a^0(\varepsilon)A_{aa}^0\bar{x}_a^0 + \bar{L}_{id}(\varepsilon)\bar{z}_d + \bar{E}_i(\varepsilon)w, \tag{11.2.56}$$

$$\bar{z}_i = z_i - [F_{adi}^+, F_{bdi}]\bar{x}_{ab}^+ - F_{adi}^0\bar{x}_a^0 = C_{q_i}\bar{x}_i,$$
$$i = 1, 2, \cdots, m_d, \tag{11.2.57}$$

$$\bar{z}_d = [\bar{z}_1, \bar{z}_2, \cdots, \bar{z}_{m_d}]', \tag{11.2.58}$$

where A_{aab}^-, A_{aab}^0, B_a^0 and L_{abd}^+ are as defined in Step 4.1 of the state feedback control law design algorithm, and $L_{iab}^+(\varepsilon)$, $L_{ia}^{01}(\varepsilon)$, $L_{ia}^{02}(\varepsilon)$, $\bar{L}_{id}(\varepsilon)$ and $\bar{E}_i(\varepsilon)$ are defined in an obvious way and, by (11.2.22) satisfy

$$|L_{iab}^+(\varepsilon)| \le l_{iab}^+, \quad |L_{ia}^{01}(\varepsilon)| \le l_{ia}^{01}, \quad |L_{ia}^{02}(\varepsilon)| \le l_{ia}^{02}, \quad |\bar{L}_{id}(\varepsilon)| \le \bar{l}_{id},$$
$$|\bar{E}_i(\varepsilon)| \le \bar{e}_i, \quad \varepsilon \in (0, 1], \tag{11.2.59}$$

for some nonnegative constants l_{iab}^+, l_{ia}^{01}, l_{ia}^{02}, $\bar{l}_{id}$, and $\bar{e}_i$ independent of ε.

We now define the following second state transformation on the closed-loop system,

$$\tilde{x}_a^- = \bar{x}_a^-, \quad \tilde{x}_{ab}^+ = \bar{x}_{ab}^+, \tag{11.2.60}$$

$$\tilde{x}_a^0 = [(\tilde{x}_{a1}^0)', (\tilde{x}_{a2}^0)', \cdots, (\tilde{x}_{al}^0)']' = S_a(\varepsilon)Q_a^{-1}(\varepsilon)(\Gamma_{sa}^0)^{-1}\bar{x}_a^0, \tag{11.2.61}$$

$$S_a(\varepsilon) = \text{blkdiag}\{S_{a1}(\varepsilon), S_{a2}(\varepsilon), \cdots, S_{al}(\varepsilon)\},$$
$$Q_a(\varepsilon) = \text{blkdiag}\{Q_{a1}(\varepsilon), Q_{a2}(\varepsilon), \cdots, Q_{al}(\varepsilon)\},$$

$$\tilde{x}_c = \varepsilon\bar{x}_c, \tag{11.2.62}$$

$$\tilde{x}_d = [\tilde{x}_1', \tilde{x}_2', \cdots, \tilde{x}_{m_d}']', \quad \tilde{x}_i = S_i(\varepsilon)\bar{x}_i, \ i = 1, 2, \cdots, m_d, \tag{11.2.63}$$

where $Q_{ai}(\varepsilon)$ and $S_{ai}(\varepsilon)$ are the $Q(\varepsilon)$ and $S(\varepsilon)$ of Lemmas 2.2.2 and 2.2.3 for the triple $(A_i, B_i, F_i(\varepsilon))$. Hence, Lemmas 2.2.2-2.2.5 all apply. In these new state variables, the closed-loop system becomes,

$$\dot{\tilde{x}}_a^- = A_{aa}^-\tilde{x}_a^- + A_{aab}^-(\varepsilon)\tilde{x}_{ab}^+ + A_{aa}^{-0}(\varepsilon)\tilde{x}_a^0 + L_{ad}^-\bar{z}_d + E_a^- w, \tag{11.2.64}$$

$$\dot{\tilde{x}}_{ab}^+ = A_{ab}^{+c}\tilde{x}_{ab}^+ + A_{aba}^{+0}(\varepsilon)\tilde{x}_a^0 + L_{abd}^+\bar{z}_d, \quad z_b = [0_{m_b \times n_a^+}, C_b]\tilde{x}_{ab}^+, \tag{11.2.65}$$

$$\dot{\tilde{x}}_a^0 = \tilde{J}_a(\varepsilon)\tilde{x}_a^0 + \tilde{B}_a(\varepsilon)\tilde{x}_a^0 + \tilde{L}_{ad}^0(\varepsilon)\bar{z}_d + \tilde{E}_a^0(\varepsilon)w, \tag{11.2.66}$$

$$\dot{\tilde{x}}_c = A_{cc}^c\tilde{x}_c + \varepsilon[A_{cab}^+\tilde{x}_{ab}^+ + A_{ca}^0(\varepsilon)\tilde{x}_a^0 + L_{cd}\bar{z}_d + E_c w], \tag{11.2.67}$$

$$z_0 = [F_{a0}^+, F_{b0}]\tilde{x}_{ab}^+ + \tilde{F}_{a0}^0(\varepsilon)\tilde{x}_a^0, \tag{11.2.68}$$

$$\varepsilon \dot{\tilde{x}}_i = (A_{q_i} + B_{q_i} F_i)\tilde{x}_i + \varepsilon \tilde{L}_{iab}^+(\varepsilon)\tilde{x}_{ab}^+ + \varepsilon \tilde{L}_{ia}^0(\varepsilon)\tilde{x}_a^0 + \varepsilon \tilde{L}_{id}(\varepsilon)\tilde{z}_d$$
$$+ \varepsilon \tilde{E}_i(\varepsilon)w,$$

$$\tilde{z}_i = \bar{z}_i = z_i - [F_{adi}^+, F_{bdi}]\tilde{x}_{ab}^+ - \tilde{F}_{adi}^0(\varepsilon)\tilde{x}_a^0 = C_{q_i}\tilde{x}_i, \qquad (11.2.69)$$

$$\tilde{z}_d = [\tilde{z}_1, \tilde{z}_2, \cdots, \tilde{z}_{m_d}]', \qquad (11.2.70)$$

where

$$A_{aa}^{-0}(\varepsilon) = [B_{0a}^+, L_{ad}^-]F_a^0(\varepsilon)\Gamma_{sa}^0 Q_a(\varepsilon)S_a^{-1}(\varepsilon), \qquad (11.2.71)$$

$$A_{aba}^{+0}(\varepsilon) = [B_{0ab}^+, L_{abd}^+]F_a^0(\varepsilon)\Gamma_{sa}^0 Q_a(\varepsilon)S_a^{-1}(\varepsilon), \qquad (11.2.72)$$

$$\tilde{J}_a(\varepsilon) = \text{blkdiag}\left\{\varepsilon \tilde{J}_{a1}(\varepsilon), \varepsilon \tilde{J}_{a2}(\varepsilon), \cdots, \varepsilon \tilde{J}_{al}(\varepsilon)\right\}, \qquad (11.2.73)$$

$$\tilde{B}_a(\varepsilon) = \begin{bmatrix} 0 & \tilde{B}_{12}(\varepsilon) & \tilde{B}_{13}(\varepsilon) & \cdots & \tilde{B}_{1l}(\varepsilon) \\ 0 & 0 & \tilde{B}_{23}(\varepsilon) & \cdots & \tilde{B}_{2l}(\varepsilon) \\ \vdots & \vdots & \vdots & \ddots & \vdots \\ 0 & 0 & 0 & \cdots & 0 \end{bmatrix}, \qquad (11.2.74)$$

$$\tilde{B}_{ij}(\varepsilon) = S_{ai}(\varepsilon)Q_{ai}^{-1}(\varepsilon)B_{ij}F_j(\varepsilon)Q_{aj}(\varepsilon)S_{aj}^{-1}(\varepsilon),$$
$$i = 1, 2, \cdots, l, j = i+1, i+2, \cdots, l,$$

$$\tilde{L}_{ad}^0(\varepsilon) = S_a(\varepsilon)Q_a^{-1}(\varepsilon)(\Gamma_{sa}^0)^{-1}(L_{ad}^0 + T_a^0 L_{abd}^+), \qquad (11.2.75)$$

$$\tilde{E}_a^0(\varepsilon) = S_a(\varepsilon)Q_a^{-1}(\varepsilon)(\Gamma_{sa}^0)^{-1}E_a^0,$$
$$\tilde{E}_a^0(\varepsilon) = [(\tilde{E}_{a1}^0(\varepsilon))' \quad (\tilde{E}_{a2}^0(\varepsilon))' \quad \cdots \quad (\tilde{E}_{al}^0(\varepsilon))']', \qquad (11.2.76)$$

$$A_{cab}^+ = [L_{cb}[0, C_b] - [B_{0c}, L_{cd}]F_{ab}^+, \qquad (11.2.77)$$

$$A_{ca}^0(\varepsilon) = [B_{0c}, L_{cd}]F_a^0(\varepsilon)\Gamma_{sa}^0 Q_a(\varepsilon)S_a^{-1}(\varepsilon), \qquad (11.2.78)$$

$$\tilde{F}_{a0}^0(\varepsilon) = -F_{a0}^0(\varepsilon)S_a^{-1}(\varepsilon)Q_a(\varepsilon)\Gamma_{sa}^0, \qquad (11.2.79)$$

$$\tilde{L}_{iab}^+(\varepsilon) = S_i(\varepsilon)L_{iab}^+(\varepsilon), \qquad (11.2.80)$$

$$\tilde{L}_{ia}^0(\varepsilon) = -S_i(\varepsilon)[L_{ia}^{01}(\varepsilon)F_a^0(\varepsilon) + L_{ia}^{02}(\varepsilon)F_a^0(\varepsilon)A_{aa}^0]$$
$$\times \Gamma_{sa}^0 Q_a(\varepsilon)S_a^{-1}(\varepsilon), \qquad (11.2.81)$$

$$\tilde{L}_{id}(\varepsilon) = S_i(\varepsilon)\bar{L}_{id}(\varepsilon), \qquad (11.2.82)$$

$$\tilde{E}_i(\varepsilon) = S_i(\varepsilon)\bar{E}_i(\varepsilon), \qquad (11.2.83)$$

$$\tilde{F}_{adi}^0(\varepsilon) = -F_{adi}^0(\varepsilon)\Gamma_{sa}^0 Q_a(\varepsilon)S_a^{-1}(\varepsilon), \qquad (11.2.84)$$

and where, for $i = 1$ to l, $\tilde{J}_{ai}(\varepsilon)$ is as defined in Lemma 2.2.2.

By (11.2.22), (11.2.59), Lemmas 2.2.4 and 2.2.5, we have that, for all $\varepsilon \in (0, 1]$,

$$|A_{aab}^-(\varepsilon)| \le a_{aab}^-, \quad |\tilde{L}_{ad}^0(\varepsilon)| \le \tilde{l}_{ad}^0, \quad |A_{cab}^+| \le a_{cab}^+, \qquad (11.2.85)$$

$$|A_{aa}^{-0}(\varepsilon)| \le a_{aa}^{-0}\varepsilon, \quad |A_{aba}^{+0}(\varepsilon)| \le a_{aa}^{+0}\varepsilon, \quad |A_{ca}^0(\varepsilon)| \le a_{ca}^0\varepsilon, \quad |\tilde{F}_{a0}^0(\varepsilon)| \le \tilde{f}_{a0}^0\varepsilon,$$
$$(11.2.86)$$

for $i = 1$ to m_d,

$$|\tilde{L}_{iab}^+(\varepsilon)| \le \tilde{l}_{ab}^+, \quad |\tilde{L}_{ia}^0(\varepsilon)| \le \tilde{l}_a^0 \varepsilon, \quad |\tilde{L}_{id}(\varepsilon)| \le \tilde{l}_d, \quad |\tilde{F}_{adi}^0(\varepsilon)| \le \tilde{f}_{ad}^0 \varepsilon, \quad |\tilde{E}_i(\varepsilon)| \le \tilde{e},$$
(11.2.87)

for $i = 1$ to l,

$$|\tilde{E}_{ai}^0(\varepsilon)| \le \tilde{e}_a^0 \varepsilon,$$
(11.2.88)

and finally, for $i = 1$ to l, $j = i + 1$ to l,

$$|\tilde{B}_{ij}(\varepsilon)| \le \tilde{b}_{ij}\varepsilon,$$
(11.2.89)

where a_{aab}^-, $\tilde{l}_{ad}^0$, a_{cab}^+, a_{aa}^{-0}, a_{aa}^{+0}, $\tilde{e}_a^0$, a_{ca}^0, $\tilde{f}_{a0}^0$, $\tilde{l}_{ab}^+$, $\tilde{l}_a^0$, $\tilde{l}_d$, $\tilde{f}_{ad}^0$, $\tilde{b}_{ij}$ and $\tilde{e}$ are some positive constants, independent of ε.

We next construct a Lyapunov function for the closed-loop system (11.2.64)-(11.2.70). We do this by composing Lyapunov functions for the subsystems. For the subsystem of $\tilde{x}_a^-$, we choose a Lyapunov function,

$$V_a^-(\tilde{x}_a^-) = (\tilde{x}_a^-)' P_a^- \tilde{x}_a^-,$$
(11.2.90)

where $P_a^- > 0$ is the unique solution to the Lyapunov equation

$$(A_{aa}^-)' P_a^- + P_a^- A_{aa}^- = -I,$$
(11.2.91)

and for the subsystem of $\tilde{x}_{ab}^+$, choose a Lyapunov function,

$$V_{ab}^+(\tilde{x}_{ab}^+) = (\tilde{x}_{ab}^+)' P_{ab}^+ \tilde{x}_{ab}^+,$$
(11.2.92)

where $P_{ab}^+ > 0$ is the unique solution to the Lyapunov equation

$$(A_{ab}^{+c})' P_{ab}^+ + P_{ab}^+ A_{ab}^{+c} = -I.$$
(11.2.93)

The existence of such P_a^- and P_{ab}^+ is guaranteed by the fact that both A_{aa}^- and A_{ab}^{+c} are asymptotically stable. For the subsystem of $\tilde{x}_a^0 = [(\tilde{x}_{a1}^0)', (\tilde{x}_{a2}^0)', \cdots, (\tilde{x}_{al}^0)']'$, we choose a Lyapunov function,

$$V_a^0(\tilde{x}_a^0) = \sum_{i=1}^{l} \frac{(\kappa_a^0)^{i-1}}{\varepsilon} (\tilde{x}_{ai}^0)' P_{ai}^0 \tilde{x}_{ai}^0,$$
(11.2.94)

where κ_a^0 is a positive scalar, whose value is to be determined later, and each P_{ai}^0 is the unique solution to the Lyapunov equation,

$$\tilde{J}_{ai}(\varepsilon)' P_{ai}^0 + P_{ai}^0 \tilde{J}_{ai}(\varepsilon) = -I,$$
(11.2.95)

which, by Lemma 2.2.3, is independent of ε. Similarly, for the subsystem $\tilde{x}_c$, choose a Lyapunov function,

$$V_c(\tilde{x}_c) = \tilde{x}_c' P_c \tilde{x}_c,$$
(11.2.96)

where $P_c > 0$ is the unique solution to the Lyapunov equation,

$$(A_{cc}^c)'P_c + P_c A_{cc}^c = -I. \tag{11.2.97}$$

The existence of such a P_c is again guaranteed by the fact that A_{cc}^c is asymptotically stable. Finally, for the subsystem of $\tilde{x}_d$, choose a Lyapunov function

$$V_d(\tilde{x}_d) = \sum_{i=1}^{m_d} \tilde{x}_i' P_i \tilde{x}_i, \tag{11.2.98}$$

where each P_i is the unique solution to the Lyapunov equation

$$(A_{q_i} + B_{q_i} F_i)'P_i + P_i(A_{q_i} + B_{q_i} F_i) = -I. \tag{11.2.99}$$

Once again, the existence of such P_i is due to the fact that $A_{q_i} + B_{q_i} F_i$ is asymptotically stable.

We now construct a Lyapunov function for the closed-loop system (11.2.64)-(11.2.70) as follows,

$$V(\tilde{x}_a^-, \tilde{x}_{ab}^+, \tilde{x}_a^0, \tilde{x}_c, \tilde{x}_d) = V_a^-(\tilde{x}_a^-) + \kappa_{ab}^+ V_{ab}^+(\tilde{x}_{ab}^+) + V_a^0(\tilde{x}_a^0) + V_c(\tilde{x}_c) + \kappa_d V_d(\tilde{x}_d), \tag{11.2.100}$$

where $\kappa_{ab}^+ = 2|P_a^-|^2(a_{aab}^-)^2 + 1$ and the value of κ_d is to be determined.

Let us first consider the derivative of $V_a^0(\tilde{x}_a^0)$ along the trajectories of the subsystem $\tilde{x}_a^0$ and obtain that,

$$\dot{V}_a^0(\tilde{x}_a^0) = \sum_{i=1}^{l} \left[-(\kappa_a^0)^{i-1}(\tilde{x}_{ai}^0)'\tilde{x}_{ai}^0 + 2 \sum_{j=i+1}^{l} \frac{(\kappa_a^0)^{i-1}}{\varepsilon}(\tilde{x}_{ai}^0)' P_{ai}^0 \tilde{B}_{ij}(\varepsilon)\tilde{x}_{aj}^0 \right]$$
$$+2\sum_{i=1}^{l} \frac{(\kappa_a^0)^{i-1}}{\varepsilon} \left[(\tilde{x}_{ai}^0)' P_{ai}^0 \tilde{L}_{ad}^0(\varepsilon)\tilde{z}_d + (\tilde{x}_{ai}^0)' P_{ai}^0 \tilde{E}_a^0(\varepsilon)w \right]. \tag{11.2.101}$$

Using (11.2.89) it is straightforward to show that, there exists a $\kappa_a^0 > 0$ such that,

$$\dot{V}_a^0(\tilde{x}_a^0) \le -\frac{3}{4}|\tilde{x}_a^0|^2 + \frac{\alpha_1}{\varepsilon}|\tilde{x}_a^0||\tilde{z}_d| + \alpha_2|w|^2, \tag{11.2.102}$$

for some nonnegative constants α_1 and α_2, independent of ε.

In view of (11.2.102), the derivative of V along the trajectory of the closed-loop system (11.2.64)-(11.2.70) can be evaluated as follows,

$$\dot{V} \le -(\tilde{x}_a^-)'\tilde{x}_a^- + 2(\tilde{x}_a^-)'P_a^- A_{aab}^-(\varepsilon)\tilde{x}_{ab}^+ + 2(\tilde{x}_a^-)'P_a^- A_{aa}^{-0}(\varepsilon)\tilde{x}_a^0 + 2(\tilde{x}_a^-)'P_a^- L_{ad}^- \tilde{z}_d$$
$$+2(\tilde{x}_a^-)'P_a^- E_a^- w$$
$$-\kappa_{ab}^+(\tilde{x}_{ab}^+)'\tilde{x}_{ab}^+ + 2\kappa_{ab}^+(x_{ab}^+)'P_{ab}^+ A_{aba}^{+0}(\varepsilon)\tilde{x}_a^0 + 2\kappa_{ab}^+(x_{ab}^+)'P_{ab}^+ L_{abd}^+ \tilde{z}_d$$

$$-\frac{3}{4}|\tilde{x}_a^0|^2 + \frac{\alpha_1}{\varepsilon}|\tilde{x}_a^0|\,|\tilde{z}_d| + \alpha_2|w|^2$$

$$-\tilde{x}_c'\tilde{x}_c + 2\varepsilon\tilde{x}_c'P_c[A_{cab}^+\tilde{x}_{ab}^+ + A_{ca}^0(\varepsilon)\tilde{x}_a^0 + L_{cd}\tilde{z}_d + E_c w]$$

$$+\kappa_d\sum_{i=1}^{m_d}\left[-\frac{1}{\varepsilon}\tilde{x}_i'\tilde{x}_i + 2\tilde{x}_i'P_i\tilde{L}_{iab}^+(\varepsilon)\tilde{x}_{ab}^+ + 2\tilde{x}_i'P_i\tilde{L}_{ia}^0(\varepsilon)\tilde{x}_a^0 + 2\tilde{x}_i'P_i\tilde{L}_{id}(\varepsilon)\tilde{z}_d\right.$$

$$\left.+ 2\tilde{x}_i'P_i\tilde{E}_i(\varepsilon)w\right]. \tag{11.2.103}$$

Using the majorizations (11.2.85)-(11.2.88) and noting the definition of κ_{ab}^+ in (11.2.100), we can easily verify that, there exist a $\kappa_d > 0$ and an $\varepsilon_1^* \in (0,1]$ such that, for all $\varepsilon \in (0,\varepsilon_1^*]$,

$$\dot{V} \le -\frac{1}{2}|\tilde{x}_a^-|^2 - \frac{1}{2}|\tilde{x}_{ab}^+|^2 - \frac{1}{2}|\tilde{x}_a^0|^2 - \frac{1}{2}|\tilde{x}_c|^2$$

$$-\frac{1}{2\varepsilon}|\tilde{x}_d|^2 + \alpha_3|w|^2, \tag{11.2.104}$$

for some positive constant α_3, independent of ε.

From (11.2.104), it follows that the closed-loop system in the absence of disturbance w is asymptotically stable. It remains to show that, for any given $\gamma > 0$, there exists an $\varepsilon^* \in (0,\varepsilon_1^*]$ such that, for all $\varepsilon \in (0,\varepsilon^*]$,

$$\|z\|_{L_2} \le \gamma\|w\|_{L_2}. \tag{11.2.105}$$

To this end, we integrate both sides of (11.2.104) from 0 to ∞. Noting that $V \ge 0$ and $V(t) = 0$ at $t = 0$, we have,

$$\|\tilde{z}_d\|_{L_2} \le \left(\sqrt{2\alpha_3\varepsilon}\right)\|w\|_{L_2}, \tag{11.2.106}$$

which, when used in (11.2.102), results in,

$$\|\tilde{x}_a^0\|_{L_2} \le 2\sqrt{\frac{4\alpha_1^2\alpha_3}{\varepsilon} + \alpha_2}\|w\|_{L_2}. \tag{11.2.107}$$

Viewing $\tilde{z}_d$ as disturbance to the dynamics of $\tilde{x}_{ab}^+$ also results in,

$$\|\tilde{x}_{ab}^+\|_{L_2} \le \alpha_4\sqrt{\varepsilon}\|w\|_{L_2}, \tag{11.2.108}$$

for some positive constant α_4, independent of ε.

Finally, recalling that

$$z = \Gamma_{\mathrm{OP}}\left[z_0', \left(\tilde{z}_d + F_{ab}^+\tilde{x}_{ab}^+ + \tilde{F}_{ad}^0(\varepsilon)\tilde{x}_a^0\right)', z_b'\right]', \tag{11.2.109}$$

where

$$\tilde{F}_{ad}^0(\varepsilon) = \begin{bmatrix} \tilde{F}_{ad1}^0(\varepsilon) \\ \tilde{F}_{ad2}^0(\varepsilon) \\ \vdots \\ F_{adm_d}^0(\varepsilon) \end{bmatrix}, \tag{11.2.110}$$

with each $\tilde{F}_{adi}(\varepsilon)$ satisfying (11.2.87), we have,

$$\|z\|_{L_2} \le |\Gamma_{\text{OP}}| \left(\sqrt{2\alpha_3\varepsilon} + \alpha_5\sqrt{\varepsilon} + \alpha_6\sqrt{4\alpha_1^2\alpha_3\varepsilon + \alpha_2\varepsilon^2} \right) \|w\|_{L_2}, \tag{11.2.111}$$

for some positive constants α_5 and α_6, both independent of ε.

To complete the proof, we choose $\varepsilon^* \in (0, \varepsilon_1^*]$ such that,

$$|\Gamma_{\text{OP}}| \left(\sqrt{2\alpha_3\varepsilon} + \alpha_5\sqrt{\varepsilon} + \alpha_6\sqrt{4\alpha_1^2\alpha_3\varepsilon + \alpha_2\varepsilon^2} \right) \le \gamma. \tag{11.2.112}$$

⊠

11.3. Discrete-Time Linear Systems

11.3.1. Problem Statement

Consider the following general discrete-time linear system,

$$\Sigma : \begin{cases} x^+ = A\,x + B\,u + E\,w, \\ y = C_1\,x \qquad\quad + D_1\,w, \\ z = C_2\,x + D_2\,u + D_{22}\,w, \end{cases} \tag{11.3.1}$$

where $x \in \mathbb{R}^n$ is the state, $u \in \mathbb{R}^m$ is the control input, $y \in \mathbb{R}^\ell$ is the measurement, $w \in \mathbb{R}^q$ is the disturbance and $z \in \mathbb{R}^p$ is the output to be controlled. A, B, E, C_1, C_2, D_1, D_2, and D_{22} are constant matrices of appropriate dimensions. For easy reference in future development, throughout this section, we define Σ_{P} to be the subsystem characterized by the matrix quadruple (A, B, C_2, D_2) and Σ_{Q} to be the subsystem characterized by the matrix quadruple (A, E, C_1, D_1). The following dynamic feedback control laws are investigated:

$$\Sigma_c : \begin{cases} x_c^+ = A_c\,x_c + B_c\,y, \\ u = C_c\,x_c + D_c\,y. \end{cases} \tag{11.3.2}$$

The controller Σ_c of (11.3.2) is said to be internally stabilizing when applied to the system Σ, if the following matrix is asymptotically stable:

$$A_{\text{cl}} = \begin{bmatrix} A + BD_cC_1 & BC_c \\ B_cC_1 & A_c \end{bmatrix}, \tag{11.3.3}$$

i.e., all its eigenvalues lie inside the open unit disc. Denote by T_{zw} the corresponding closed-loop transfer matrix from the disturbance w to the output to be controlled z, i.e.,

$$T_{zw}(z) = [C_2 + D_2 D_c C_1 \quad D_2 C_c] \left(zI - \begin{bmatrix} A + BD_c C_1 & BC_c \\ B_c C_1 & A_c \end{bmatrix} \right)^{-1}$$

$$\times \begin{bmatrix} E + BD_c D_1 \\ B_c D_1 \end{bmatrix} + D_2 D_c D_1 + D_{22}. \tag{11.3.4}$$

where the z in the subscript of $T_{zw}(z)$ is the controlled output and the z inside ( ) is the z-transform variables. Since both are standard notations and their distinction can easily be made, we have not changed one of them to an alternative notation.

The H_∞-norm of the transfer matrix T_{zw} is given by

$$\|T_{zw}\|_\infty = \sup_{\omega \in [0,\pi]} \sigma_{\max}[T_{zw}(e^{j\omega})], \tag{11.3.5}$$

where $\sigma_{\max}[\cdot]$ denotes the maximum singular value. Then the general H_∞-ADDPMS for the the given discrete-time system Σ of (11.3.1) can be formally defined as follows.

Problem 11.3.1. *The general H_∞ almost disturbance decoupling problem with measurement feedback and internal stability (the general H_∞-ADDPMS) for (11.3.1) is defined as follows. For any given positive scalar $\gamma > 0$, find a controller of the form (11.3.2) such that,*

1. *in the absence of disturbance, the closed-loop system comprising the system (11.3.1) and the controller (11.3.2) is asymptotically stable, i.e., the matrix A_{cl} as given by (11.3.3) is asymptotically stable;*

2. *the closed-loop system has an L_2-gain, from the disturbance w to the controlled output z, that is less than or equal to γ, i.e.,*

$$\|z\|_{L_2} \le \gamma \|w\|_{L_2}, \ \forall w \in L_2 \text{ and for } (x(0), x_c(0)) = (0,0). \tag{11.3.6}$$

Equivalently, the H_∞-norm of the closed-loop transfer matrix from w to z, T_{zw}, is less than or equal to γ, i.e., $\|T_{zw}\|_\infty \le \gamma$. ▣

A set of necessary and sufficient conditions under which the above problems was recently given in [11]. To state these conditions, we need to define the following geometric subspaces.

Definition 11.3.1. *Consider a linear time-invariant system* Σ_* *characterized by a matrix quadruple* (A, B, C, D). *The weakly unobservable subspaces of* Σ_*, $\mathcal{V}^{\times}$, *and the strongly controllable subspaces of* Σ_*, $\mathcal{S}^{\times}$, *are defined as follows:*

1. $\mathcal{V}^{\times}(\Sigma_*)$ *is the maximal subspace of* $\mathbb{R}^n$ *which is* $(A + BF)$-*invariant and contained in* $\mathrm{Ker}\,(C + DF)$ *such that the eigenvalues of* $(A + BF)|\mathcal{V}^{\times}$ *are contained in* $\mathbb{C}^{\times} \subseteq \mathbb{C}$ *for some constant matrix* F.

2. $\mathcal{S}^{\times}(\Sigma_*)$ *is the minimal* $(A + LC)$-*invariant subspace of* $\mathbb{R}^n$ *containing* $\mathrm{Im}\,(B + LD)$ *such that the eigenvalues of the map which is induced by* $(A + LC)$ *on the factor space* $\mathbb{R}^n/\mathcal{S}^{\times}$ *are contained in* $\mathbb{C}^{\times} \subseteq \mathbb{C}$ *for some constant matrix* L.

Furthermore, we denote $\mathcal{V}^{\circ} = \mathcal{V}^{\times}$ *and* $\mathcal{S}^{\circ} = \mathcal{S}^{\times}$, *if* $\mathbb{C}^{\times} = \mathbb{C}^{\circ} \cup \mathbb{C}^{\circ}$; $\mathcal{V}^{\circledast} = \mathcal{V}^{\times}$ *and* $\mathcal{S}^{\circledast} = \mathcal{S}^{\times}$, *if* $\mathbb{C}^{\times} = \mathbb{C}^{\circledast}$; *and finally* $\mathcal{V}^* = \mathcal{V}^{\times}$ *and* $\mathcal{S}^* = \mathcal{S}^{\times}$, *if* $\mathbb{C}^{\times} = \mathbb{C}$. ▣

Definition 11.3.2. *Consider a linear system* Σ_* *characterized by a quadruple* (A, B, C, D). *For any* $\lambda \in \mathbb{C}$, *we define*

$$\mathcal{S}_{\lambda}(\Sigma_*) = \left\{ x \in \mathbb{C}^n \,\middle|\, \exists u \in \mathbb{C}^{n+m} \,:\, \begin{pmatrix} x \\ 0 \end{pmatrix} = \begin{bmatrix} A - \lambda I & B \\ C & D \end{bmatrix} u \right\} \qquad (11.3.7)$$

and

$$\mathcal{V}_{\lambda}(\Sigma_*) = \left\{ x \in \mathbb{C}^n \,\middle|\, \exists u \in \mathbb{C}^m \,:\, 0 = \begin{bmatrix} A - \lambda I & B \\ C & D \end{bmatrix} \begin{pmatrix} x \\ u \end{pmatrix} \right\}. \qquad (11.3.8)$$

$\mathcal{V}_{\lambda}(\Sigma_*)$ *and* $\mathcal{S}_{\lambda}(\Sigma_*)$ *are associated with the so-called state zero directions of* Σ_* *if* λ *is an invariant zero of* Σ_*. ▣

The following results concerning the solvability of the discrete-time general H_{∞}-ADDPFIS and the general H_{∞}-ADDPMS are recalled from [11].

Theorem 11.3.1. *Consider the given discrete-time linear time-invariant system* Σ *as given by (11.3.1) with the measurement output being*

$$y = \begin{pmatrix} x \\ w \end{pmatrix}, \quad \text{or} \quad C_1 = \begin{pmatrix} I \\ 0 \end{pmatrix}, \quad D_1 = \begin{pmatrix} 0 \\ I \end{pmatrix} \qquad (11.3.9)$$

(i.e., all state variables and disturbances (full information) are measurable and available for feedback). The general H_{∞} *almost disturbance decoupling problem with full information feedback and with internal stability (the general* H_{∞}-*ADDPFIS) is solvable if and only if the following conditions are satisfied:*

(a) (A, B) *is stabilizable.*

(b) $\mathrm{Im}\,(D_{22}) \subset \mathrm{Im}\,(D_2)$, i.e., $D_{22}+D_2S = 0$, with $S = -(D_2'D_2)^\dagger D_2'D_{22}$.

(c) $\mathrm{Im}\,(E + BS) \subset \{\mathcal{V}^\circ(\Sigma_{\mathrm{P}}) + B\mathrm{Ker}\,(D_2)\} \cap \left\{\bigcap_{|\lambda|=1} \mathcal{S}_\lambda(\Sigma_{\mathrm{P}})\right\}.$ ⊤

Theorem 11.3.2. *Consider the given discrete-time linear time-invariant system Σ as given by (11.3.1). The general H_∞ almost disturbance decoupling problem with measurement feedback and with internal stability (the general H_∞-ADDPMS) is solvable if and only if the following conditions are satisfied:*

(a) (A, B) *is stabilizable.*

(b) (A, C_1) *is detectable.*

(c) $D_{22} + D_2SD_1 = 0$, *where* $S = -(D_2'D_2)^\dagger D_2'D_{22}D_1'(D_1D_1')^\dagger.$

(d) $\mathrm{Im}\,(E + BSD_1) \subset \{\mathcal{V}^\circ(\Sigma_{\mathrm{P}}) + B\mathrm{Ker}\,(D_2)\} \cap \left\{\bigcap_{|\lambda|=1} \mathcal{S}_\lambda(\Sigma_{\mathrm{P}})\right\}.$

(e) $\mathrm{Ker}\,(C_2 + D_2SC_1) \supset \{\mathcal{S}^\circ(\Sigma_{\mathrm{Q}}) \cap C_1^{-1}\{\mathrm{Im}\,(D_1)\}\} \cup \left\{\bigcup_{|\lambda|=1} \mathcal{V}_\lambda(\Sigma_{\mathrm{Q}})\right\}.$

(f) $\mathcal{S}^\circ(\Sigma_{\mathrm{Q}}) \subset \mathcal{V}^\circ(\Sigma_{\mathrm{P}}).$ ⊤

The following remark concerns the full state feedback case.

Remark 11.3.1. *For special case when all the states of the system (11.3.1) are measurable and available for feedback, i.e., $y = x$, it can be easily derived from Theorem 11.3.2 that the H_∞ almost disturbance decoupling problem with full state feedback and with internal stability for the given system is solvable if and only if the following conditions are satisfied:*

(a) (A, B) *is stabilizable.*

(b) $D_{22} = 0$.

(c) $\mathrm{Im}\,(E) \subset \mathcal{V}^\circ(\Sigma_{\mathrm{P}}) \cap \left\{\bigcap_{|\lambda|=1} \mathcal{S}_\lambda(\Sigma_{\mathrm{P}})\right\}.$ ⓡ

The objective of this section is to demonstrate how the eigenstructure assignment based low gain feedback design technique of Section 2.3.1 can be employed to construct families of feedback control laws of the form (11.3.2), parameterized in a single parameter, say ε, that, under the necessary and sufficient conditions of [11], solve the above defined general H_∞-ADDPMS, H_∞-ADDFIS and H_∞-ADDPS for general systems whose subsystems Σ_{P} and Σ_{Q} may have invariant zeros on the unit circle.

11.3.2. Solution to the General H_∞-ADDPS

The general H_∞-ADDPMS is solved by explicit construction of feedback laws. The feedback laws we are to construct are observer-based. A family of static state feedback control laws parameterized in a single parameter is first constructed to solve the general H_∞-ADDPS. A class of observers parameterized in the same parameter ε is then constructed to implement the state feedback control laws and thus obtain a family of dynamic measurement feedback control laws parameterized in a single parameter ε that solve the general H_∞-ADDPMS. The observer gains for this family of observers are constructed by applying the algorithm for constructing state feedback gains to the dual system. Besides the low gain feedback design technique of Chapter 2, another basic tool we use in the construction of such families of feedback control laws is the special coordinate basis (SCB) [92,95], in which a linear system is decomposed into several subsystems corresponding to its finite and infinite zero structures as well as its invertibility structures. A summary of SCB and its properties is given in Appendix A.

Since our objective here is to demonstrate how the low gain feedback design technique can be used in the construction of observer based feedback laws that solve the general H_∞-ADDPMS and since both the state feedback gain matrix and the observer gain matrix can be constructed using the same algorithm, we will only present the algorithm for explicitly constructing a family of state feedback laws that solves the general H_∞-ADDPS. The solution to the general H_∞-ADDPMS can be found in [50]. In [50] it is also shown how the H_∞-ADDPFIS is converted into an H_∞-ADDPS and solved.

More specifically, we will present a design procedure that constructs a family of parameterized static state feedback control laws,

$$u = F(\varepsilon)x, \qquad (11.3.10)$$

that solves the general H_∞-ADDPMS for the following system,

$$\begin{cases} x^+ = A\,x + B\,u + E\,w, \\ y = x, \\ z = C_2\,x + D_2\,u + D_{22}\,w. \end{cases} \qquad (11.3.11)$$

That is, under this family of state feedback control laws, the resulting closed-loop system is asymptotically stable for sufficiently small ε and the H_∞-norm of the closed-loop transfer matrix from w to z, $T_{zw}(z,\varepsilon)$, tends to zero as ε goes to zero, where

$$T_{zw}(z,\varepsilon) = [C_2 + D_2 F(\varepsilon)][zI - A - BF(\varepsilon)]^{-1}E + D_{22}. \qquad (11.3.12)$$

Clearly, $D_{22} = 0$ is a necessary condition for the solvability of the general H_∞-ADDPS.

The algorithm for constructing a family of state feedback laws that solves the H_∞-ADDPS consists of the following five steps.

Step 1: Decomposition of Σ_P. Transform the subsystem Σ_P, i.e., the matrix quadruple (A, B, C_2, D_2), into the special coordinate basis (SCB) as given by Theorem A.1.1. Denote the state, output and input transformation matrices as Γ_{SP}, Γ_{OP} and Γ_{IP}, respectively.

Step 2: Gain matrix for the subsystem associated with $\mathcal{X}_c$. Let F_c be any constant matrix subject to the constraint that

$$A_{cc}^c = A_{cc} + B_c F_c, \qquad (11.3.13)$$

is an asymptotically stable matrix. Note that the existence of such an F_c is guaranteed by the property of the special coordinate basis, i.e., (A_{cc}, B_c) is controllable.

Step 3: Gain matrix for the subsystem associated with $\mathcal{X}_a^+$, $\mathcal{X}_b$ and $\mathcal{X}_d$. Let

$$F_{abd} = \begin{bmatrix} 0 & 0 & F_{a0}^+ & F_{b0} & F_{d0} \\ E_{da}^- & E_{da}^0 & F_{ad}^+ & F_{bd} & F_{dd} \end{bmatrix}, \qquad (11.3.14)$$

where

$$F_{abd}^+ = \begin{bmatrix} F_{a0}^+ & F_{b0} & F_{d0} \\ F_{ad}^+ & F_{bd} & F_{dd} \end{bmatrix}, \qquad (11.3.15)$$

is any constant matrix subject to the constraint that

$$A_{abd}^{+c} = \begin{bmatrix} A_{aa}^+ & L_{ab}^+ C_b & L_{ad}^+ C_d \\ 0 & A_{bb} & L_{bd} C_d \\ B_d E_{da}^+ & B_d E_{db} & A_{dd} \end{bmatrix} + \begin{bmatrix} B_{0a}^+ & 0 \\ B_{0b} & 0 \\ B_{0d} & B_d \end{bmatrix} F_{abd}^+ \qquad (11.3.16)$$

is an asymptotically stable matrix. Again, the existence of such an F_{abd}^+ is guaranteed by the property of the special coordinate basis.

Step 4: Gain matrix for the subsystem associated with A_{aa}^0. The construction of this gain matrix is carried out in the following four sub-steps.

Step 4.1: Preliminary coordinate transformation. Noting that

$$A_{abd} = \begin{bmatrix} A_{aa} & L_{ab} C_b & L_{ad} C_d \\ 0 & A_{bb} & L_{bd} C_d \\ B_d E_{da} & B_d E_{db} & A_{dd} \end{bmatrix}, \quad B_{abd} = \begin{bmatrix} B_{0a} & 0 \\ B_{0b} & 0 \\ B_{0d} & B_d \end{bmatrix},$$

we have

$$A_{abd} + B_{abd}F_{abd} = \begin{bmatrix} A_{aa}^- & 0 & A_{abd}^- \\ 0 & A_{aa}^0 & A_{abd}^0 \\ 0 & 0 & A_{abd}^{+c} \end{bmatrix}, \quad B_{abd} = \begin{bmatrix} B_{0a}^- & 0 \\ B_{0a}^0 & 0 \\ B_{0abd}^+ & B_{abd}^+ \end{bmatrix},$$

(11.3.17)

where

$$B_{0abd}^+ = \begin{bmatrix} B_{0a}^+ \\ B_{0b} \\ B_{0d} \end{bmatrix}, \quad B_{abd}^+ = \begin{bmatrix} 0 \\ 0 \\ B_d \end{bmatrix},$$

(11.3.18)

$$A_{abd}^0 = [\, 0 \quad L_{ab}^0 C_b \quad L_{ad}^0 C_d \,] + [\, B_{0a}^0 \quad 0\,] F_{abd}^+,$$

(11.3.19)

and

$$A_{abd}^- = [\, 0 \quad L_{ab}^- C_b \quad L_{ad}^- C_d \,] + [\, B_{0a}^- \quad 0\,] F_{abd}^+.$$

(11.3.20)

Clearly, the pair $(A_{abd} + B_{abd}F_{abd}, B_{abd})$ remains stabilizable. Construct the following nonsingular transformation matrix,

$$\Gamma_{abd} = \begin{bmatrix} I_{n_a^-} & 0 & 0 \\ 0 & 0 & I_{n_a^+ + n_b + n_d} \\ 0 & I_{n_a^0} & T_a^0 \end{bmatrix}^{-1},$$

(11.3.21)

where T_a^0 is the unique solution to the following Lyapunov equation,

$$A_{aa}^0 T_a^0 - T_a^0 A_{abd}^{+c} = A_{abd}^0.$$

(11.3.22)

We note here that such a unique solution to the above Lyapunov equation always exists since all the eigenvalues of A_{aa}^0 are on the unit circle and all the eigenvalues of A_{abd}^{+c} are on the open unit disc. It is now easy to verify that

$$\Gamma_{abd}^{-1}(A_{abd} + B_{abd}F_{abd})\Gamma_{abd} = \begin{bmatrix} A_{aa}^- & A_{abd}^- & 0 \\ 0 & A_{abd}^{+c} & 0 \\ 0 & 0 & A_{aa}^0 \end{bmatrix},$$

(11.3.23)

and

$$\Gamma_{abd}^{-1}B_{abd} = \begin{bmatrix} B_{0a}^- & 0 \\ B_{0abd}^+ & B_{abd}^+ \\ B_{0a}^0 + T_a^0 B_{0abd}^+ & T_a^0 B_{abd}^+ \end{bmatrix}.$$

(11.3.24)

Hence, the matrix pair (A_{aa}^0, B_a^0) is controllable, where

$$B_a^0 = [\, B_{0a}^0 + T_a^0 B_{0abd}^+ \quad T_a^0 B_{abd}^+ \,].$$

(11.3.25)

Step 4.2: Further coordinate transformation. Find the nonsingular transformation matrices Γ_{sa}^0 and Γ_{Ia}^0 to transform the pair (A_{aa}^0, B_a^0) into the block diagonal control canonical form,

$$(\Gamma_{sa}^0)^{-1} A_{aa}^0 \Gamma_{sa}^0 = \begin{bmatrix} A_1 & 0 & \cdots & 0 \\ 0 & A_2 & \cdots & 0 \\ \vdots & \vdots & \ddots & \vdots \\ 0 & 0 & \cdots & A_l \end{bmatrix}, \qquad (11.3.26)$$

and

$$(\Gamma_{sa}^0)^{-1} B_a^0 \Gamma_{Ia}^0 = \begin{bmatrix} B_1 & B_{12} & \cdots & B_{1l} & \star \\ 0 & B_2 & \cdots & B_{2l} & \star \\ \vdots & \vdots & \ddots & \vdots & \vdots \\ 0 & 0 & \cdots & B_l & \star \end{bmatrix}, \qquad (11.3.27)$$

where l is an integer and for $i = 1, 2, \cdots, l$,

$$A_i = \begin{bmatrix} 0 & 1 & 0 & \cdots & 0 \\ 0 & 0 & 1 & \cdots & 0 \\ \vdots & \vdots & \vdots & \ddots & \vdots \\ 0 & 0 & 0 & \cdots & 1 \\ -a_{n_i}^i & -a_{n_i-1}^i & -a_{n_i-2}^i & \cdots & -a_1^i \end{bmatrix}, \quad B_i = \begin{bmatrix} 0 \\ 0 \\ \vdots \\ 0 \\ 1 \end{bmatrix}.$$

We note that all the eigenvalues of A_i are on the unit circle. Here, the $\star$'s represent sub-matrices of less interest. We also note that the existence of the above canonical form was shown in Wonham [126] while its software realization can be found in Chen [9].

Step 4.3: Subsystem design. For each (A_i, B_i), let $F_i(\varepsilon) \in \mathbb{R}^{1 \times n_i}$ be the state feedback gain such that

$$\lambda(A_i + B_i F_i(\varepsilon)) = (1 - \varepsilon)\lambda(A_i), \qquad (11.3.28)$$

Clearly, all the eigenvalues of $A_i + B_i F_i(\varepsilon)$ are strictly inside the unit circle and $F_i(\varepsilon)$ is unique.

Step 4.4: Composition of gain matrix for subsystem associated with $\mathcal{X}_a^0$. Let

$$F_a^0(\varepsilon) = \Gamma_{Ia}^0 \begin{bmatrix} F_1(\varepsilon) & 0 & \cdots & 0 & 0 \\ 0 & F_2(\varepsilon) & \cdots & 0 & 0 \\ \vdots & \vdots & \ddots & \vdots & \vdots \\ 0 & 0 & \cdots & F_{l-1}(\varepsilon) & 0 \\ 0 & 0 & \cdots & 0 & F_l(\varepsilon) \\ 0 & 0 & \cdots & 0 & 0 \end{bmatrix} (\Gamma_{sa}^0)^{-1},$$

$$(11.3.29)$$

where $\varepsilon \in (0,1]$ is a design parameter whose value is to be specified later.

We recall that the above Steps 4.2-4.4 are the application to the pair (A_a^0, B_a^0) of the eigenstructure assignment based low gain design procedure of Section 2.3.1. Clearly, we have

$$|F_a^0(\varepsilon)| \le f_a^0 \varepsilon, \ \ \varepsilon \in (0,1], \tag{11.3.30}$$

for some positive constant f_a^0, independent of ε. For future use, we partition

$$F_a^0(\varepsilon) = \begin{bmatrix} F_{a0}^0(\varepsilon) \\ F_{ad}^0(\varepsilon) \end{bmatrix}, \tag{11.3.31}$$

and

$$F_a^0(\varepsilon)T_a^0 = \begin{bmatrix} F_{a0+}^0(\varepsilon) & F_{a0b}^0(\varepsilon) & F_{a0d}^0(\varepsilon) \\ F_{ad+}^0(\varepsilon) & F_{adb}^0(\varepsilon) & F_{add}^0(\varepsilon) \end{bmatrix}. \tag{11.3.32}$$

Step 5: Composition of parameterized gain matrix $F(\varepsilon)$. In this step, various gains calculated in Steps 2 to 4 are put together to form a composite state feedback gain matrix $F(\varepsilon)$. It is given by

$$F(\varepsilon) = \Gamma_{\text{IP}}\left[F_0 + F_\star(\varepsilon)\right]\Gamma_{\text{SP}}^{-1}, \tag{11.3.33}$$

where

$$F_0 = -\begin{bmatrix} C_{0a}^- & C_{0a}^0 & C_{0a}^+ - F_{a0}^+ & C_{0b} - F_{b0} & C_{0c} & C_{0d} - F_{d0} \\ E_{da}^- & E_{da}^0 & -F_{ad}^+ & -F_{bd} & E_{dc} & -F_{dd} \\ E_{ca}^- & E_{ca}^0 & E_{ca}^+ & 0 & -F_c & 0 \end{bmatrix}, \tag{11.3.34}$$

and

$$F_\star(\varepsilon) = \begin{bmatrix} 0 & F_{a0}^0(\varepsilon) & F_{a0+}^0(\varepsilon) & F_{a0b}^0(\varepsilon) & 0 & F_{a0d}^0(\varepsilon) \\ 0 & F_{ad}^0(\varepsilon) & F_{ad+}^0(\varepsilon) & F_{adb}^0(\varepsilon) & 0 & F_{add}^0(\varepsilon) \\ 0 & 0 & 0 & 0 & 0 & 0 \end{bmatrix}. \tag{11.3.35}$$

This completes the construction of the parameterized state feedback gain matrix $F(\varepsilon)$. ▣

The following theorem then shows that the above algorithm indeed yields a family of state feedback laws that solves the general H_∞-ADDPS.

Theorem 11.3.3. *Consider the given system (11.3.11) in which all the states are available for feedback. Assume that the problem of H_∞ almost disturbance* decoupling with internal stability for (11.3.11) is solvable, i.e., the solvability conditions of Remark 11.3.1 are satisfied. Then, the closed-loop system comprising (11.3.11) and the full state feedback control law

$$u = F(\varepsilon)x, \tag{11.3.36}$$

with $F(\varepsilon)$ given by (11.3.33), has the following properties: For any given $\gamma > 0$, there exists a positive scalar $\varepsilon^* > 0$ such that for all $\varepsilon \in (0, \varepsilon^*]$,

1. the closed-loop system is asymptotically stable, i.e., $\lambda(A + BF(\varepsilon))$ are on the open unit disc; and

2. the H_∞-norm of the closed-loop transfer matrix from the disturbance w to the controlled output z is less than or equal to γ, i.e., $\|T_{zw}(z, \varepsilon)\|_\infty \leq \gamma$.

Hence, the family of control laws as given by (11.3.36) solves the H_∞-ADDPMS for (11.3.11).   ⊞

Proof of Theorem 11.3.3. Under the feedback control law $u = F(\varepsilon)x$, the closed-loop system on the special coordinate basis can be written as follows,

$$x_a^{-+} = A_{aa}^- x_a^- + B_{0a}^- z_0 + L_{ad}^- z_d + L_{ab}^- z_b + E_a^- w, \tag{11.3.37}$$

$$(x_a^0)^+ = A_{aa}^0 x_a^0 + B_{0a}^0 z_0 + L_{ad}^0 z_d + L_{ab}^0 z_b + E_a^0 w, \tag{11.3.38}$$

$$(x_{abd}^+)^+ = A_{abd}^{+c} x_{abd}^+ + [B_{0abd}^+, B_{abd}^+] F_a^0(\varepsilon)[x_a^0 + T_a^0 x_{abd}^+] + E_{abd}^+ w, \tag{11.3.39}$$

$$x_c^+ = A_{cc}^c x_c + B_{0c} z_0 + L_{cb} z_b + L_{cd} z_d + E_c w, \tag{11.3.40}$$

$$z_0 = [F_{a0}^+, F_{b0}, F_{d0}] x_{abd}^+ + F_{a0}^0(\varepsilon)(x_a^0 + T_a^0 x_{abd}^+), \tag{11.3.41}$$

$$z_b = [0_{m_b \times n_a^+}, C_b, 0_{m_b \times n_d}] x_{abd}^+, \tag{11.3.42}$$

$$z_d = [0_{m_b \times n_a^+}, 0_{m_b \times n_b}, C_d] x_{abd}^+, \tag{11.3.43}$$

where x_a^-, x_a^0, $x_{abd}^+ = [(x_a^+)', x_b', x_d']'$ and x_c are the state variables with the time index k suppressed, $(x_a^-)^+$, $(x_a^0)^+$, $(x_{abd}^+)^+$ and x_c^+ in the left hand side of the equations denote respectively $x_a^-(k+1)$, $x_a^0(k+1)$, $x_{abd}^+(k+1)$ and $x_c(k+1)$, and B_{0abd}^+ is as defined in Step 4.1 of the state feedback design algorithm. We have also used Condition (b) of Remark 11.3.1, i.e., $D_{22} = 0$, and E_a^-, E_a^0, E_{abd}^+, E_b and E_c are defined as follows,

$$\Gamma_{\mathrm{SP}}^{-1} E = [(E_a^-)' \quad (E_a^0)' \quad (E_{ab}^+)' \quad E_c' \quad E_d']', \quad E_{abd}^+ = [(E_{ab}^+)' \quad E_d']'. \tag{11.3.44}$$

Condition (c) of Remark 11.3.1 then implies that

$$E_{abd}^+ = 0, \tag{11.3.45}$$

and

$$\text{Im}(E_a^0) \subset \mathcal{S}(A_{aa}^0) = \cap_{\omega \in \lambda(A_{aa}^0)} \text{Im}\{\omega I - A_{aa}^0\}. \tag{11.3.46}$$

To complete the proof, we will make two state transformations on the closed-loop system (11.3.37)-(11.3.43). The first state transformation is given as follows,

$$\bar{x}_{abd} = \Gamma_{abd}^{-1} x_{abd}, \quad \bar{x}_c = x_c, \tag{11.3.47}$$

where $x_{abd} = [(x_a^-)', (x_a^0)', (x_{abd}^+)']'$ and $\bar{x}_{abd} = [(\bar{x}_a^-)', (\bar{x}_{abd}^+)', (\bar{x}_a^0)']'$. In the new state variables (11.3.47), the closed-loop system becomes,

$$(\bar{x}_a^-)^+ = A_{aa}^- \bar{x}_a^- + A_{aabd+}^- \bar{x}_{abd}^+ + B_{0a}^- F_{a0}^0(\varepsilon)\bar{x}_a^0 + E_a^- w, \tag{11.3.48}$$

$$(\bar{x}_{abd}^+)^+ = A_{abd}^{+c} \bar{x}_{abd}^+ + [B_{0abd}^+, B_{abd}^+]F_a^0(\varepsilon)\bar{x}_a^0, \tag{11.3.49}$$

$$(\bar{x}_a^0)^+ = (A_{aa}^0 + B_a^0 F_a^0(\varepsilon))\bar{x}_a^0 + E_a^0 w, \tag{11.3.50}$$

$$\bar{x}_c^+ = A_{cc}^c \bar{x}_c + A_{cabd+} \bar{x}_{abd}^+ + B_{0c} F_{a0}^0(\varepsilon)\bar{x}_a^0 + E_c w, \tag{11.3.51}$$

$$z_0 = [F_{a0}^+, F_{b0}, F_{d0}]x_{abd}^+ + F_{a0}^0(\varepsilon)\bar{x}_a^0, \tag{11.3.52}$$

$$z_b = [0_{m_b \times n_a^+}, C_b, 0_{m_b \times n_d}]x_{abd}^+, \tag{11.3.53}$$

$$z_d = [0_{m_b \times n_a^+}, 0_{m_b \times n_b}, C_d]x_{abd}^+, \tag{11.3.54}$$

where

$$A_{aabd+}^- = B_{0a}^-[F_{a0}^+ \quad F_{b0} \quad F_{d0}] + L_{ad}^-[0 \quad 0 \quad C_d] + L_{ab}^-[0 \quad C_b \quad 0] \tag{11.3.55}$$

and

$$A_{cabd+} = B_{0c}[F_{a0}^+ \quad F_{b0} \quad F_{d0}] + L_{cb}[0 \quad C_b \quad 0] + L_{cd}[0 \quad 0 \quad C_d]. \tag{11.3.56}$$

We now define the following second state transformation on the closed-loop system,

$$\tilde{x}_a^- = \bar{x}_a^-, \quad \tilde{x}_{abd}^+ = \bar{x}_{abd}^+, \tag{11.3.57}$$

$$\tilde{x}_a^0 = [(\tilde{x}_{a1}^0)', (\tilde{x}_{a2}^0)', \cdots, (\tilde{x}_{al}^0)']' = S_a(\varepsilon)Q_a^{-1}(\varepsilon)(\Gamma_{sa}^0)^{-1}\bar{x}_a^0, \tag{11.3.58}$$

$$S_a(\varepsilon) = \text{blkdiag}\{S_{a1}(\varepsilon), S_{a2}(\varepsilon), \cdots, S_{al}(\varepsilon)\},$$

$$Q_a(\varepsilon) = \text{blkdiag}\{Q_{a1}(\varepsilon), Q_{a2}(\varepsilon), \cdots, Q_{al}(\varepsilon)\},$$

$$\tilde{x}_c = \varepsilon\bar{x}_c, \tag{11.3.59}$$

where $Q_{ai}(\varepsilon)$ and $S_{ai}(\varepsilon)$ are the $Q(\varepsilon)$ and $S(\varepsilon)$ of Lemmas 2.3.2 and 2.3.3 for the triple $(A_i, B_i, F_i(\varepsilon))$. Hence, Lemmas 2.3.2-2.3.5 all apply. In these new state variables, the closed-loop system becomes,

$$(\tilde{x}_a^-)^+ = A_{aa}^- \tilde{x}_a^- + A_{aabd+}^- \tilde{x}_{abd}^+$$

$$+B_{0a}^- F_{a0}^0(\varepsilon)\Gamma_{sa}^0 Q_a(\varepsilon)S_a^{-1}(\varepsilon)\tilde{x}_a^0 + E_a^- w, \tag{11.3.60}$$

$$(\tilde{x}_{abd}^+)^+ = A_{abd}^{+c}\tilde{x}_{abd}^+ + [B_{0abd}^+, B_{abd}^+]F_a^0(\varepsilon)\Gamma_{sa}^0 Q_a(\varepsilon)S_a^{-1}(\varepsilon)\tilde{x}_a^0, \tag{11.3.61}$$

$$(\tilde{x}_a^0)^+ = \tilde{J}_a(\varepsilon)\tilde{x}_a^0 + \tilde{B}_a(\varepsilon)\tilde{x}_a^0 + \tilde{E}_a^0(\varepsilon)w, \tag{11.3.62}$$

$$\tilde{x}_c^+ = A_{cc}^c\tilde{x}_c + \varepsilon[A_{cabd}+\tilde{x}_{abd}^+$$

$$+B_{0c}F_{a0}^0(\varepsilon)\Gamma_{sa}^0 Q_a(\varepsilon)S_a^{-1}(\varepsilon)\tilde{x}_a^0 + E_c w], \tag{11.3.63}$$

$$z_0 = [F_{a0}^+, F_{b0}, F_{d0}]x_{abd}^+ + F_{a0}^0(\varepsilon)\Gamma_{sa}^0 Q_a(\varepsilon)S_a^{-1}(\varepsilon)\tilde{x}_a^0, \tag{11.3.64}$$

$$z_b = [0_{m_b\times n_a^+}, C_b, 0_{m_b\times n_d}]x_{abd}^+, \tag{11.3.65}$$

$$z_d = [0_{m_b\times n_a^+}, 0_{m_b\times n_b}, C_d]x_{abd}^+, \tag{11.3.66}$$

where

$$\tilde{J}_a(\varepsilon) = \text{blkdiag}\{\tilde{J}_{a1}(\varepsilon), \tilde{J}_{a2}(\varepsilon), \cdots, \tilde{J}_{al}(\varepsilon)\}, \tag{11.3.67}$$

$$\tilde{B}_a(\varepsilon) = \begin{bmatrix} 0 & \tilde{B}_{12}(\varepsilon) & \tilde{B}_{13}(\varepsilon) & \cdots & \tilde{B}_{1l}(\varepsilon) \\ 0 & 0 & \tilde{B}_{23}(\varepsilon) & \cdots & \tilde{B}_{2l}(\varepsilon) \\ \vdots & \vdots & \vdots & \ddots & \vdots \\ 0 & 0 & 0 & \cdots & 0 \end{bmatrix}, \tag{11.3.68}$$

$$\tilde{B}_{ij}(\varepsilon) = S_{ai}(\varepsilon)Q_{ai}^{-1}(\varepsilon)B_{ij}F_j(\varepsilon)Q_{aj}(\varepsilon)S_{aj}^{-1}(\varepsilon),$$

$$i = 1, 2, \cdots, l, j = i+1, i+2, \cdots, l, \tag{11.3.69}$$

$$\tilde{E}_a^0(\varepsilon) = S_a(\varepsilon)Q_a^{-1}(\varepsilon)(\Gamma_{sa}^0)^{-1}E_a^0,$$

$$\tilde{E}_a^0(\varepsilon) = [(\tilde{E}_{a1}^0(\varepsilon))' \quad (\tilde{E}_{a2}^0(\varepsilon))' \quad \cdots \quad (\tilde{E}_{al}^0(\varepsilon))']', \tag{11.3.70}$$

and where, for $i = 1$ to l, $\tilde{J}_{ai}(\varepsilon)$ is as defined in Lemma 2.3.2.

By Lemmas 2.2.4 and 2.2.5, we have that, for all $\varepsilon \in (0, 1]$,

$$|F_a^0(\varepsilon)\Gamma_{sa}^0 Q_a(\varepsilon)S_a^{-1}(\varepsilon)| \leq \tilde{f}_{a0}^0\varepsilon, \tag{11.3.71}$$

for $i = 1$ to l,

$$|\tilde{E}_{ai}^0(\varepsilon)| \leq \tilde{e}_a^0\varepsilon, \tag{11.3.72}$$

and finally, for $i = 1$ to l, $j = i+1$ to l,

$$|\tilde{B}_{ij}(\varepsilon)| \leq \tilde{b}_{ij}\varepsilon, \tag{11.3.73}$$

where $\tilde{f}_{a0}^0$, $\tilde{e}_a^0$, and $\tilde{b}_{ij}$ are some positive constants, independent of ε.

We next construct a Lyapunov function for the closed loop system (11.3.60)-(11.3.66). We do this by composing Lyapunov functions for the subsystems. For the subsystem of $\tilde{x}_a^-$, we choose a Lyapunov function,

$$V_a^-(\tilde{x}_a^-) = (\tilde{x}_a^-)'P_a^-\tilde{x}_a^-, \tag{11.3.74}$$

where $P_a^- > 0$ is the unique solution to the Lyapunov equation,

$$(A_{aa}^-)'P_a^- A_{aa}^- - P_a^- = -I, \tag{11.3.75}$$

and for the subsystem of $\tilde{x}_{abd}^+$, choose a Lyapunov function,

$$V_{abd}^+(\tilde{x}_{abd}^+) = (\tilde{x}_{abd}^+)'P_{abd}^+ \tilde{x}_{abd}^+, \tag{11.3.76}$$

where $P_{abd}^+ > 0$ is the unique solution to the Lyapunov equation,

$$(A_{abd}^{+c})'P_{abd}^+ A_{abd}^{+c} - P_{abd}^+ = -I. \tag{11.3.77}$$

The existence of such P_a^- and P_{ab}^+ is guaranteed by the fact that both A_{aa}^- and A_{abd}^{+c} are asymptotically stable. For the subsystem of $\tilde{x}_a^0 = [(\tilde{x}_{a1}^0)', (\tilde{x}_{a2}^0)', \cdots,$ $(\tilde{x}_{al}^0)']'$, we choose a Lyapunov function,

$$V_a^0(\tilde{x}_a^0) = \sum_{i=1}^{l} \frac{(\kappa_a^0)^{i-1}}{\varepsilon}(\tilde{x}_{ai}^0)'P_{ai}^0(\varepsilon)\tilde{x}_{ai}^0, \tag{11.3.78}$$

where κ_a^0 is a positive scalar, whose value is to be determined later, and each $P_{ai}^0(\varepsilon)$ is the unique solution to the Lyapunov equation,

$$\tilde{J}_{ai}(\varepsilon)'P_{ai}^0 \tilde{J}_{ai}(\varepsilon) - P_{ai}^0 = -\varepsilon I, \tag{11.3.79}$$

which, by Lemma 2.3.3, satisfies,

$$P_{ai}(\varepsilon) \le P_{ai2}, \quad \forall \varepsilon \in (0, \varepsilon_0^*], \tag{11.3.80}$$

for some $\varepsilon_0^* \in (0, 1]$ and some positive definite P_{ai2} independent of ε. Similarly, for the subsystem $\tilde{x}_c$, choose a Lyapunov function,

$$V_c(\tilde{x}_c) = \tilde{x}_c' P_c \tilde{x}_c, \tag{11.3.81}$$

where $P_c > 0$ is the unique solution to the Lyapunov equation,

$$(A_{cc}^c)'P_c A_{cc}^c - P_c = -I. \tag{11.3.82}$$

The existence of such a P_c is again guaranteed by the fact that A_{cc}^c is asymptotically stable.

We now construct a Lyapunov function for the closed-loop system (11.3.60)-(11.3.66) as follows,

$$V(\tilde{x}_a^-, \tilde{x}_{abd}^+, \tilde{x}_a^0, \tilde{x}_c) = V_a^-(\tilde{x}_a^-) + \kappa_{abd}^+ V_{abd}^+(\tilde{x}_{abd}^+) + V_a^0(\tilde{x}_a^0) + V_c(\tilde{x}_c), \tag{11.3.83}$$

where $\kappa_{abd}^+ = 2|P_a^{-1}|^2|A_{aa}^-|^2 + 1$.

Let us first consider the difference of $V_a^0(\tilde{x}_a^0)$ along the trajectories of the subsystem $\tilde{x}_a^0$ and obtain that,

$$\Delta V_a^0 = \sum_{i=1}^{l} \left[-(\kappa_a^0)^{i-1}(\tilde{x}_{ai}^0)'\tilde{x}_{ai}^0 + 2 \sum_{j=i+1}^{l} \frac{(\kappa_a^0)^{i-1}}{\varepsilon}(\tilde{x}_{ai}^0)'\tilde{J}_{ai}'(\varepsilon)P_{ai}^0(\varepsilon) \right.$$
$$\times \left[\tilde{B}_{ij}(\varepsilon)\tilde{x}_{aj}^0 + \tilde{E}_{ai}^0(\varepsilon)w \right]$$
$$+ \frac{(\kappa_a^0)^{i-1}}{\varepsilon}\left(\sum_{j=i+1}^{l} \tilde{B}_{ij}(\varepsilon)\tilde{x}_{aj}^0(\varepsilon) + \tilde{E}_{ai}^0(\varepsilon)w \right)' P_{ai}^0(\varepsilon)$$
$$\left. \times \left(\sum_{j=i+1}^{l} \tilde{B}_{ij}(\varepsilon)\tilde{x}_{aj}^0(\varepsilon) + \tilde{E}_{ai}^0(\varepsilon)w \right) \right] \tag{11.3.84}$$

Using (11.3.72), (11.3.73) and Lemma 2.3.3, it is straightforward to show that, there exists a $\kappa_a^0 > 0$ such that,

$$\Delta V_a^0 \le -\frac{3}{4}|\tilde{x}_a^0|^2 + \alpha_1|w|^2, \tag{11.3.85}$$

for some nonnegative constants α_1, independent of ε.

In view of (11.3.85), the difference of V along the trajectory of the closed-loop system (11.3.60)-(11.3.66) can be evaluated as follows,

$$\Delta V \le -(\tilde{x}_a^-)'\tilde{x}_a^- + 2(\tilde{x}_a^-)'(A_{aa}^-)'P_a^-$$
$$\times[A_{aabd}^-(\varepsilon)\tilde{x}_{abd}^+ + B_{0a}^- F_{a0}^0(\varepsilon)\Gamma_{sa}^0 Q_a(\varepsilon)S_a^{-1}(\varepsilon)\tilde{x}_a^0 + E_a^- w]$$
$$-\kappa_{abd}^+(\tilde{x}_{abd}^+)'\tilde{x}_{abd}^+ + 2\kappa_{abd}^+(\tilde{x}_{abd}^+)'(A_{abd}^{+c})'P_{abd}^+$$
$$\times[B_{0abd}^+, B_{abd}^+]F_a^0(\varepsilon)\Gamma_{sa}^0 Q_a(\varepsilon)S_a^{-1}(\varepsilon)\tilde{x}_a^0$$
$$-\frac{3}{4}|\tilde{x}_a^0|^2 + \alpha_1|w|^2 - \tilde{x}_c'\tilde{x}_c + 2\varepsilon\tilde{x}_c'(A_{cc}^{+c})'P_c[A_{cabd+}\tilde{x}_{abd}^+$$
$$+B_{0c}F_{a0}^0(\varepsilon)\Gamma_{sa}^0 Q_a(\varepsilon)S_a^{-1}(\varepsilon)\tilde{x}_a^0 + E_c w]. \tag{11.3.86}$$

Using (11.3.71) and noting the definition of κ_{abd}^+, we can easily verify that, there exists an $\varepsilon_1^* \in (0, \varepsilon_0^*]$ such that, for all $\varepsilon \in (0, \varepsilon_1^*]$,

$$\dot{V} \le -\frac{1}{2}|\tilde{x}_a^-|^2 - \frac{1}{2}|\tilde{x}_{ab}^+|^2 - \frac{1}{2}|\tilde{x}_a^0|^2 - \frac{1}{2}|\tilde{x}_c|^2 + \alpha_2|w|^2, \tag{11.3.87}$$

for some positive constant α_2, independent of ε.

From (11.3.87), it follows that the closed-loop system in the absence of disturbance w is asymptotically stable. It remains to show that, for any given $\gamma > 0$, there exists an $\varepsilon^* \in (0, \varepsilon_1^*]$ such that, for all $\varepsilon \in (0, \varepsilon^*]$,

$$\|z\|_{l_2} \le \gamma\|w\|_{l_2}. \tag{11.3.88}$$

To this end, we sum both sides of (11.3.87) from 0 to ∞. Noting that $V \geq 0$ and $V(k) = 0$ at $k = 0$, we have,

$$\|\tilde{x}_a^0\|_{l_2} \leq (\sqrt{2\alpha_3}) \|w\|_{l_2}, \tag{11.3.89}$$

which, when used together with (11.3.71) in (11.3.61), results in,

$$\|\tilde{x}_{abd}^+\|_{l_2} \leq \alpha_3 \varepsilon \|w\|_{l_2}. \tag{11.3.90}$$

for some positive constant α_3, independent of ε.

Finally, recalling that

$$z = \Gamma_{\mathrm{OP}}[z_0', z_d', z_b']', \tag{11.3.91}$$

where z_0, z_d and z_b are as defined in the closed-loop system (11.3.60)-(11.3.66), we have,

$$\|z\|_{l_2} \leq \alpha_4 |\Gamma_{\mathrm{OP}}| \varepsilon \|w\|_{l_2}, \tag{11.3.92}$$

for some positive constant α_4 independent of ε.

To complete the proof, we choose $\varepsilon^* \in (0, \varepsilon_1^*]$ such that,

$$\alpha_4 |\Gamma_{\mathrm{OP}}| \varepsilon \leq \gamma. \tag{11.3.93}$$

⊠

11.4. Nonlinear Systems

11.4.1. Problem Statement

We consider the problem of global almost disturbance decoupling with stability for nonlinear systems of the form,

$$\dot{x} = f(x) + g(x)u + p(x)w, \quad y = h(x), \tag{11.4.1}$$

where $x \in \mathbb{R}^n$ is the state, $u \in \mathbb{R}$ is the control input, $w \in \mathbb{R}$ is the disturbance, $y \in \mathbb{R}$ is the regulated output, f, g, and p are smooth vector fields with $f(0) = 0$, and h is a smooth function with $h(0) = 0$. The problem of almost disturbance decoupling with stability was originally formulated for linear systems by Willems in [124] (see also Sections 11.2 and 11.3). Since then various generalizations to nonlinear systems have been made (see, for example, Chapters 6 and 7 and [25,26,74,75,90] and the references therein). The problem we are to consider in this section is formulated as follows.

Problem 11.4.1. (L_2 almost disturbance decoupling with global asymptotic stability) *The problem of L_2 almost disturbance decoupling with global asymptotic stability is said to be solvable for system (11.4.1) if, for any given $\gamma > 0$,* there is a smooth feedback law $u = u(x; \gamma)$ with $u(0, \gamma) = 0$, such that the corresponding closed-loop system

(a) has a globally asymptotically stable equilibrium at $x = 0$;

(b) has an L_2 gain, from the disturbance input w to the regulated output y, that is less than or equal to γ, i.e.,

$$\int_0^\infty y^2(t)dt \le \gamma^2 \int_0^\infty w^2(t)dt, \quad \forall w \in L_2 \text{ and for } x(0) = 0. \tag{11.4.2}$$

P

In a recent series of papers ([25,26,75]), some elegant results were obtained for system (11.4.1) in the following special form,

$$
\begin{aligned}
\dot{z} &= f_0(z, \xi_1) + p_0(z, \xi_1)w, \\
\dot{\xi}_1 &= \xi_2 + p_1(z, \xi_1)w, \\
\dot{\xi}_2 &= \xi_3 + p_2(z, \xi_1, \xi_2)w, \\
&\vdots \\
\dot{\xi}_{r-1} &= \xi_r + p_{r-1}(z, \xi_1, \xi_2, \cdots, \xi_{r-1})w, \\
\dot{\xi}_r &= u + p_r(z, \xi_1, \xi_2, \cdots, \xi_{r-1}, \xi_r)w, \\
y &= \xi_1,
\end{aligned}
\tag{11.4.3}
$$

in which it is assumed that $f_0(0,0) = 0$ and the dynamics

$$\dot{z} = f_0(z, 0) \tag{11.4.4}$$

is referred to as the zero dynamics of the system (11.4.3) [24]. Throughout this section, we will also, by somewhat abuse of terminology, refer to the first equation of (11.4.3) as the zero dynamics equation. More specifically, it is shown in [75] that the problem of L_2 almost disturbance decoupling with global asymptotic stability is solvable if

(i) the equilibrium $z = 0$ of the zero dynamics (11.4.4) is globally asymptotically stable; and

(ii) $p_0(z, 0) = 0$.

and, under these conditions, feedback laws of high gain type that solve the problem are also explicitly constructed. This result of [75] was recently generalized in [25] in the sense that Condition (ii) is replaced by a weaker one. The requirement that the zero dynamics be globally asymptotically stable (Condition (i)), however, remains un-relaxed.

More recently, the results of [25,75] are further generalized in [26] to allow part of the zero dynamics to be unstable as long as it satisfies certain stabilizability condition and its corresponding zero dynamics equation *is unaffected* by the disturbance w. The zero dynamics equation (recall that the first equation of (11.4.1) is referred to as the zero dynamics equation) considered in [26] takes the following cascade-connected form with two subsystems,

$$
\begin{aligned}
\dot{z}_a &= f_a(z_a, z_c, \xi_1) + p_a(z_a, z_c, \xi_1)w, \\
\dot{z}_c &= f_c(z_c, \xi_1),
\end{aligned}
\tag{11.4.5}
$$

where the first one characterizes a "stable part" of the zero dynamics (more precisely, $z_a = 0$ is a globally asymptotically stable equilibrium of $\dot{z}_a = f_a(z_a, 0, 0)$), and the second one characterizes a possibly unstable but stabilizable and disturbance unaffected part of the zero dynamics. The conditions needed on both subsystems for solving the problem of L_2 almost disturbance decoupling with globally asymptotic stability can be made more precise by recalling the following result from [26].

Theorem 11.4.1. *Suppose that*

(i) there exists a smooth real-valued function $V_a(z_a)$, which is positive definite and proper, such that

$$
\frac{\partial V_a}{\partial z_a}\left[f_a(z_a, z_c, \xi_1) + p_a(z_a, z_c, \xi_1)w\right] \leq -\alpha_a(|z_a|) + \gamma_0^2|w|^2 + \gamma_0^2|z_c|^2 + \gamma_0^2|\xi_1|^2,
\tag{11.4.6}
$$

for some $\mathcal{K}_\infty$ function α_a and some positive real number γ_0, and

(ii) there exists a smooth real-valued function $v_c(z_c)$, with $v_c(0) = 0$, and a smooth real-valued function $V_c(z_c)$, which is positive definite and proper, such that

$$
\frac{\partial V_c}{\partial z_c} f_c(z_c, v_c(z_c)) + |v_c(z_c)|^2 \leq -\alpha_c(|z_c|),
\tag{11.4.7}
$$

for some $\mathcal{K}_\infty$ function α_c.

Then, the problem of L_2 almost disturbance decoupling with global asymptotic stability is solvable for system (11.4.3) with its first equation in the form of (11.4.5). ⊡

For use in the proof of our main result, we also recall the following observation from [26].

Observation 11.4.1. *We note here the emphasis on that (11.4.6) be true for some positive real number γ_0. Indeed what is needed in the proof is that for the arbitrary positive number γ there exist V_a and α_a such that (11.4.6) hold. One observes that the former is sufficient to guarantee the latter, for one can multiply both sides of (11.4.6) by γ^2/γ_0^2 and redefine V_1 and α_a accordingly.* ◻

The objective of this section is to show that a certain class of unstable zero dynamics is actually allowed to be affected by the disturbance in solving the problem of L_2 almost disturbance decoupling with global asymptotic stability. This would complement the result of [26], where unstable zero dynamics is not allowed to be affected by the disturbance. The disturbance affected unstable part of the zero dynamics we consider contains a chain of integrators of arbitrary length with every integator except the last one affected by the disturbance. The key to arriving at this result is again to use low gain feedback to stabilize this disturbance affected unstable part of the zero dynamics.

11.4.2. Solution of the H_∞-ADDPS for a Class of Nonlinear Systems

Consider the system (11.4.3) with its first equation in the form of,

$$
\begin{aligned}
\dot{z}_a &= f_a(z_a, z_b, \xi_1) + p_a(z_a, z_b, \xi_1)w, \\
\dot{z}_{b1} &= z_{b2} + p_{b1}(z_a, z_b, \xi_1)w, \\
\dot{z}_{b2} &= z_{b3} + p_{b2}(z_a, z_b, \xi_1)w, \\
&\quad\vdots \\
\dot{z}_{bq-1} &= z_{bq} + p_{bq-1}(z_a, z_b, \xi_1)w, \\
\dot{z}_{bq} &= \xi_1,
\end{aligned}
\tag{11.4.8}
$$

where $z_b = [z_{b1}, z_{b2}, \cdots, z_{bq}]'$. We note here that the dynamics of z_b contains a chain of integrators of length q and is unstable for $q \geq 2$ and stable but not asymptotically stable for $q = 1$.

Our main result is presented in the following theorem.

Theorem 11.4.2. *Consider the system (11.4.3) with its first equation in the form of (11.4.8). Suppose that*

(i) there exists a smooth real-valued function $V_a(z_a)$, which is positive definite and proper, such that

$$
\frac{\partial V_a}{\partial z_a}\left[f_a(z_a, z_b, \xi_1) + p_a(z_a, z_b, \xi_1)w\right] \leq -\alpha_a(|z_a|) + \gamma_0^2 |w|^2 + \gamma_0^2 |z_b|^2 + \gamma_0^2 |\xi_1|^2,
\tag{11.4.9}
$$

for some $\mathcal{K}_\infty$ function α_a and some positive real number γ_0, and

(ii) there exists a constant number $\delta \geq 0$ such that

$$
|p_{bi}(z_a, z_b, \xi_1)| \leq \delta, \ \ for \ i = 1 \ to \ q - 1.
\tag{11.4.10}
$$

Then, the problem of L_2 almost disturbance decoupling with global asymptotic stability is solvable.   ⊡

Proof of Theorem 11.4.2. We begin the proof by defining a low gain state feedback

$$
u_b(z_b) = -\varepsilon^q k_{b1} z_{b1} - \varepsilon^{q-1} k_{b2} z_{b2} - \cdots - \varepsilon k_{bq} z_{bq},
\tag{11.4.11}
$$

where the vector $k_b = [k_{b1}, k_{b2}, \cdots, k_{bq}]'$ is chosen in such a way that the polynomial

$$s^q + k_{bq}s^{q-1} + k_{bq-1}s^{q-2} + \cdots + k_{b1} \tag{11.4.12}$$

is Hurwitz and $\varepsilon \in (0,1]$ is a design parameter to be specified later. We next rename the output of the system as

$$\tilde{y} = \xi_1 - u_b(z_b) = \xi_1 + \varepsilon^q k_{b1}z_{b1} + \varepsilon^{q-1}k_{b2}z_{b2} + \cdots + \varepsilon k_{bq}z_{bq}. \tag{11.4.13}$$

With this new output $\tilde{y}$, we define a new set of state variables for the system, $\tilde{z}_a, \tilde{z}_b, \tilde{\xi}_1, \tilde{\xi}_2, \cdots, \tilde{\xi}_r$, as,

$$\tilde{z}_a = z_a,$$
$$\tilde{z}_b = S_b(\varepsilon)z_b, \quad S_b(\varepsilon) = \text{diag}\{\varepsilon^{q-1}, \varepsilon^{q-2}, \cdots, \varepsilon, 1\},$$
$$\tilde{\xi}_1 = \tilde{y} = \xi_1 + \varepsilon^q k_{b1}z_{b1} + \varepsilon^{q-1}k_{b2}z_{b2} + \cdots + \varepsilon^2 k_{bq-1}z_{bq-1} + \varepsilon k_{bq}z_{bq},$$
$$\tilde{\xi}_2 = \xi_2 + \varepsilon^q k_{b1}z_{b2} + \varepsilon^{q-1}k_{b2}z_{b3} + \cdots + \varepsilon^2 k_{bq-1}z_{bq} + \varepsilon k_{bq}z_{bq+1},$$
$$\vdots$$
$$\tilde{\xi}_r = \xi_r + \varepsilon^q k_{b1}z_{br} + \varepsilon^{q-1}k_{b2}z_{br+1} + \cdots + \varepsilon^2 k_{bq-1}z_{bq+r-2} + \varepsilon k_{bq}z_{bq+r-1},$$
$$\tag{11.4.14}$$

where we have denoted $z_{bi} = \xi_{i-q}$ for $i \geq q + 1$.

We also choose a pre-feedback law as,

$$u = -\varepsilon^q k_{b1}z_{br+1} - \varepsilon^{q-1}k_{b2}z_{br+2} - \cdots - \varepsilon^2 k_{bq-1}z_{bq+r-1} - \varepsilon k_{bq}z_{bq+r} + \tilde{u}. \tag{11.4.15}$$

Under this pre-feedback law, the closed-loop system in the new state variables can be rewritten as follows,

$$\begin{aligned}
\dot{z}_a &= f_a(\tilde{z}_a, S_b^{-1}(\varepsilon)\tilde{z}_b, \tilde{x}_1 + u_b(S_b^{-1}(\varepsilon)\tilde{z}_b)) \\
&\quad + p_a(\tilde{z}_a, S_b^{-1}(\varepsilon)\tilde{z}_b, \tilde{x}_1 + u_b(S_b^{-1}(\varepsilon)\tilde{z}_b))w, \\
\dot{\tilde{z}}_b &= \varepsilon A_b\tilde{z}_b + B_b\tilde{x}_1 + \varepsilon\tilde{p}_b(\tilde{z}_a, \tilde{z}_b, \tilde{x}_1)w, \\
\dot{\tilde{\xi}}_1 &= \tilde{x}_2 + \tilde{p}_1(\tilde{z}_a, \tilde{z}_b, \tilde{x}_1)w, \\
\dot{\tilde{\xi}}_2 &= \tilde{x}_3 + \tilde{p}_2(\tilde{z}_a, \tilde{z}_b, \tilde{x}_1, \tilde{x}_2)w, \\
&\quad\vdots \\
\dot{\tilde{\xi}}_{r-1} &= \tilde{x}_q + \tilde{p}_{r-1}(\tilde{z}_a, \tilde{z}_b, \tilde{x}_1, \tilde{x}_2, \cdots, \tilde{x}_{r-1})w, \\
\dot{\tilde{\xi}}_r &= \tilde{u} + \tilde{p}_r(\tilde{z}_a, \tilde{z}_b, \tilde{x}_1, \tilde{x}_2, \cdots, \tilde{x}_r)w,
\end{aligned} \tag{11.4.16}$$

where

$$A_b = \begin{bmatrix} 0 & 1 & 0 & \cdots & 0 \\ 0 & 0 & 1 & \cdots & 0 \\ \vdots & \vdots & \vdots & \ddots & \vdots \\ 0 & 0 & 0 & \cdots & 1 \\ -k_{b1} & -k_{b2} & -k_{b3} & \cdots & -k_{bq} \end{bmatrix}, \quad B_b = \begin{bmatrix} 0 \\ 0 \\ \vdots \\ 0 \\ 1 \end{bmatrix}, \tag{11.4.17}$$

$$\tilde{p}_b(\tilde{z}_a, \tilde{z}_b, \tilde{x}_1) = \begin{bmatrix} \varepsilon^{q-2} p_{b1}(\tilde{z}_a, S_b^{-1}(\varepsilon)\tilde{z}_b, \tilde{x}_1 + u_b(S_b^{-1}(\varepsilon)\tilde{z}_b)) \\ \varepsilon^{q-3} p_{b2}(\tilde{z}_a, S_b^{-1}(\varepsilon)\tilde{z}_b, \tilde{x}_1 + u_b(S_b^{-1}(\varepsilon)\tilde{z}_b)) \\ \vdots \\ \varepsilon p_{bq-2}(\tilde{z}_a, S_b^{-1}(\varepsilon)\tilde{z}_b, \tilde{x}_1 + u_b(S_b^{-1}(\varepsilon)\tilde{z}_b)) \\ p_{bq-1}(\tilde{z}_a, S_b^{-1}(\varepsilon)\tilde{z}_b, \tilde{x}_1 + u_b(S_b^{-1}(\varepsilon)\tilde{z}_b)) \\ 0 \end{bmatrix}, \tag{11.4.18}$$

and $\tilde{p}_i(\tilde{z}_a, \tilde{z}_b, \tilde{x}_1, \tilde{x}_2, \cdots, \tilde{x}_i)$, for $i = 1$ to r, are defined in a straightforward way.

We now observe that system (11.4.16) is in the form of (11.4.3) and (11.4.5) with the first equation of (11.4.5) corresponding to the dynamics of $\tilde{z}_a$ and $\tilde{z}_b$ and the second equation of (11.4.5) non-existent. We hence can apply Theorem 11.4.1 to system (11.4.16). Condition (ii) of Theorem 11.4.1 is automatically satisfied. To verify Condition (i) of Theorem 11.4.1, we will show that there exists an $\varepsilon^* \in (0, 1]$ such that for each $\varepsilon \in (0, \varepsilon^*]$, there exists a $V_{ab}(\tilde{z}_a, \tilde{z}_b)$ and α_{ab}, the following inequality corresponding to (11.4.6) holds,

$$\begin{aligned} \frac{\partial V_{ab}}{\partial \tilde{z}_a} [\ &f_a(\tilde{z}_a, S_b^{-1}(\varepsilon)\tilde{z}_b, \tilde{x}_1 + u_b(S_b^{-1}(\varepsilon)\tilde{z}_b)) \\ &+ p_a(\tilde{z}_a, S_b^{-1}(\varepsilon)\tilde{z}_b, \tilde{x}_1 + u_b(S_b^{-1}(\varepsilon)\tilde{z}_b))w] \\ &+ \frac{\partial V_{ab}}{\partial z_b}[\varepsilon A_b \tilde{z}_b + B_b \tilde{x}_1 + \varepsilon \tilde{p}_b(\tilde{z}_a, \tilde{z}_b, \tilde{x}_1)w] \\ \leq\ &-\alpha_{ab}(|[\tilde{z}_a', \tilde{z}_b']'|) + \varepsilon^2 |w|^2 + \varepsilon^2 |\tilde{x}_1|^2. \end{aligned} \tag{11.4.19}$$

Let us choose,

$$V_{ab}(\tilde{z}_a, \tilde{z}_b) = \varepsilon^{2q+6} V_a(\tilde{z}_a) + \varepsilon^5 \tilde{z}_b' P_b \tilde{z}_b, \tag{11.4.20}$$

where the function V_a is as given by Condition (i) of the theorem, P_b is the positive definite solution of the following Lyapunov function,

$$A_b' P_b + P_b A_b = -I. \tag{11.4.21}$$

Such a solution exists since the matrix A_b is asymptotically stable. Noting that $|\tilde{p}_b(\tilde{z}_a, \tilde{z}_b, \tilde{x}_1)| \leq (q-1)\delta$ for all $\varepsilon \in (0, 1]$ and $u_b(S_b^{-1}(\varepsilon)\tilde{z}_b) = -\varepsilon k_b \tilde{z}_b$, it follows from (11.4.9) and (11.4.21) that,

$$\begin{aligned} \frac{\partial V_{ab}}{\partial \tilde{z}_a} [\ &f_a(\tilde{z}_a, S_b^{-1}(\varepsilon)\tilde{z}_b, \tilde{x}_1 + u_b(S_b^{-1}(\varepsilon)\tilde{z}_b)) \\ &+ p_a(\tilde{z}_a, S_b^{-1}(\varepsilon)\tilde{z}_b, \tilde{x}_1 + u_b(S_b^{-1}(\varepsilon)\tilde{z}_b))w] \\ &+ \frac{\partial V_{ab}}{\partial z_b}[\varepsilon A_b \tilde{z}_b + B_b \tilde{x}_1 + \varepsilon \tilde{p}_b(\tilde{z}_a, \tilde{z}_b, \tilde{x}_1)w] \\ \leq\ &-\varepsilon^{2q+6}\alpha_a(|\tilde{z}_a|) + \varepsilon^{2q+6}\gamma_0^2 |w|^2 + \varepsilon^{2q+6}\gamma_0^2 |S_b^{-1}(\varepsilon)|^2 |\tilde{z}_b|^2 \\ &+ \varepsilon^{2q+6}\gamma_0^2 (2|\tilde{x}_1|^2 + 2\varepsilon^2 |k_b|^2 |\tilde{z}_b|^2) \end{aligned}$$

$$-\varepsilon^6|\tilde{z}_b|^2 + 2\varepsilon^5\tilde{z}_b'P_bB_b\tilde{x}_1 + 2\varepsilon^6\tilde{z}_b'P_b\tilde{p}_b(\tilde{z}_a,\tilde{z}_b,\tilde{x}_1)w$$

$$\leq -\varepsilon^{2q+6}\alpha_a(|\tilde{z}_a|) - \left[\varepsilon^6 - \varepsilon^7 - \varepsilon^8 - \varepsilon^8\gamma_0^2 - 2\varepsilon^{2q+8}\gamma_0^2|k_b|^2\right]|\tilde{z}_b|^2$$

$$+[\varepsilon^{2q+6}\gamma_0^2 + \varepsilon^4(q-1)^2\delta^2|P_b|^2]|w|^2$$

$$+[2\varepsilon^{2q+6}\gamma_0^2 + \varepsilon^3|P_b|^2]|\tilde{x}_1|^2. \tag{11.4.22}$$

It is straightforward to verify that there exists an $\varepsilon^* \in (0,1]$ such that for all $\varepsilon \in (0,\varepsilon^*]$,

$$\varepsilon^6 - \varepsilon^7 - \varepsilon^8 - \varepsilon^8\gamma_0^2 - 2\varepsilon^{2q+6}\gamma_0^2|k_b|^2 \geq \varepsilon^6/2,$$
$$\varepsilon^{2q+6}\gamma_0^2 + \varepsilon^4(q-1)^2\delta^2|P_b|^2 \leq \varepsilon^2, \tag{11.4.23}$$
$$2\varepsilon^{2q+6}\gamma_0^2 + \varepsilon^3|P_b|^2 \leq \varepsilon^2.$$

Also note that, for every $\varepsilon \in (0,\varepsilon^*]$, the function $W(\tilde{z}_a,\tilde{z}_b) = \varepsilon^{2q+6}\alpha_a(|z_a|) + \frac{1}{2}\varepsilon^6|\tilde{z}_b|^2$ is continuous positive definite and is radially unbounded. It follows from [29, Lemma 3.5, p. 138] that there exists a $\mathcal{K}_\infty$ function α_{ab} such that $W(\tilde{z}_a,\tilde{z}_b) \geq \alpha_{ab}(|[\tilde{z}_a', \tilde{z}_b']'|)$. Thus, with this choice of α_{ab}, (11.4.19) is satisfied for every $\varepsilon \in (0,\varepsilon^*]$.

We now apply Theorem 11.4.1 to system (11.4.16) and obtain that, for every $\varepsilon \in (0,\varepsilon^*]$, there exists a smooth state feedback $\tilde{u}(\tilde{z}_a,\tilde{z}_b,\tilde{x}_1,\tilde{x}_2,\cdots,\tilde{x}_r;\varepsilon)$ such that the closed-loop system consisting of system (11.4.16) and this feedback law

(a) has a globally asymptotically stable equilibrium at the origin;

(b) has an L_2 gain, from the disturbance w to the renamed output $\tilde{y} = \tilde{x}_1$, that is less than or equal to ε, i.e.,

$$\int_0^\infty \tilde{y}^2(t)dt = \int_0^\infty \tilde{x}_1^2(t)dt \leq \varepsilon^2 \int_0^\infty w^2(t)dt. \tag{11.4.24}$$

To obtain the L_2 gain from the disturbance w to the regulated output y, we examine the second equation of (11.4.16) with $\tilde{x}_1$ viewed as a disturbance. For the Lyapunov function $V_b(\tilde{z}_b) = \tilde{z}_b'P_b\tilde{z}_b$, we have,

$$\frac{\partial V_b}{\partial \tilde{z}_b}[\varepsilon A_b\tilde{z}_b + B_b\tilde{x}_1 + \varepsilon\tilde{p}_b(\tilde{z}_a,\tilde{z}_b,\tilde{x}_1)w]$$

$$\leq -\varepsilon|\tilde{z}_b|^2 + 2\tilde{z}_b'P_bB_b\tilde{x}_1 + 2\varepsilon\tilde{z}_b'P_b\tilde{p}_b(\tilde{z}_a,\tilde{z}_b,\tilde{x}_1)w$$

$$\leq -\frac{\varepsilon}{2}|\tilde{z}_b|^2 + \frac{4}{\varepsilon}|P_b|^2|\tilde{x}_1|^2 + 4\varepsilon(q-1)^2\delta^2|P_b|^2w^2. \tag{11.4.25}$$

Integrating both sides of the above inequality and using $V_b(0) = 0$ and (11.4.24), we obtain

$$\int_0^\infty |\tilde{z}_b(t)|^2dt \leq 8[1 + (q-1)^2\delta^2]|P_b|^2 \int_0^\infty w^2(t)dt. \tag{11.4.26}$$

Recalling that $y = \xi_1 = \tilde{x}_1 - \varepsilon k_b \tilde{z}_b$, it follows from (11.4.24) and (11.4.26) that,

$$\int_0^\infty y^2(t)dt \leq \int_0^\infty (2\tilde{x}_1^2(t) + 2\varepsilon^2 |k_b|^2 \tilde{z}_b^2(t))dt$$

$$\leq [2 + 16|k_b|^2(1 + (q-1)^2\delta^2)|P_b|^2]\varepsilon^2 \int_0^\infty w^2(t)dt. \quad (11.4.27)$$

Finally, for any given $\gamma > 0$, let $\varepsilon \in (0, \varepsilon^*]$ be such that

$$\left[2 + 16|k_b|^2(1 + (q-1)^2\delta^2)|P_b|^2\right]\varepsilon^2 \leq \gamma^2 \qquad (11.4.28)$$

to complete the proof.   ⊠

11.5. Concluding Remarks

In this chapter we have demonstrated how low gain feedback can be utilized to solve the general H_∞ almost disturbance decoupling problem with internal stability. The role low gain feedback plays here is the treatment of $j\omega$-axis (unit circle) invariant zeros. The major challenge in explicit construction of feedback laws for solving the ADDPMS comes from the presence of $j\omega$ (unit circle) zeros. In the literature on the explicit construction of these feedback laws, $j\omega$ (unit circle) invariant zeros have always been excluded from consideration. We have also demonstrated how low gain feedback can be utilized to solve the H_∞-ADDPS for a class of nonlinear systems. Our results on the nonlinear systems H_∞-ADDPS complements some recent results on the topic.

Chapter 12

Robust Stabilization of an Inverted Pendulum on a Carriage with Restricted Travel

12.1. Introduction and Problem Statement

Balancing an inverted pendulum has been a benchmark example in demonstrating and motivating various control design techniques. For example, much of the material presented in [37] was illustrated by an inverted pendulum with its pivot mounted on a carriage, which is in turn driven by a horizontal force (Fig. 12.1.1). Recently, this same example was again examined in [121] where the physical limitations impose a constraint on the maximum allowable motion of the carriage. As a result, nonlinear controllers were constructed which successfully balance the inverted pendulum under the maximum allowable motion constraint. The purpose of this chapter is to explore the possibility of using the idea of combining low gain and high gain feedback to construct robust *linear* controllers that balance the pendulum without violating the maximum allowable motion constraint.

To facilitate our presentation, we denote the displacement of the carriage at time t by $s(t)$, while the angular rotation of the pendulum at time t is denoted by $\theta(t)$. The pendulum consists of a weightless rod of length L with a mass m attached to its tip. The moment of inertia with respect to the center of gravity (the tip) is J. The carriage has a mass M. The friction coefficient between the carriage and the floor is F. The horizontal force exerted on the carriage at time

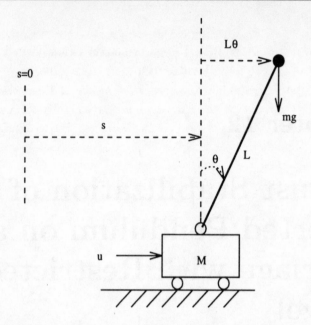

Figure 12.1.1: Inverted Pendulum on a Carriage

t is $u(t)$. We also assume that m is small with respect to M and J is small with respect to mL^2. Under these assumptions, a nonlinear model for this inverted pendulum on a carriage system was derived in [37] as given below,

$$\begin{cases} M\ddot{s} = u - F\dot{s}, \\ \ddot{\phi} = \frac{g}{L}\sin\phi - \frac{1}{L}\ddot{s}\cos\phi. \end{cases} \qquad (12.1.1)$$

Selecting the state variables $x_1 = s$, $x_2 = \dot{s}$, $x_3 = s + L\theta$ and $x_4 = \dot{s} + L\dot{\theta}$, a state space representation of (12.1.1) is given as,

$$\begin{cases} \dot{x}_1 = x_2, \\ \dot{x}_2 = -\dfrac{F}{M}x_2 + \dfrac{1}{M}u, \\ \dot{x}_3 = x_4, \\ \dot{x}_4 = g\sin\dfrac{x_3 - x_1}{L} - \dfrac{F}{M}\left(1 - \cos\dfrac{x_3 - x_1}{L}\right)x_2 \\ \qquad + \dfrac{1}{M}\left(1 - \cos\dfrac{x_3 - x_1}{L}\right)u. \end{cases} \qquad (12.1.2)$$

Linearizing (12.1.2) at the origin of the state space yields

$$\begin{cases} \dot{x}_1 = x_2, \\ \dot{x}_2 = -\dfrac{F}{M}x_2 + \dfrac{1}{M}u, \\ \dot{x}_3 = x_4, \\ \dot{x}_4 = \dfrac{g}{L}x_3 - \dfrac{g}{L}x_1. \end{cases} \qquad (12.1.3)$$

Assuming that the pendulum and the carriage are not in motion before the driving force is exerted, the initial conditions for the system (12.1.3) are then given by $x_1(0) = s(0)$, $x_2(0) = 0$, $x_3(0) = s(0) + L\theta(0)$ and $x_4(0) = 0$. The design objective is to stabilize the system by means of linear state feedback under the constraints that the carriage remains within a certain maximum allowable distance from the origin ($s = 0$). Moreover, the feedback controllers thus obtained should possess a certain degree of robustness. More specifically, our design objectives can be precisely stated as follows.

1. For any *a priori* given (arbitrarily small) numbers η_1 and η_2, find a linear state feedback law that stabilizes the system subject to the restriction that

$$\begin{cases} 0 \leq x_1(t) \leq (1+\eta_1)\,(s(0)+L\theta(0)) + \eta_2, & \text{if } \theta(0) \geq 0,\ s(0) + L\theta(0) \geq 0, \\ 0 \geq x_1(t) \geq (1+\eta_1)\,(s(0)+L\theta(0)) - \eta_2, & \text{if } \theta(0) \leq 0,\ s(0) + L\theta(0) \leq 0; \end{cases} \tag{12.1.4}$$

2. The closed-loop system has an infinite amount of gain margin in the sense that, if the feedback gain is perturbed by any multiplying factor greater than one, the controller will still balance the pendulum without requiring greater traveling distance than the maximum allowable.

We note that $s(0) + L\theta(0)$ is a linearized approximation of the projection of the pendulum tip on the floor at time $t = 0$ and $x_1(t)$ is the displacement of the carriage at time t. Hence, equation (12.1.4) in the first design objective sets the maximum allowable travel for the carriage, which becomes the initial projection of the pendulum tip on the floor as both η_1 and η_2 approach zero. We also note that the the gain margin in the second design objective is different from the traditional notion of stability gain margin in that here not only stability but also performance are maintained as the gain is increased.

Our design algorithm utilizes both the idea of low gain state feedback and high gain state feedback. Throughout this monograph, the mixture of low gain and high gain feedbacks has already proven to be a powerful design technique.

12.2. Linear Low-and-High Gain Design

In this section, we first present a design algorithm which leads to a linear low-and-high gain state feedback law and then show that such a linear high-and-low gain state feedback law would indeed achieve our design objectives.

The design algorithm is given in the following three steps.

Step 1. Taking $y = x_1$ as the output, the system has two invariant zeros at

$$\left\{ \sqrt{\frac{g}{L}}, \ -\sqrt{\frac{g}{L}} \right\},$$

and has the following zero dynamics (the dynamics of (12.1.3) when the output $y = x_1$ is set to zero by state feedback and appropriate choice of initial conditions),

$$\begin{cases} \dot{x}_3 = x_4, \\ \dot{x}_4 = \dfrac{g}{L} x_3. \end{cases} \tag{12.2.1}$$

By choosing

$$u_0 = -\left(\frac{g}{L} + \varepsilon \sqrt{\frac{g}{L}} \right) x_3 - \left(\sqrt{\frac{g}{L}} + \varepsilon \right) x_4$$

and renaming the output as

$$\tilde{y} = \tilde{x}_1 = x_1 + \frac{L}{g} u_0, \tag{12.2.2}$$

we place the poles of the zero dynamics at

$$\left\{ -\varepsilon, \ -\sqrt{\frac{g}{L}} \right\},$$

where ε is a positive scalar satisfying

$$\varepsilon \leq \min \left\{ \frac{1}{2}, \ \frac{\eta_1}{2 + 3\sqrt{\frac{L}{g}}}, \ \frac{\eta_2}{\pi L \left(2\sqrt{\frac{g}{L}} + 1 \right)} \right\}. \tag{12.2.3}$$

We note here that the low gain component in the renamed output $\tilde{y}$ moves an invariant zero from $s = 0$ to $s = -\varepsilon$.

Step 2. With $\tilde{x}_1$ as the new output, we rewrite the system (12.1.3) as

$$\begin{cases} \dot{\tilde{x}}_1 = -\dfrac{1}{\varepsilon} \tilde{x}_1 + \tilde{x}_2, \\[2mm] \dot{\tilde{x}}_2 = \dfrac{1}{\varepsilon} x_2 + \dfrac{L}{\varepsilon g} \dot{u}_0 + \dfrac{L}{g} \ddot{u}_0 - \dfrac{F}{M} x_2 + \dfrac{1}{M} u, \\[2mm] \dot{x}_3 = x_4, \\[2mm] \dot{x}_4 = -\varepsilon \sqrt{\dfrac{g}{L}} x_3 - \left(\sqrt{\dfrac{g}{L}} + \varepsilon \right) x_4 - \dfrac{g}{L} \tilde{x}_1, \end{cases} \tag{12.2.4}$$

where

$$\tilde{x}_2 = x_2 + \frac{1}{\varepsilon} x_1 + \frac{L}{\varepsilon g} u_0 + \frac{L}{g} \dot{u}_0, \tag{12.2.5}$$

and where

$$\begin{cases} u_0 = -\left(\sqrt{\frac{g}{L}} + \varepsilon\right)\left(\sqrt{\frac{g}{L}}x_3 + x_4\right), \\ \dot{u}_0 = -\left(\sqrt{\frac{g}{L}} + \varepsilon\right)\left(\sqrt{\frac{g}{L}}x_4 + \frac{g}{L}x_3 - \frac{g}{L}x_1\right), \\ \ddot{u}_0 = -\left(\sqrt{\frac{g}{L}} + \varepsilon\right)\left(\sqrt{\frac{g}{L}}\left(\frac{g}{L}x_3 - \frac{g}{L}x_1\right) + \frac{g}{L}x_4 - \frac{g}{L}x_2\right). \end{cases} \tag{12.2.6}$$

Step 3. Choose the linear high gain state feedback law as

$$u = -\frac{M}{\mu}\tilde{x}_2, \tag{12.2.7}$$

where μ is a positive scalar whose value is to be chosen later. Ⓐ

The theorem below shows that the linear feedback law as given by (12.2.7) indeed achieves our design objectives.

Theorem 12.2.1. *Consider the closed-loop system consisting of the system (12.1.3) and the linear state feedback law (12.2.7). Then, there exists a $\mu^* > 0$ such that for each $\mu \in (0, \mu^*]$, the closed-loop system is stable with (12.1.4) satisfied. This, in turn, also shows that the linear state feedback law (12.2.7) possesses an infinite gain margin in the sense that, if the feedback gain is perturbed by a multiplying factor greater than one, the controller still stabilizes the system without requiring greater traveling distance than the maximum allowable.* Ⓣ

Proof of Theorem 12.2.1 With the state feedback law (12.2.7), the closed-loop system can be written as,

$$\begin{cases} \dot{\tilde{x}}_1 = -\frac{1}{\varepsilon}\tilde{x}_1 + \tilde{x}_2, \\ \dot{\tilde{x}}_2 = \frac{1}{\varepsilon}x_2 + \frac{L}{\varepsilon g}\dot{u}_0 + \frac{L}{g}\ddot{u}_0 - \frac{F}{M}x_2 - \frac{1}{\mu}\tilde{x}_2, \\ \dot{x}_3 = x_4, \\ \dot{x}_4 = -\varepsilon\sqrt{\frac{g}{L}}x_3 - \left(\sqrt{\frac{g}{L}} + \varepsilon\right)x_4 - \frac{g}{L}\tilde{x}_1, \end{cases} \tag{12.2.8}$$

with initial conditions given by

$$
\begin{cases}
\tilde{x}_1(0) = -L\theta(0) - \sqrt{\dfrac{L}{g}}\varepsilon(s(0) + L\theta(0)), \\[2mm]
\tilde{x}_2(0) = -\left(\dfrac{1}{\varepsilon} + \sqrt{\dfrac{g}{L}} + \varepsilon\right) L\theta(0) - \sqrt{\dfrac{L}{g}}(s(0) + L\theta(0)), \\[2mm]
x_3(0) = s(0) + L\theta(0), \\[2mm]
x_4(0) = 0.
\end{cases}
\tag{12.2.9}
$$

Clearly, the closed-loop system is in the standard singular perturbation form with $\tilde{x}_2$ as the fast variable and the rest as the slow ones. Letting $\mu = 0$, we obtain that $\tilde{x}_2 = 0$ and hence the stable reduced system,

$$
\begin{cases}
\dot{\tilde{x}}_{1s} = -\dfrac{1}{\varepsilon}\tilde{x}_{1s}, \ \tilde{x}_{1s}(0) = \tilde{x}_1(0), \\[2mm]
\dot{x}_{3s} = x_{4s}, \ x_{3s}(0) = x_3(0), \\[2mm]
\dot{x}_{4s} = -\varepsilon\sqrt{\dfrac{g}{L}}x_{3s} - \left(\sqrt{\dfrac{g}{L}} + \varepsilon\right)x_{4s} - \dfrac{g}{L}\tilde{x}_{1s}, \ x_{4s}(0) = x_4(0) = 0.
\end{cases}
\tag{12.2.10}
$$

From the first equation of the reduced system (12.2.10), we have that,

$$
\tilde{x}_{1s}(t) = \tilde{x}_1(0)e^{-\frac{1}{\varepsilon}t}.
\tag{12.2.11}
$$

Viewing $\tilde{x}_{1s}$ as an input signal to the dynamics of x_{3s} and $\tilde{x}_{4s}$, we solve the last two equations of the reduced system (12.2.10) and obtain

$$
\sqrt{\dfrac{g}{L}}x_{3s} + x_{4s} = \sqrt{\dfrac{g}{L}}x_3(0)e^{-\varepsilon t} - \dfrac{\varepsilon}{1-\varepsilon^2}\dfrac{g}{L}\tilde{x}_1(0)\left(e^{-\varepsilon t} - e^{-\frac{1}{\varepsilon}t}\right).
\tag{12.2.12}
$$

Now standard singular perturbation arguments show that there exists a $\mu_1^* > 0$ such that for each $\mu \in (0, \mu_1^*]$ the closed-loop system is stable and

$$
\sqrt{\dfrac{g}{L}}x_3 + x_4 = \sqrt{\dfrac{g}{L}}x_{3s} + x_{4s} + O(\mu)\tilde{x}_1(0)
$$
$$
+ O(\mu)\tilde{x}_2(0) + O(\mu)x_3(0),
\tag{12.2.13}
$$
$$
\tilde{x}_1 = \tilde{x}_{1s} + O(\mu)\tilde{x}_1(0) + O(\mu)\tilde{x}_2(0) + O(\mu)x_3(0),
\tag{12.2.14}
$$

which, together with (12.2.12), show that

$$
u_0(t) = -\left(\sqrt{\dfrac{g}{L}} + \varepsilon\right)\left(\sqrt{\dfrac{g}{L}}x_3 + x_4\right)
$$
$$
= -\left(\sqrt{\dfrac{g}{L}} + \varepsilon\right)\left(\sqrt{\dfrac{g}{L}}x_3(0)e^{-\varepsilon t} - \dfrac{\varepsilon}{1-\varepsilon^2}\dfrac{g}{L}\tilde{x}_1(0)\left(e^{-\varepsilon t} - e^{-\frac{1}{\varepsilon}t}\right)\right)
$$
$$
+ O(\mu)\tilde{x}_1(0) + O(\mu)\tilde{x}_2(0) + O(\mu)x_3(0).
\tag{12.2.15}
$$

Hence, from (12.2.14) and (12.2.15) we have

$$x_1(t) = \tilde{x}_1(t) - \frac{L}{g} u_0(t)$$

$$= \left[e^{-\frac{1}{\varepsilon}t} - \left(\sqrt{\frac{g}{L}} + \varepsilon \right) \frac{\varepsilon}{1 - \varepsilon^2} \left(e^{-\varepsilon t} - e^{-\frac{1}{\varepsilon}t} \right) + O(\mu) \right] \tilde{x}_1(0)$$

$$+ \left[\left(\sqrt{\frac{g}{L}} + \varepsilon \right) \sqrt{\frac{L}{g}} e^{-\varepsilon t} + O(\mu) \right] x_3(0) + O(\mu). \qquad (12.2.16)$$

In the case that $\theta(0) \geq 0$ and $s(0) + L\theta(0) \geq 0$, we observe that $\tilde{x}_1(0) \leq 0$ and $x_3(0) \geq 0$. Hence, assuming that $\theta(0) \leq \pi/2$, it follows from (12.2.16) and (12.2.9) that

$$x_1(t) \leq \left[1 + \varepsilon \left(1 + \frac{3}{2} \sqrt{\frac{L}{g}} \right) + O(\mu) \right] (s(0) + L\theta(0))$$

$$+ \pi L \left(\frac{1}{2} + \sqrt{\frac{g}{L}} \right) \varepsilon + O(\mu)$$

$$\leq \left[1 + \frac{1}{2}\eta_1 + O(\mu) \right] (s(0) + L\theta(0)) + \frac{\eta_2}{2} + O(\mu), \qquad (12.2.17)$$

and

$$x_1(t) \geq - \left[L\theta(0) + \sqrt{\frac{L}{g}} \varepsilon(s(0) + L\theta(0)) \right] e^{-\frac{1}{\varepsilon}t}$$

$$+ \left[\left(\sqrt{\frac{g}{L}} + \varepsilon \right) \sqrt{\frac{L}{g}} e^{-\varepsilon t} + O(\mu) \right] (s(0) + L\theta(0)) + O(\mu)$$

$$\geq s(0) e^{-\varepsilon t} + O(\mu). \qquad (12.2.18)$$

Noting that $x_1(t)$ decays exponentially with a slowest term $\varepsilon^{-\varepsilon t}$, it is now clear that there exists a $\mu_2^* \in (0, \mu_1^*]$ such that for all $\mu \in (0, \mu_2^*]$,

$$0 \leq x_1(t) \leq (1 + \eta_1)(s(0) + L\theta(0)) + \eta_2, \qquad (12.2.19)$$

which is the first item of (12.1.4).

Similarly, in the case that $\theta(0) \leq 0$ and $s(0) + L\theta(0) \leq 0$, we observe that $\tilde{x}_1(0) \geq 0$ and $x_3(0) \leq 0$. Hence, assuming that $\theta(0) \geq -\pi/2$, it follows from (12.2.16) that

$$x_1(t) \geq \left[1 + \varepsilon \left(1 + \frac{3}{2} \sqrt{\frac{L}{g}} \right) + O(\mu) \right] (s(0) + L\theta(0))$$

$$-\pi L \left(\frac{1}{2} + \sqrt{\frac{g}{L}} \right) \varepsilon + O(\mu)$$

$$\geq \left[1 + \frac{1}{2}\eta_1 + O(\mu) \right] (s(0) + L\theta(0)) - \frac{\eta_2}{2} + O(\mu), \qquad (12.2.20)$$

and

$$x_1(t) \leq -s(0)e^{-\varepsilon t} + O(\mu). \qquad (12.2.21)$$

Again, it is clear that there exists a $\mu_3^* \in (0, \mu_1^*]$ such that for all $\mu \in (0, \mu_3^*]$,

$$0 \geq x_1(t) \geq (1 + \eta_1)(s(0) + L\theta(0)) - \eta_2, \qquad (12.2.22)$$

which is the second equation of (12.1.4).

Finally, taking $\mu^* = \min\{\mu_2^*, \mu_3^*\}$, the proof is completed.   ⊠

12.3. Simulations

To demonstrate our design algorithm, we take the numerical values for the system parameters as follows,

$$\frac{F}{M} = 1s^{-1}, \; \frac{1}{M} = 1kg^{-1}, \; \frac{g}{L} = 16s^{-1}, \; L = 0.613m. \qquad (12.3.1)$$

Let $\eta_1 = 0.1$ and $\eta_2 = 0.1m$, we choose $\varepsilon = 0.03$. With these numerical values, the linear feedback law (12.2.7) is given by

$$u = -\frac{1}{\mu}(37.36x_1 + x_2 - 37.56x_3 - 9.39x_4).$$

We simulate the above control law with both the nonlinear model (12.1.2) and the linearized model (12.1.3). Extensive simulation shows that the feedback law design on the basis of the linearized model works satisfactorily when applied to the original nonlinear model. In fact, simulation shows the performance difference between the linearized model and the nonlinear model is almost unnoticeable. We believe that this is due to the two time scale nature of our control law.

Figs. 12.3.1 and 12.3.2 are simulation results for the initial conditions $s(0) = 0.1$, $\theta(0) = 0.1$. In the figures, we have plotted only the first ten seconds of the state transience for better visualization of the early fast responses due to the high gain action. Plots for a longer time period show the slow convergence due to the low gain action. We also note that with these initial conditions, $(1 + \eta_1)(s(0) + L\theta(0)) + \eta_2 = 0.28$.

12.4. Conclusions

We have presented a linear state feedback law that successfully balances an inverted pendulum on a carriage which has limited travel. The design once again demonstrates the usefulness of the technique of combining low gain and high gain feedback.

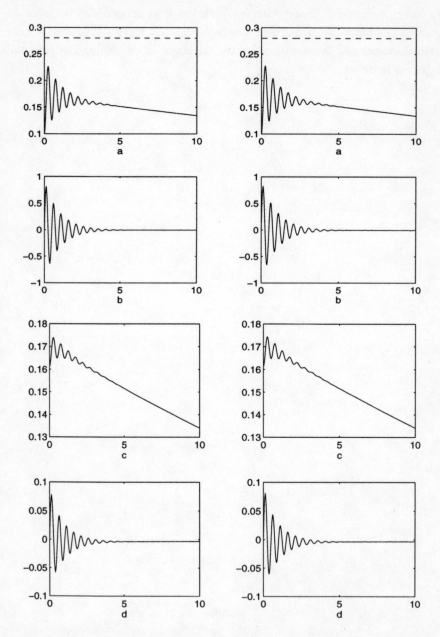

Figure 12.3.1: $\mu = 0.2$. left column: linear model; right column: nonlinear model. a) $x_1(t)$;   b) $x_2(t)$;   c) $x_3(t)$;   d) $x_4(t)$.

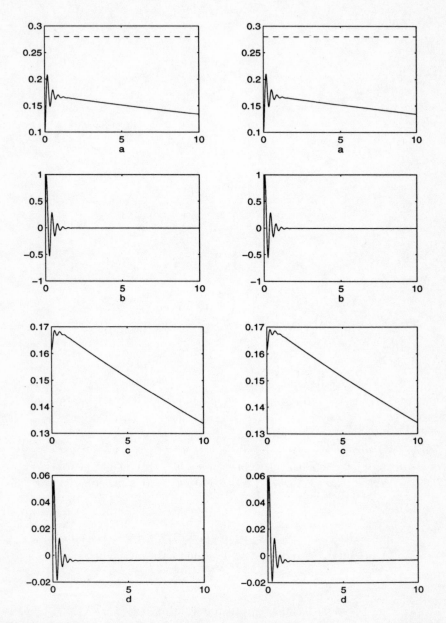

Figure 12.3.2: $\mu = 0.1$. left column: linear model; right column: nonlinear model. a) $x_1(t)$;   b) $x_2(t)$;   c) $x_3(t)$;   d) $x_4(t)$.

Chapter 13

Feedback Design for an F-16 Fighter Aircraft with Rate Limited Deflector

13.1. Introduction

In the past few years there has been much interest concerning stabilization of linear systems with magnitude saturating actuators, resulting in several promising design techniques. Chapters 2-8 described some of these techniques. Additional design and analysis techniques can be found in [3].

The objective of this chapter is to show how the available design techniques for linear systems with saturating actuators can be utilized to design robust tracking controllers for an F-16 fighter aircraft with rate limited deflector. More specifically, we will combine the low-and-high gain (LHG) design technique of Chapter 4, and the piecewise linear LQ control (PLC) design technique of [127], both originally proposed for dealing with actuator magnitude saturation, to arrive at feedback laws that control linear systems subject to actuator rate saturation.

The resulting combined PLC/LHG state feedback design method inherits the advantages of both design techniques, while avoiding their disadvantages. In particular, in the LHG design, a low gain feedback law is first designed in such a way that the actuator does not saturate in magnitude and the closed-loop system remains linear. The gain is chosen low to enlarge the region in which the closed-loop system remains linear and hence to enlarge the basin of attraction of the closed-loop system. Then, utilizing an appropriate Lyapunov function for the closed-loop system under this low gain feedback control law, a linear high gain feedback control law is constructed and added to the low gain feedback control to form the final LHG feedback control law. Such a linear low-and-high gain feedback control law speeds up the transient response for the state in a certain subspace of the state space and is able to stabilize the system in the presence of input-additive plant uncertainties and to reject arbitrarily large bounded input-additive disturbances. The disadvantage of this control law is that the transient response for the state outside that subspace of the state space remains that of the low gain feedback, which is typically sluggish (due to low feedback gain for a large basin of attraction). On the other hand, the aim of the PLC [127] (or HPB control [78]) is to increase the feedback gain piecewisely while adhering to the input bound as the trajectories converge toward the origin. Such a design results in fast transience for all states. However, it lacks robustness to large uncertainties and the ability to reject disturbances.

In this chapter, we will use the combined PLC/LHG design technique to arrive at feedback laws that achieve robust tracking performance. The signals to be tracked are modeled as the state of a reference system and the tracking control problem is formulated as an output regulation problem ([16] and Chapter 5). In the case that only the output is available for feedback, the performance of the state feedback laws is preserved by the use of a fast observer. We note here that the PLC design method as originally proposed in [127] is only for state feedback. Although the LHG output feedback design is available, its design and analysis do not carry through in the combined PLC/LHG design, since the closed-loop system under the combined PLC/LHG state feedback law is discontinuous. Indeed, discontinuity in the PLC/LHG state feedback laws is the primary source of difficulty in its implementation with observer state. Our output feedback design also provides an output feedback implementation of the PLC controllers as a special case.

The remainder of the chapter is organized as follows. In Section 13.2, we precisely formulate the two problems, the state feedback robust tracking problem and the output feedback robust tracking problem. Section 13.3 describes the combined PLC/LHG state feedback design algorithm for obtaining stabilizing state feedback laws. This design algorithm is used in Section 13.4 to construct feedback laws that solve the state feedback robust tracking control problem as formulated in Section 13.2. Section 13.5 deals with the case that only output is available for feedback. In this case, a fast observer is used to implement the state feedback laws constructed in Section 13.4 and shown to preserve the performance of the state feedback laws. In Section 13.6, the proposed combined PLC/LHG design method is applied to an F-16 fighter aircraft.

Simulation results demonstrate its effectiveness. Concluding remarks are made in Section 13.7.

13.2. Problem Statement

Consider the linear dynamical system

$$
\begin{cases}
\dot{x} = Ax + Bv, \quad x(0) \in \mathcal{X} \subset \mathbb{R}^n, \\
\dot{v} = \text{sat}_\Delta(-Tv + Tu + d), \quad v(0) \in \mathcal{V} \subset \mathbb{R}^m, \\
\dot{w} = Sw, \quad w(0) \in \mathcal{W} \subset \mathbb{R}^s, \\
e = Cx + Qw, \\
y = Ex,
\end{cases}
\tag{13.2.1}
$$

where the first equation describes the plant, with state $x \in \mathbb{R}^n$, input $u \in \mathbb{R}^m$ and output $y \in \mathbb{R}^q$, the second equation represents actuator dynamics with state $v \in \mathbb{R}^m$ and is subject to disturbance $d \in \mathbb{R}^m$ and rate saturation $\text{sat}_\Delta(\cdot)$, the third equation defines the reference signal $-Qw \in \mathbb{R}^p$ to be tracked by the plant output $Cx \in \mathbb{R}^p$. For $\Delta = (\Delta_1, \Delta_2, \cdots, \Delta_m)$, $\Delta_i > 0$, the actuator rate saturation function $\text{sat}_\Delta : \mathbb{R}^m \to \mathbb{R}^m$ is the standard saturation function, i.e., $\text{sat}_\Delta(v) = [\text{sat}_{\Delta_1}(v_1), \text{sat}_{\Delta_2}(v_2), \cdots, \text{sat}_{\Delta_m}(v_m)]'$, $\text{sat}_{\Delta_i}(v_i) = \text{sign}(v_i) \min\{\Delta_i, |v_i|\}$, the positive definite diagonal matrices $T = \text{diag}(\tau_1, \tau_2, \cdots, \tau_m)$ represents the "time constants" of the actuators. Finally, all three sets $\mathcal{X}$, $\mathcal{V}$ and $\mathcal{W}$ are bounded.

We also make the following assumptions on the system.

Assumption 13.2.1. *The pair (A, B) is stabilizable.* ⊞

Assumption 13.2.2. *The eigenvalues of S have nonnegative real parts.* ⊞

Assumption 13.2.3. *The disturbance is uniformly bounded by a known (arbitrarily large) constant D, i.e., $|d(t)| \leq D$, $\forall t \geq 0$.* ⊞

Remark 13.2.1. *We note that Assumption 13.2.1 is obviously necessary, Assumption 13.2.2 does not involve any loss of generality since asymptotically stable modes of the reference model do not affect the tracking performance of the plant, and Assumption 13.2.3 is satisfied by essentially all practical disturbances.* ℝ

Before stating the problem, we make the following preliminary definition.

Definition 13.2.1. *The data $(D, \mathcal{Z}_\infty, \mathcal{E}_\infty)$ is said to be admissible for state feedback [output feedback] if D is a nonnegative real number, and $\mathcal{Z}_\infty$ and $\mathcal{E}_\infty$ are respectively a subset of $\mathbb{R}^{n+m}$ [$\mathbb{R}^{2n+m}$] and $\mathbb{R}^p$ both containing the origin as an interior point.* Ⅾ

The problems we are to consider in this chapter are the following,

Problem 13.2.1. *Given the data* $(D, \mathcal{Z}_\infty, \mathcal{E}_\infty)$, *admissible for state feedback, the problem of state feedback robust tracking problem is to find a state feedback law* $u = F(x, v, w)$, *such that the closed-loop system satisfies,*

1. *Every trajectory of the system*

$$\begin{cases} \dot{x} = Ax + Bv, \\ \dot{v} = \mathrm{sat}_\Delta(-Tv + TF(x, v, 0) + d) \end{cases} \tag{13.2.2}$$

   starting from $\mathcal{X} \times \mathcal{V}$ *enters* $\mathcal{Z}_\infty$ *in a finite time and remains in it thereafter;*

2. *For all* $x(0) \in \mathcal{X}$, $v(0) \in \mathcal{V}$ *and* $w(0) \in \mathcal{W}$, *the solution of the closed-loop system is such that* $e(t)$ *enters the set* $\mathcal{E}_\infty$ *in a finite time and remains in it thereafter.* ☐

Problem 13.2.2. *Given the data* $(D, \mathcal{Z}_\infty, \mathcal{E}_\infty)$, *admissible for output feedback, the problem of output feedback robust tracking is to find a feedback law*

$$\begin{cases} \dot{\psi} = \alpha(\psi, y), \quad \psi(0) \in \mathcal{X} \subset \mathbb{R}^n, \\ u = F(\psi, v, w), \end{cases}$$

such that the closed-loop system satisfies

1. *Every trajectory of the system*

$$\begin{cases} \dot{x} = Ax + Bv, \\ \dot{v} = \mathrm{sat}_\Delta(-Tv + TF(\psi, v, 0) + d), \\ \dot{\psi} = \alpha(\psi, Ex), \end{cases}$$

   starting from $\mathcal{X} \times \mathcal{V} \times \mathcal{X}$ *enters* $\mathcal{Z}_\infty$ *in a finite time and remains in it thereafter;*

2. *For all* $x(0), \psi(0) \in \mathcal{X}$, $v(0) \in \mathcal{V}$ *and* $w(0) \in \mathcal{W}$, *the solution of the closed-loop system is such that* $e(t)$ *enters the set* $\mathcal{E}_\infty$ *in a finite time and remains in it thereafter.* ☐

13.3. The Combined PLC/LHG Design Algorithm

In this section, we describe a design algorithm that is a combination of the piecewise linear LQ control [127] and the low-and-high gain feedback design technique of Chapter 4. For completeness, we organize this section as follows. Sections 13.3.1 and 13.3.2 respectively recapitulate the PLC and the LHG design techniques. Section 13.3.3 presents the combined PLC/LHG design algorithm.

13.3.1. Piecewise Linear LQ Control Design (PLC)

Consider the linear dynamical system subject to actuator magnitude saturation,

$$\dot{x} = Ax + B\text{sat}_\Delta(u), \quad x(0) \in \mathcal{X} \subset \mathbf{R}^n, u \in \mathbf{R}^m, \tag{13.3.1}$$

where the saturation function $\text{sat}_\Delta : \mathbf{R}^m \to \mathbf{R}^m$ is as defined in Section 13.2, the pair (A, B) is assumed to be stabilizable, and $\mathcal{X}$ is bounded.

The PLC design is based on the following LQ algebraic Riccati equation (ARE),

$$A'P + PA - PBR^{-1}B'P + I = 0, \tag{13.3.2}$$

where $R = \text{diag}(\epsilon) = \text{diag}(\epsilon_1, \epsilon_2, \cdots, \epsilon_m)$, $\epsilon_i > 0$, are the design parameters to be chosen later.

Key to the PLC scheme is the notion of invariant sets. A nonempty subset of ε in $\mathbf{R}^n$ is positively invariant if for a dynamical system and for any initial condition $x(0) \in \varepsilon$, $x(t) \in \varepsilon$ for all $t \geq 0$. For the closed-loop system comprising system (13.3.1) and the LQ control $u = -R^{-1}B'Px$, simple Lyapunov analysis shows that the Lyapunov level set

$$\varepsilon(P, \rho) = \{x : x'Px \leq \rho\}, \ \forall \rho > 0$$

is an invariant set, provided that saturation does not occur for all $x \in \varepsilon(P, \rho)$. To avoid the saturation from occurring, while fully utilizing the available control capacity, for a given ρ, $\epsilon = (\epsilon_1, \epsilon_2, \cdots, \epsilon_m)$ will be chosen to be the smallest such that

$$|u_i| = \left| \frac{1}{\epsilon_i} B_i'Px \right| \leq \Delta_i, \ \forall x \in \varepsilon(P, \rho),$$

where B_i is the ith column of matrix B and u_i is the ith element of u. The existence and uniqueness of such an ϵ are established, and an algorithm for computing such an ϵ is also given, in [127]. More specifically, it is shown through the existence of a unique fixed point that the following iteration converges from any initial value to the desired value of ϵ,

$$\epsilon_{n+1} = \sqrt{\rho}\Phi(\epsilon_n), \tag{13.3.3}$$

where

$$\Phi(\epsilon) = [\phi_1(\epsilon), \phi_2(\epsilon), \cdots, \phi_m(\epsilon)]',$$

and for each $i = 1$ to m,

$$\phi_i(\epsilon) = \frac{1}{\Delta_i}\sqrt{B_i'P(\epsilon)B_i}.$$

The aim of the PLC scheme is to increase the state feedback gain piecewisely while adhering to actuator bounds as the trajectories converge towards the origin. This is achieved by constructing nested level sets, $\varepsilon_0, \varepsilon_1, \cdots, \varepsilon_N$, in such a way that the trajectories traverse successively the surface of each ε_i and the control law is switched to higher and higher gains as each surface is crossed.

The procedure for designing a PLC law is as follows. Given the set of initial conditions $\mathcal{X} \subset \mathbb{R}^n$, choose an initial level set ε_0 as,

$$\varepsilon_0 = \inf_\rho \{\varepsilon(P, \rho) : \mathcal{X} \subset \varepsilon(P, \rho)\}. \tag{13.3.4}$$

We denote the value of ρ associated with ε_0 as ρ_0, and the corresponding values of ϵ, R and P as ϵ_0, R_0 and P_0 respectively. A simple approach to determining ε_0 and ρ_0 can also be found in [127]. More specifically, it is shown that the size of ε_0 grows monotonically as the parameter ρ grows. Hence, ε_0 and ρ_0 can be determined by a simple iteration procedure. Here we would like to note that, as explained in [127], increasing ρ indefinitely for exponentially unstable A will not result in an ε_0 that grows without bound.

To determine the inner level sets ε_i's, choose successively smaller ρ_i where $\rho_{i+1} < \rho_i$ for each $i = 1, 2, \cdots, N$. A simple choice of such ρ_i's is the geometric sequence of the form

$$\rho_i = \rho_0 (\Delta\rho)^i, \quad i = 0, 1, 2, \cdots, N,$$

where the ρ-reduction factor $\Delta\rho \in (0, 1)$. (Consequently, the values of ϵ, R and P associated with each of these ρ_i's are denoted as ϵ_i, R_i and P_i respectively.) For a discussion on the choice of N and $\Delta\rho$, see [127].

As shown in [127], a critical property of such a sequence of level sets ε_i is that they are strictly nested in the sense that ε_{i+1} is strictly inside ε_i for each $i = 0$ to $N - 1$. Here and throughout this chapter, by saying set S_2 is strictly inside set S_1 we mean that $S_2 \subset S_1$ and their boundaries do not overlap.

13.3.2. Low-and-High Gain Feedback Design (LHG)

Consider the linear system subject to actuator magnitude saturation, input additive disturbances and uncertainties,

$$\dot{x} = Ax + B\text{sat}_\Delta (u + f(x) + d), \quad x(0) \in \mathcal{X} \subset \mathbb{R}^n, u \in \mathbb{R}^m, \tag{13.3.5}$$

where the saturation function $\text{sat}_\Delta : \mathbb{R}^m \to \mathbb{R}^m$ is as defined in Section 13.2, the locally Lipschitz function $f : \mathbb{R}^n \to \mathbb{R}^m$ represents the input additive plant uncertainties and d the input-additive disturbance. The LHG feedback design for this system is given as follows. First, the level set ε_0 is determined as in the PLC design. Correspondingly, a state feedback law with possibly low feedback gain is determined as,

$$u_L = -R_0^{-1}B'P_0x. \tag{13.3.6}$$

A high gain state feedback is then constructed as,

$$u_H = -kR_0^{-1}B'P_0x, \quad k \geq 0. \tag{13.3.7}$$

The final low-and-high gain state feedback is then given by a simple addition of the low and high gain feedbacks u_L and u_H, viz.,

$$u = -(1+k)R_0^{-1}B'P_0x, \quad k \geq 0. \tag{13.3.8}$$

Here the design parameter k is referred to as the high gain parameter. As demonstrated in Chapter 4, the freedom in choosing the value of this high gain parameter can be utilized to achieve robust stabilization in the presence of input additive plant uncertainties $f(x)$ and input-additive disturbance rejection. Moreover, the transient speed for the states not in the range space of $B'P_0$ will increase as the value of k increases. To see this, let us consider the following Lyapunov function,

$$V_0(x) = x'P_0x. \tag{13.3.9}$$

The evaluation of $\dot{V}$ along the trajectories of the closed-loop system in the absence of uncertainties and disturbances gives,

$$\begin{aligned}
\dot{V} &= -x'x - x'P_0BR_0^{-1}B'P_0x \\
&\quad +2x'P_0B[\text{sat}_\Delta(-(k+1)R_0^{-1}B'P_0x) + R_0^{-1}B'P_0x] \\
&= -x'x - x'P_0BR_0^{-1}B'P_0x \\
&\quad -2\sum_{i=1}^{m} v_i[\text{sat}_{\Delta_i}((k+1)v_i) - v_i], \tag{13.3.10}
\end{aligned}$$

where we have denoted the ith element of $v = -R_0^{-1}B'P_0x$ as v_i. By the choice of P_0, it is clear that $|v_i| \leq \Delta_i$ and hence $-v_i[\text{sat}_{\Delta_i}((k+1)v_i) - v_i] \leq 0$, for each $i = 1$ to m. If x is not in the range space of $B'P_0$, that is $B'P_0x \neq 0$, then, for any i such that $v_i \neq 0$, $-v_i[\text{sat}_{\Delta_i}((k_2+1)v_i) - v_i] \leq -v_i[\text{sat}_{\Delta_i}((k_1+1)v_i) - v_i]$ whenever $k_2 \geq k_1$. However, for any x in the null space of $B'P_0$, $-v_i[\text{sat}_{\Delta_i}((k+1)v_i) - v_i] = 0$ for any value of k.

13.3.3. Combined PLC/LHG Feedback Design

In this subsection, we present the proposed combined PLC/LHG state feedback design for the linear system subject to actuator rate saturation (13.2.1).

Step 1. Choose a pre-feedback

$$u = v + \bar{u}. \tag{13.3.11}$$

Let $\tilde{x} = [x', v']'$. Then the system (13.2.1) under the above pre-feedback is given by,

$$\dot{\tilde{x}} = \tilde{A}\tilde{x} + \tilde{B}\mathrm{sat}_\Delta(T\bar{u} + d), \quad \tilde{x}(0) \in \mathcal{X} \times \mathcal{V} \subset \mathbb{R}^{n+m}, \tag{13.3.12}$$

where

$$\tilde{A} = \begin{bmatrix} A & B \\ 0 & 0 \end{bmatrix}, \quad \tilde{B} = \begin{bmatrix} 0 \\ I \end{bmatrix}. \tag{13.3.13}$$

Assumption 13.2.1, i.e., the pair (A, B) is stabilizable, implies that $(\tilde{A}, \tilde{B})$ is stabilizable.

Step 2. Apply the PLC state feedback design algorithm to system (13.3.12), and obtain a sequence of nested level sets $\varepsilon_0, \varepsilon_1, \cdots, \varepsilon_N$ (and correspondingly, the parameters $\epsilon_0, \epsilon_1, \cdots, \epsilon_N$) and a piecewise linear state feedback law,

$$\bar{u} = \begin{cases} \bar{u}_i = -T^{-1}\tilde{R}_i^{-1}\tilde{B}'\tilde{P}_i\tilde{x} & \text{for } \tilde{x} \in \varepsilon_i \setminus \varepsilon_{i+1}, \, i = 0, \text{ to } N-1, \\ \\ \bar{u}_N = -T^{-1}\tilde{R}_N^{-1}\tilde{B}'\tilde{P}_N\tilde{x} & \text{for } \tilde{x} \in \varepsilon_N. \end{cases} \tag{13.3.14}$$

Step 3. Design the LHG state feedback based on the PLC feedback law (13.3.14) and obtain the following combined final PLC/LHG feedback law,

$$u = \begin{cases} u_i = v - (k+1)T^{-1}\tilde{R}_i^{-1}\tilde{B}'\tilde{P}_i\tilde{x} & \text{for } \tilde{x} \in \varepsilon_i \setminus \varepsilon_{i+1}, \, i = 0 \text{ to } N-1, \\ \\ u_N = v - (k+1)T^{-1}\tilde{R}_N^{-1}\tilde{B}'\tilde{P}_N\tilde{x} & \text{for } \tilde{x} \in \varepsilon_N, \end{cases} \tag{13.3.15}$$

where $k \geq 0$ is a design parameter to be specified later. $\boxed{\text{A}}$

13.4. Robust Tracking via State Feedback

In this section, we will show that the combined PLC/LHG design technique as described in the previous section can be used to construct feedback laws that solve the state feedback robust tracking problem (i.e., Problem 13.2.1). The state feedback results are presented in a theorem as follows.

Theorem 13.4.1. *Let Assumptions 13.2.1-13.2.3 hold, and given the data* $(D, \mathcal{W}_\infty, \mathcal{E}_\infty)$, *admissible for state feedback, then Problem 13.2.1 is solvable if there exist matrices* Π *and* Γ *such that*

1. *they satisfy the linear matrix equations,*

$$\begin{cases} \Pi S = \tilde{A}\Pi + \tilde{B}\Gamma, \\ \tilde{C}\Pi + Q = 0, \end{cases} \tag{13.4.1}$$

where $\tilde{A}$ and $\tilde{B}$ are as given by (13.3.13), and $\tilde{C} = [C \quad 0_{p \times m}]$;

2. *there exists a $\delta = [\delta_1, \delta_2, \cdots, \delta_m]$, $\delta_i > 0$, such that $|\Gamma_i w(t)| \leq \Delta_i - \delta_i$ for all $w(0) \in \mathcal{W}$ and all $t \geq 0$, where Γ_i is the ith row of Γ.*

Moreover, the feedback laws that solve Problem 13.2.1 can be explicitly constructed using the combined PLC/LHG design technique as described in the previous section. $\boxed{\text{T}}$

Remark 13.4.1. *It follows from [16] that Condition 1 is necessary for solving Problem 1, while, similar arguments as in [70] show that, under some extra mild conditions on the plant and the reference model, Condition 2 is also necessary.* $\boxed{\text{R}}$

Proof of Theorem 13.4.1. We will prove this theorem by explicitly constructing a family of combined PLC/LHG state feedback laws, parameterized in a parameter k and showing that there exists a $k^* > 0$ such that for any $k \geq k^*$, the feedback law solves Problem 13.2.1.

Construction of Parameterized State Feedback Laws:

Step 1. Carry out Step 1 of the combined PLC/LHG feedback design of Section 13.3 as follows. Choose a pre-feedback

$$u = v + \bar{u}. \tag{13.4.2}$$

Let $\tilde{x} = [x', v']'$, Then system (13.2.1) under the above pre-feedback becomes,

$$\dot{\tilde{x}} = \tilde{A}\tilde{x} + \tilde{B}\text{sat}_\Delta(T\bar{u} + d), \tag{13.4.3}$$

where $\tilde{A}$ and $\tilde{B}$ are given by (13.3.13). Also let $\tilde{x}(0) \in \Xi = (\mathcal{X} \times \mathcal{V}) \cup \{\tilde{x} - \Pi w : \tilde{x} \in \mathcal{X} \times \mathcal{V}, w \in \mathcal{W}\}$.

Step 2. Carry out Step 2 of the combined PLC/LHG feedback design of Section 13.3 using δ_i instead of Δ_i and Ξ instead of $\mathcal{X} \times \mathcal{V}$, and obtain a sequence of nested level sets $\varepsilon_0, \varepsilon_1, \cdots, \varepsilon_N$ (and correspondingly, the parameters $\epsilon_0, \epsilon_1, \cdots, \epsilon_N$) and a piecewise linear feedback law,

$$\bar{u} = \begin{cases} \bar{u}_i = -T^{-1}\tilde{R}_i^{-1}\tilde{B}'\tilde{P}_i\tilde{x} & \text{for } \tilde{x} \in \varepsilon_i \setminus \varepsilon_{i+1}, \ i = 0 \text{ to } N-1, \\ \bar{u}_N = -T^{-1}\tilde{R}_N^{-1}\tilde{B}'\tilde{P}_N\tilde{x} & \text{for } \tilde{x} \in \varepsilon_N. \end{cases} \tag{13.4.4}$$

Step 3. Design the final combined PLC/LHG tracking control laws as,

$$u = \begin{cases} u_i = v - (k+1)T^{-1}\tilde{R}_i^{-1}\tilde{B}'\tilde{P}_i\tilde{x} + T^{-1}[(k+1)\tilde{B}'\tilde{R}_i^{-1}\tilde{P}_i\Pi + \Gamma]w, \\ \qquad \text{for } \tilde{x} \in \varepsilon_i \setminus \varepsilon_{i+1}, \ i = 0 \text{ to } N-1, \\ \\ u_N = v - (k+1)T^{-1}\tilde{R}_N^{-1}\tilde{B}'\tilde{P}_N\tilde{x} + T^{-1}[(k+1)\tilde{B}'\tilde{R}_N^{-1}\tilde{P}_N\Pi + \Gamma]w \\ \qquad \text{for } \tilde{x} \in \varepsilon_N, \end{cases}$$

$$(13.4.5)$$

where $k \geq 0$ is a design parameter to be specified later.   Ⓐ

We now proceed to show that,

Point 1. there exists a $k_1^* > 0$ such that, for all $k \geq k_1^*$, Item 1 of Problem 13.2.1 holds;

Point 2. there exists a $k_2^* > 0$ such that, for all $k \geq k_2^*$, Item 2 of Problem 13.2.1 holds,

from which the result of Theorem 13.4.1 follows with $k^* = \max\{k_1^*, k_2^*\}$.

Let us first show the existence of $k_2^* > 0$. To do so, let us introduce an invertible, triangular coordinate change $\xi = \tilde{x} - \Pi w$. Using Condition 1 of the theorem, we have,

$$\dot{\xi} = \tilde{A}\xi + \tilde{B}[\text{sat}_\Delta(-Tv + Tu + d) - \Gamma w], \tag{13.4.6}$$

where u is given by (13.4.5). Let $\Xi_\infty \subset \mathbb{R}^{n+m}$ be a set that contains the origin as an interior point and be such that $\xi \in \Xi_\infty$ implies that $C\xi \in \mathcal{E}_\infty$. We next show that there exists a $k_2^* > 0$ such that for any $k \geq k_2^*$ the solution of (13.4.6) starting from any $\xi(0) \in \Xi$ enters the set Ξ_∞ in a finite time and remains in it thereafter. This can be done in two steps. In the first step, we show that, for each $i = 0$ to $N - 1$, there exits a $k_{2i}^* > 0$, such that for all $k \geq k_{2i}^*$, in the presence of any d satisfying Assumption 13.2.3, all trajectories starting from $\varepsilon_i \setminus \varepsilon_{i+1}$ will remain in ε_i and enter into the inner level set ε_{i+1} in a finite time. This in turn implies that, for any $k \geq \max\{k_{20}^*, k_{21}^*, \cdots, k_{2N-1}^*\}$, all the trajectories of the closed-loop system starting from $\Xi \subset \varepsilon_0$ will enter the inner-most level set ε_N in a finite time. The second step of the proof is to show that there exists a $k_{2N}^* > 0$ such that, for all $k \geq k_{2N}^*$, all the trajectories of the closed-loop system starting from ε_N will remain in it and enter and remain in the set Ξ_∞ in a finite time. Once these two steps are completed, the proof of Point 2 is then completed by taking $k_2^*(D, \Xi_\infty) = \max\{k_{20}^*, k_{21}^*, \cdots, k_{2N}^*\}$.

We start by considering the closed-loop system (13.4.6) for $\xi \in \varepsilon_i \setminus \varepsilon_{i+1}$, $i = 0$ to N,

$$\dot{\xi} = \tilde{A}\xi + \tilde{B}[\text{sat}(-(k+1)\tilde{R}_i^{-1}\tilde{B}'\tilde{P}_i\xi + \Gamma w + d) - \Gamma w]$$

$$= (\tilde{A} - \tilde{B}\tilde{R}_i^{-1}\tilde{B}'\tilde{P}_i)\xi + \tilde{B}[\text{sat}_\Delta(-(k+1)\tilde{R}_i^{-1}\tilde{B}'\tilde{P}_i\xi + \Gamma w + d)$$
$$+\tilde{R}_i^{-1}\tilde{B}'\tilde{P}_i\xi - \Gamma w], \tag{13.4.7}$$

where $\varepsilon_{N+1} = \emptyset$.

Pick the Lyapunov function,

$$V_i = \xi'\tilde{P}_i\xi. \tag{13.4.8}$$

The evaluation of $\dot{V}_i$ along the trajectories of the closed-loop system (13.4.7) gives,

$$\dot{V}_i = -\xi'\xi - \xi'\tilde{P}_i\tilde{B}\tilde{R}_i^{-1}\tilde{B}'\tilde{P}_i\xi$$
$$+2\xi'\tilde{P}_i\tilde{B}[\text{sat}_\Delta(-(k+1)\tilde{R}_i^{-1}\tilde{B}'\tilde{P}_i\xi + \Gamma w + d) + \tilde{R}_i^{-1}\tilde{B}'\tilde{P}_i\xi - \Gamma w]$$
$$\le -\xi'\xi - 2\sum_{i=1}^{m} v_i[\text{sat}_{\Delta_i}((k+1)v_i + \theta_i + d_i) - v_i - \theta_i], \tag{13.4.9}$$

where we have denoted the ith elements of $v = -\tilde{R}_i^{-1}\tilde{B}'\tilde{P}_i\xi$, Γw and d respectively as v_i, θ_i and d_i.

By the construction of ε_i, it is clear that $|v_i + \theta_i| \le \Delta_i$ for all $\xi \in \varepsilon_i$. Hence we have,

$$|kv_i| \ge |d_i| \implies v_i[\text{sat}_{\Delta_i}((k+1)v_i+\theta_i+d_i)-v_i-\theta_i] \ge v_i[\text{sat}_{\Delta_i}(v_i+\theta_i)-v_i-\theta_i] = 0,$$

and,

$$|kv_i| < |d_i| \implies v_i[\text{sat}_{\Delta_i}((k+1)v_i + \theta_i + d_i) - v_i - \theta_i] \le \frac{2|d_i|^2}{k},$$

where we have used the fact that,

$$|\text{sat}_{\Delta_i}(v_1) - \text{sat}_{\Delta_i}(v_2)| \le |v_1 - v_2|, \quad \forall v_1, v_2 \in \mathbf{R}.$$

Hence, we can conclude that, for all $\xi \in \varepsilon_i \setminus \varepsilon_{i+1}$, $i = 0$ to $N - 1$,

$$\dot{V}_i \le -\xi'\xi + \frac{4|d|^2}{k} \le -\xi'\xi + \frac{4D^2}{k}. \tag{13.4.10}$$

To complete the first step of the proof, for each $i = 0$ to $N - 1$, we let

$$k_{2i}^*(D) = \frac{5\lambda_{\max}(\tilde{P}_{i+1})D^2}{\rho_{i+1}}.$$

We then have that, for all $k \ge k_i^*$,

$$\dot{V}_i \le -\frac{1}{\lambda_{\max}(\tilde{P}_{i+1})}\left(V_{i+1} - \frac{4\lambda_{\max}(\tilde{P}_{i+1})D^2}{k}\right), \tag{13.4.11}$$

and hence,

$$\dot{V}_i < 0, \quad \forall \tilde{x} \in \varepsilon_i \setminus \varepsilon_{i+1}, \tag{13.4.12}$$

which, together with the fact that ε_{i+1} is strictly inside ε_i, show that all the trajectories of the closed-loop system starting from $\varepsilon_i \setminus \Xi_\infty$ will remain in ε_i and enter the level set ε_{i+1} in a finite time.

For the second step of the proof, let $\rho_{N+1} \in (0, \rho_N]$ be such that $\varepsilon(\tilde{P}_N, \rho_{N+1})$ $\subset \Xi_\infty$. The existence of such a ρ_{N+1} is due to the fact that Ξ_∞ contains the origin of the state space as an interior point. Choose $k_{2N}^*(D, \Xi_\infty)$ as follows,

$$k_{2N}^*(D, \Xi_\infty) = \frac{5\lambda_{\max}(\tilde{P}_N)D^2}{\rho_{N+1}}.$$

We will also have that, for all $k \geq k_{2N}^*$,

$$\dot{V}_N \leq -\frac{1}{\lambda_{\max}(\tilde{P}_N)} \left(V_N - \frac{4\lambda_{\max}(\tilde{P}_N)D^2}{k} \right), \tag{13.4.13}$$

and hence,

$$\dot{V}_N < 0, \quad \forall \xi \in \varepsilon_N \setminus \varepsilon(\tilde{P}_N, \rho_{N+1}), \tag{13.4.14}$$

which shows that all trajectories of the closed-loop system starting from ε_N will remain in ε_N and enter the set $\varepsilon(\tilde{P}_N, \rho_{N+1}) \subset \Xi_\infty$ in a finite time and remain in it thereafter. This completes the proof of the existence of k_2^*.

The existence of κ_1^* can be shown in a similar way. In particular, in the proof of the existence of κ_2^*, let $\mathcal{W} = \{0\}$ and hence $\xi = x$. Then the existence of k_1^* follows. This concludes the proof of Theorem 13.4.1. ⊠

13.5. Robust Tracking via Output Feedback

In this section, we consider the case that only the output is available for feedback, i.e., the output feedback robust tracking problem. We will use a fast observer to preserve the performance of the combined PLC/LHG state feedback laws proposed in the previous section. In order to build the fast observer, we make the following assumption,

Assumption 13.5.1. *The pair (A, E) is observable.* ⊞

Our output feedback result is given as follows.

Theorem 13.5.1. *Let Assumptions 13.2.1-13.2.3 and 13.5.1 hold, and given the data $(D, \mathcal{W}_\infty, \mathcal{E}_\infty)$, admissible for output feedback, then Problem 13.2.2 is solvable if there exist matrices Π and Γ such that*

1. they satisfy the linear matrix equations,

$$\begin{cases} \Pi S = \tilde{A}\Pi + \tilde{B}\Gamma, \\ \tilde{C}\Pi + Q = 0, \end{cases} \tag{13.5.1}$$

where $\tilde{A}$ and $\tilde{B}$ are as given by (13.3.13), and $\tilde{C} = [\, C \quad 0_{p \times m}\,]$;

2. there exists a $\delta = [\delta_1, \delta_2, \cdots, \delta_m]$, $\delta_i > 0$, such that $|\Gamma_i w(t)| \le \Delta_i - \delta_i$ for all $w(0) \in \mathcal{W}$ and all $t \ge 0$, where Γ_i is the ith row of Γ.

Moreover, the output feedback laws that solve Problem 13.2.2 can be explicitly constructed by implementing the state feedback law of the previous section with a fast observer. ▣

Proof of Theorem 13.5.1. We will prove this theorem by explicitly constructing a family of fast observer based combined PLC/LHG output feedback laws, parameterized in two parameters k and ℓ and showing that, there exists a $k^* > 0$, and for each $k \ge k^*$, there exists an $\ell^*(k) > 0$ such that, for any $\ell \ge \ell^*(k)$, $k \ge k^*$, the constructed output feedback law solves Problem 13.2.2.

Construction of Parameterized Output Feedback Laws:

Step 1: Fast Observer Design. The family of fast observers, parameterized in ℓ is given by,

$$\dot{\hat{x}} = A\hat{x} - L(\ell)(y - C\hat{x}) + Bv, \quad \hat{x}(0) \in \mathcal{X}, \tag{13.5.2}$$

where $L(\ell)$, $\ell > 0$ is chosen such that

$$\lambda(A + L(\ell)C) = \ell\Lambda, \tag{13.5.3}$$

where Λ is any set of n complex numbers, with negative real parts and close under complex conjugation.

Step 2: State Feedback Design. Carry out the state feedback design as in the proof of Theorem 13.4.1. In the design, instead of letting $\Xi = (\mathcal{X} \times \mathcal{V}) \cup \{\tilde{x} - \Pi w : \tilde{x} \in \mathcal{X} \times \mathcal{V}, w \in \mathcal{W}\}$, let Ξ be any bounded set such that $(\mathcal{X} \times \mathcal{V}) \cup \{\tilde{x} - \Pi w : \tilde{x} \in \mathcal{X} \times \mathcal{V}, w \in \mathcal{W}\}$ is strictly inside it. Let the resulting state feedback laws be given by,

$$u = \begin{cases} u_i = v - (k+1)T^{-1}\tilde{R}_i^{-1}\tilde{B}'\tilde{P}_i\tilde{x} + T^{-1}[(k+1)\tilde{B}'\tilde{R}^{-1}\tilde{P}_i\Pi + \Gamma]w, \\ \qquad \text{for } \tilde{x} \in \varepsilon_i \setminus \varepsilon_{i+1},\ i = 0, 1, \cdots, N-1, \\[2ex] u_N = v - (k+1)T^{-1}\tilde{R}_N^{-1}\tilde{B}'\tilde{P}_N\tilde{x} + T^{-1}[(k+1)\tilde{B}'\tilde{R}_N^{-1}\tilde{P}_N\Pi + \Gamma]w \\ \qquad \text{for } \tilde{x} \in \varepsilon_N, \end{cases}$$
$$\tag{13.5.4}$$

where

$$\varepsilon_i = \{\tilde{x} \in \mathbb{R}^{n+m} : \tilde{x}'\tilde{P}_i\tilde{x} \leq \rho_i\}, \, i = 0, 1, \cdots, N,$$

is the sequence of strictly nested level sets and $k \geq 0$ is a design parameter to be specified later.

Step 3: Output Feedback Laws. The final output feedback laws are obtained by implementing the state feedback law as obtained in Step 1 with the fast observer as obtained in Step 2. The resulting output feedback law is given by,

$$u = \begin{cases} u_i = v - (k+1)T^{-1}\tilde{R}_i^{-1}\tilde{B}'\tilde{P}_i\hat{\tilde{x}} + T^{-1}[(k+1)\tilde{B}'\tilde{R}^{-1}\tilde{P}_i\Pi + \Gamma]w, \\ \qquad \text{for } \hat{\tilde{x}} \in \varepsilon_i \setminus \varepsilon_{i+1}, \, i = 0, 1, \cdots, N-1, \\ \\ u_N = v - (k+1)T^{-1}\tilde{R}_N^{-1}\tilde{B}'\tilde{P}_N\hat{\tilde{x}} + T^{-1}[(k+1)\tilde{B}'\tilde{R}_N^{-1}\tilde{P}_N\tilde{P}i + \Gamma]w, \\ \qquad \text{for } \hat{\tilde{x}} \in \varepsilon_N, \end{cases}$$

$$(13.5.5)$$

where $\hat{\tilde{x}} = [\hat{x}', v']'$. Ⓐ

We next proceed to show that,

Point 1. there exists a $k_1^* > 0$, and for each $k \geq k_1^*$, there exists an $\ell_1^*(k) > 0$ such that, for all $\ell \geq \ell_1^*(k)$, $k \geq k_1^*$, Item 1 of Problem 13.2.2 holds;

Point 2. there exists a $k_2^* > 0$, and for each $k \geq k_2^*$, there exists an $\ell_2^*(k) > 0$ such that, for all $\ell \geq \ell_2^*(k)$, $k \geq k_2^*$, Item 2 of Problem 13.2.2 holds,

from which the results of Theorem 13.5.1 follows with $k^* = \max\{k_1^*, k_2^*\}$, and for each $k \geq k^*$, $\ell^*(k) = \max\{\ell_1^*(k), \ell_2^*(k)\}$.

Proof of Point 1:

To show the existence of k_1^*, and $\ell_1^*(k)$ for each $k \geq k_1^*$, let us write out the closed-loop system,

$$\begin{cases} \dot{\tilde{x}} = \tilde{A}\tilde{x} + \tilde{B}\text{sat}_\Delta(-(k+1)\tilde{R}_i^{-1}\tilde{B}'\tilde{P}_i\tilde{x} + (k+1)\tilde{R}_i^{-1}\tilde{B}'\tilde{P}_i\tilde{\phi} + d), \\ \qquad\qquad\qquad\qquad \tilde{x} \in \varepsilon_i \setminus \varepsilon_{i+1}, \, \tilde{x}(0) \in \mathcal{X} \times \mathcal{V}, \qquad (13.5.6) \\ \dot{\tilde{\phi}} = (A + L(\ell)C)\phi, \, \phi(0) \in \mathcal{X}, \end{cases}$$

where $\tilde{x} = [x', v']'$, $\tilde{\phi} = [\phi', 0]' = [x' - \hat{x}', 0]'$. Noting that $A + L(\ell)C$ is asymptotically stable for all $\ell > 0$ and hence $\lim_{t\to\infty} \phi(t) = 0$, we need only to show that, for any $k \geq k_1^*$ and $\ell \geq \ell_1^*(k)$, $(\hat{x}(t), v(t))$ enters $\mathcal{X}_\infty \times \mathcal{V}_\infty$ in a finite time and remains in it thereafter, where $\mathcal{X}_\infty$ and $\mathcal{V}_\infty$ both contain the origin as an interior point and are such that $\mathcal{X}_\infty \times \mathcal{V}_\infty \times \mathcal{X}_\infty$ is strictly inside the set $\mathcal{Z}_\infty$. To do this, we recall that $\mathcal{X} \times \mathcal{V}$ is strictly inside the set Ξ. Since the dynamics of $\tilde{x}$ is a linear system driven by a bounded input whose bound is independent of ℓ, it follows from (13.5.3) that there exist an $\ell_{1a}^* > 0$ and a $T_0 \geq 0$ such that for all $\ell \geq \ell_{1a}^*$,

$$\tilde{x}(T_0) = (x(T_0), v(T_0)) \in \Xi, \; \hat{\tilde{x}}(T_0) = (\hat{x}(T_0), v(T_0)) \in \Xi, \; \forall(x(0), v(0)) \in \mathcal{X} \times \mathcal{V}.$$

Hence $(\hat{x}(T_0), v(T_0)) \in \varepsilon_0$. We next show that there exists a $k_1^* > 0$ and for each $k \geq k_1^*$ there exists an $\ell_1^*(k) > 0$ such that, for any $\ell \geq \ell_1^*(k)$, $k \geq k_1^*$, the trajectory $(\hat{x}(t), v(t))$ will enter $\mathcal{X}_\infty \times \mathcal{V}_\infty$ in a finite time and remain in it thereafter. This can be done in two steps. In the first step, we show that, for each $i = 0$ to $N-1$, there exists a k_{1i}^* and for each $k \geq k_{1i}^*$ there exists an $\ell_{1i}^*(k) \geq 0$ such that for all $\ell \geq \ell_{1i}^*(k)$, $k \geq k_{1i}^*$, all trajectories $(\hat{x}(t), v(t))$ starting from $\varepsilon_i \setminus \varepsilon_{i+1}$ will remain in ε_i and enter into the inner level set ε_{i+1} in a finite time. This in turn implies that, for any $\ell \geq \max\{\ell_{10}^*(k), \ell_{11}^*(k), \cdots, \ell_{1N-1}^*(k)\}$, $k \geq \max\{k_{10}^*, k_{11}^*, \cdots, k_{1N-1}^*\}$, all trajectories $(\hat{x}(t), v(t))$ starting from ε_0 will enter the inner-most level set ε_N in a finite time. The second step of the proof is to show that there exists a $k_{1N}^* > 0$ and for each $k \geq k_{1N}^*$ there is an $\ell_{1N}^* \geq \ell_{1a}^*(k)$ such that, for all $\ell \geq \ell_{1N}^*(k)$, $k \geq k_{1N}^*$, all the trajectories $(\hat{x}(t), v(t))$ starting from ε_N will enter and remain in the set $\mathcal{X}_\infty \times \mathcal{V}_\infty$ in a finite time. Once these two steps are completed, the proof of Point 1 is then completed by taking $k_1^* = \max\{k_{10}^*, k_{11}^*, \cdots, k_{1N}^*\}$ and for each $k \geq k_1^*$, $\ell_1^*(k) = \max\{\ell_{10}^*(k), \ell_{11}^*(k), \cdots, \ell_{1N}^*(k)\}$.

We start by considering the closed-loop system (13.5.6) for $(\hat{x}, v) \in \varepsilon_i \setminus \varepsilon_{i+1}$ and $t \geq T_0$,

$$\begin{cases} \dot{\tilde{x}} = \tilde{A}\tilde{x} + \tilde{B}\text{sat}_\Delta(-(k+1)\tilde{R}_i^{-1}\tilde{B}'\tilde{P}_i\tilde{x} + (k+1)\tilde{R}_i^{-1}\tilde{B}'\tilde{P}_i\tilde{\phi} + d), \\ \dot{\phi} = (A + L(\ell)C)\phi, \; \phi(0) \in \mathcal{X}. \end{cases} \tag{13.5.7}$$

Pick the Lyapunov function

$$V_i(\hat{\tilde{x}}) = \hat{\tilde{x}}'\tilde{P}_i\hat{\tilde{x}}. \tag{13.5.8}$$

Recalling that

$$\hat{\tilde{x}} = \begin{bmatrix} \hat{x} \\ v \end{bmatrix} = \begin{bmatrix} x - \phi \\ v \end{bmatrix} = \tilde{x} - \tilde{\phi},$$

we have

$$V_i(\hat{\tilde{x}}) = \tilde{x}'\tilde{P}_i\tilde{x} - 2\tilde{x}'\tilde{P}_i\tilde{\phi} + \tilde{\phi}'\tilde{P}_i\tilde{\phi}. \tag{13.5.9}$$

Now for any $k > 0$, let $\ell_{1b}^*(k) \geq \ell_{1a}^*$ be such that, for any $\ell \geq \ell_{1b}^*(k)$, $k > 0$, and for all $(x(0), v(0), \hat{x}(0)) \in \mathcal{X} \times \mathcal{V} \times \mathcal{X}$,

$$\begin{cases} |(k+1)\tilde{R}_i^{-1}\tilde{B}'\tilde{P}_i\tilde{\phi}(t)| \leq D, \; \forall t \geq T_0, \\ |\tilde{x}'\tilde{P}_i[A + L(\varepsilon)C]\tilde{\phi}(t)| \leq \frac{D^2}{k}, \; \forall(\hat{\tilde{x}}, v) \in \varepsilon_0, \forall t \geq T_0, \\ |\tilde{\phi}'(t)\tilde{P}_i[\tilde{A}\tilde{x} + \tilde{B}\text{sat}_\Delta(\cdot)]| \leq \frac{D^2}{k}, \; \forall(\hat{\tilde{x}}, v) \in \varepsilon_0, \forall t \geq T_0, \\ |\tilde{\phi}'(t)\tilde{P}_i[A + L(c)C]\tilde{\phi}(t)| \leq \frac{D^2}{k}, \; \forall t \geq T_0. \end{cases} \tag{13.5.10}$$

Let $\tilde{d} = (k+1)\tilde{R}_i^{-1}\tilde{B}'\tilde{P}_i\tilde{\phi}(t) + d$. In view of Assumption 13.2.3 and (13.5.10), we have that

$$|\tilde{d}| \leq 2D, \ \forall t \geq T_0. \tag{13.5.11}$$

Viewing $\tilde{d}$ as disturbance and in view of (13.5.10), it follows from the same arguments as used for arriving at (13.4.10) that

$$\dot{V}_i \leq -\tilde{x}'\tilde{x} + \frac{22D^2}{k}$$

$$\leq -\frac{1}{\lambda_{\max}(\tilde{P}_{i+1})}\left(\tilde{x}'\tilde{P}_{i+1}\tilde{x} - \frac{22\lambda_{\max}(\tilde{P}_{i+1})D^2}{k}\right),$$

$$\forall \hat{\tilde{x}} \in \varepsilon_i \setminus \varepsilon_{i+1}. \tag{13.5.12}$$

For each k, let $\ell_{1i}^*(k) \geq \ell_{1b}^*(k)$ be such that, for all $t \geq T_0$, $\hat{\tilde{x}}'\tilde{P}_{i+1}\hat{\tilde{x}} \geq \rho_{i+1}$ implies that $\tilde{x}'\tilde{P}_{i+1}\tilde{x} \geq \frac{1}{2}\rho_{i+1}$. Also let

$$k_{1i}^*(D) = \frac{45\lambda_{\max}(\tilde{P}_{i+1})D^2}{\rho_{i+1}}. \tag{13.5.13}$$

We then have that, for all $\ell \geq \ell_{1i}^*(k)$, $k \geq k_{1i}^*$,

$$\dot{V}_i < 0, \ \forall \hat{\tilde{x}} \in \varepsilon_i \setminus \varepsilon_{i+1}, \tag{13.5.14}$$

which implies that all trajectories $(\hat{x}(t), v(t))$ of the closed-loop system starting from $\varepsilon_i \setminus \varepsilon_{i+1}$ will remain in ε_i and enter the inner set ε_{i+1} in a finite time and remain in it thereafter.

For the second step of the proof of Point 1, let $\rho_{N+1} \in (0, \rho_N]$ be such that $\varepsilon(\tilde{P}_N, \rho_{N+1}) \subset \mathcal{X}_\infty \times \mathcal{V}_\infty$. The existence of such a ρ_{N+1} is due to the fact that $\mathcal{X}_\infty \times \mathcal{V}_\infty$ contains the origin of the state space as an interior point. We will have,

$$\dot{V}_N \leq -\frac{1}{\lambda_{\max}(\tilde{P}_N)}\left(\tilde{x}\tilde{P}_N\tilde{x} - \frac{22\lambda_{\max}(\tilde{P}_N)D^2}{k}\right), \ \forall \hat{\tilde{x}} \in \varepsilon_N. \tag{13.5.15}$$

For each $k > 0$, let $\ell_{iN}^*(k) \geq \ell_{1b}^*(k)$ be such that, for all $t \geq T_0$, $\hat{\tilde{x}}'\tilde{P}_N\hat{\tilde{x}} \geq \rho_{N+1}$ implies that $\tilde{x}'\tilde{P}_N\tilde{x} \geq \frac{1}{2}\rho_{N+1}$. Also let

$$k_{iN}^*(D) = \frac{45\lambda_{\max}(\tilde{P}_N)D^2}{\rho_{N+1}}. \tag{13.5.16}$$

We then have that, for all $\ell \geq \ell_{1N}^*(k)$, $k \geq k_{iN}^*$,

$$\dot{V}_N < 0, \ \forall \hat{\tilde{x}} \in \varepsilon_N \setminus \varepsilon(\tilde{P}_N, \rho_{N+1}), \tag{13.5.17}$$

which shows that all trajectories $(\hat{x}(t), v(t))$ of the closed-loop system that start from ε_N will remain in ε_N and enter the set $\varepsilon(\tilde{P}_N, \rho_{N+1}) \subset \mathcal{X}_\infty \times \mathcal{V}_\infty$ in a finite time and remain in it thereafter.

This completes the proof of Point 1. We next proceed to prove Point 2.

Proof of Point 2:

Let us start with the introduction of an invertible, triangular coordinate change $\xi = \tilde{x} - \Pi w$. Using Condition 1 of the theorem, we can write the closed-loop system as follows,

$$\begin{cases} \dot{\xi} = \tilde{A}\xi + \tilde{B}[\text{sat}_\Delta(-(k+1)\tilde{R}_i^{-1}\tilde{B}'\tilde{P}_i\xi(k+1)\tilde{R}_i^{-1}\tilde{B}'\tilde{P}_i\tilde{\phi} + \Gamma w + d) - \Gamma w], \\ \qquad\qquad\qquad\qquad\qquad \hat{\tilde{x}} \in \varepsilon_i \setminus \varepsilon_{i+1}, \ \tilde{x}(0) \in \mathcal{X} \times \mathcal{V}, \\ \dot{\phi} = (A + L(\ell)C)\phi, \ \phi(0) \in \mathcal{X}, \end{cases}$$

$$(13.5.18)$$

which, except for the term Γw, is exactly the same as (13.5.6). The rest of the proof follows the arguments used in the proof of Point 2 of Theorem 13.4.1 and those used in the proof of Point 1 of the current theorem. ⊠

13.6. Robust Tracking of an F-16 Fighter Aircraft Aircraft

In this section, the proposed combined PLC/LHG design algorithm is applied to an F-16 fighter aircraft derivative. At the flight condition corresponding to an altitude of 10,000 feet and a Mach number of 0.7, the second order pitch plane dynamics (short period) of this aircraft is given by [84],

$$\begin{cases} \begin{bmatrix} \dot{\alpha} \\ \dot{q} \end{bmatrix} = \begin{bmatrix} -1.1500 & 0.9937 \\ 3.7240 & -1.2600 \end{bmatrix} \begin{bmatrix} \alpha \\ q \end{bmatrix} + \begin{bmatrix} -0.1770 \\ -19.5000 \end{bmatrix} \delta, \\ \qquad |\alpha(0)| \leq 0.1\text{rad}, |q(0)| \leq 0.5\text{rad/sec}, \\ \dot{\delta} = \text{sat}_1(-T\delta + Tu + d), \quad |\delta(0)| \leq 0.1\text{rad}, \\ \dot{w} = \begin{bmatrix} 0 & 1 \\ -1 & 0 \end{bmatrix} w, \ |w_i(0)| \leq 0.5\text{rad/sec}, i = 1, 2, \\ e = \begin{bmatrix} 0 & 1 \end{bmatrix} \begin{bmatrix} \alpha \\ q \end{bmatrix} + \begin{bmatrix} 0 & 1 \end{bmatrix} w, \\ y = \begin{bmatrix} 1 & 0 \end{bmatrix} \begin{bmatrix} \alpha \\ q \end{bmatrix}, \end{cases}$$

where α, q and δ are respectively the trim value of the angle of attack, pitch rate and stabilizer deflection. Here the actuator is rate limited to ±1.0 rad/sec. The constants $T = 20$ sec, which corresponds to an actuator bandwidth of 20 rad/sec.

It is straightforward to see that the above system is in the form of (13.2.1). Our goal here is to design an effective robust tracking control law that will cause the pitch rate q to track a class of sinusoidal reference signals in the presence of actuator disturbance d. The control law to be designed is the combined PLC/LHG feedback law proposed in the previous two sections.

State Feedback Design:

In the case that both states α and q are available for feedback, we will design state feedback laws as proposed in Section 13.4. We begin by checking the conditions of Theorem 13.4.1. It is straightforward to check that Condition 1 is satisfied. More specifically, the matrices

$$\Pi = \begin{bmatrix} -0.4141 & -0.4993 \\ 0 & -1 \\ -0.1304 & -0.0307 \end{bmatrix}, \quad \Gamma = [\,0.0307 \quad -0.1304\,] \tag{13.6.1}$$

solve the linear matrix equations (13.4.1). It is also straightforward to verify that $|\Gamma w(t)| \le 0.2$ for all $w(0) \in \mathcal{W}$ and for all $t \ge 0$. Hence, Condition 2 of the theorem is satisfied with $\delta = 0.8$.

Following the design procedure given in Section 13.4, we get

$$\Xi = \{(\xi_1, \xi_2, \xi_3) : |\xi_1| \le 0.5567, |\xi_2| \le 1, |\xi_3| \le 0.1805\},$$

and,

$$\rho_0 = 9.1920, \;\; \epsilon_0 = 26.8917, \;\; \tilde{P}_0 = \begin{bmatrix} 2.0409 & 1.0217 & -10.2609 \\ 1.0217 & 0.7091 & -5.7828 \\ -10.2609 & -5.7828 & 78.6731 \end{bmatrix}.$$

Choosing $N = 5$ and $\Delta\rho = 0.5$, we obtain a feedback law of the form (13.4.5) with,

$$\rho_1 = 4.5960, \;\; \epsilon_1 = 15.0397, \;\; \tilde{P}_1 = \begin{bmatrix} 1.5809 & 0.7792 & -6.9023 \\ 0.7792 & 0.5805 & -4.0412 \\ -6.9023 & -4.0412 & 49.2153 \end{bmatrix},$$

$$\rho_2 = 2.2980, \;\; \epsilon_2 = 8.4670, \;\; \tilde{P}_2 = \begin{bmatrix} 1.2568 & 0.6043 & -4.7011 \\ 0.6043 & 0.4849 & -2.8789 \\ -4.7011 & -2.8789 & 31.1964 \end{bmatrix},$$

$$\rho_3 = 1.1490, \;\; \epsilon_3 = 4.7942, \;\; \tilde{P}_3 = \begin{bmatrix} 1.0252 & 0.4753 & -3.2344 \\ 0.4753 & 0.4118 & -2.0852 \\ -3.2344 & -2.0852 & 20.0040 \end{bmatrix},$$

$$\rho_4 = 0.5745, \;\; \epsilon_4 = 2.7281, \;\; \tilde{P}_4 = \begin{bmatrix} 0.8580 & 0.3784 & -2.2435 \\ 0.3784 & 0.3541 & -1.5314 \\ -2.2435 & -1.5314 & 12.9550 \end{bmatrix},$$

$$\rho_5 = 0.2873, \quad \epsilon_5 = 1.5590, \quad \tilde{P}_5 = \begin{bmatrix} 0.7363 & 0.3045 & -1.5666 \\ 0.3045 & 0.3075 & -1.1376 \\ -1.5666 & -1.1376 & 8.4611 \end{bmatrix}.$$

We note here again that this control law reduces to an LHG feedback control law if $N = 0$ and to a PLC law if $k = 0$. Figs. 13.6.1 and 13.6.2 are simulation results. In these simulations, $(\alpha(0), q(0), \delta(0), w_1(0), w_2(0)) = (-0.1, 0.5, 0.1, 0, 0.5)$, and $d = 5\sin(4t + 3)$. They show good utilization of the available actuator rate capacity, and that the degree of tracking accuracy in the presence of actuator disturbance increases as the value of the parameter k increases.

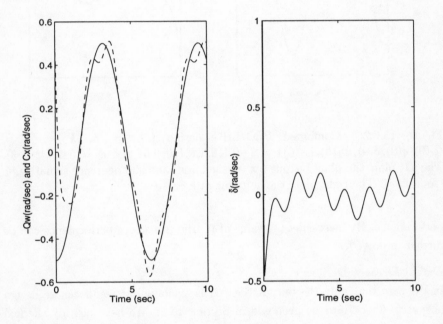

Figure 13.6.1: Combined PLC/LHG design ($N = 5$, $k = 40$). $(\alpha(0), q(0), \delta(0), w_1(0), w_2(0)) = (-0.1, 0.5, 0.1, 0, 0.5)$ and $d = 5\sin(4t + 3)$. The left plot: the plant output Cx (dashed line) and the reference signal $-Qw$ (solid line). The right plot: the actuator rate $\dot{\delta}$.

We next compare the performance of the combined PLC/LHG design with both the PLC and the LHG design. To compare with the PLC laws, we set $k = 0$ in (13.4.5). The simulation (Fig. 13.6.3) shows that in the presence of the same disturbance, the aircraft becomes unstable. To compare with the LHG feedback law, we set $N = 0$ in (13.4.5). The simulation (Fig. 13.6.4) shows that the combined PLC/LHG feedback law results in much better transient

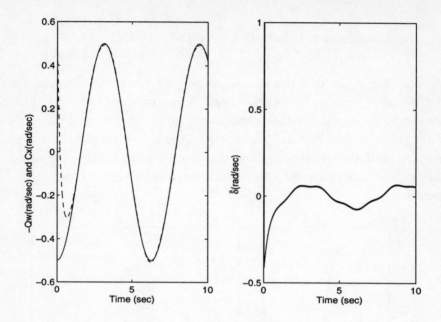

Figure 13.6.2:   Combined PLC/LHG design ($N = 5$, $k = 400$). $(\alpha(0), q(0), \delta(0), w_1(0), w_2(0)) = (-0.1, 0.5, 0.1, 0, 0.5)$ and $d = 5\sin(4t + 3)$. The left plot: the plant output Cx (dashed line) and the reference signal $-Qw$ (solid line). The right plot: the actuator rate $\dot{\delta}$.

performance. By increasing the value of N, the transience performance can be further improved.

Output Feedback Design:

In the case that only the output (i.e., α) is available for feedback, we design output feedback laws as proposed in Section 13.5. We have already checked in the state feedback case that all conditions of Theorem 13.5.1 are satisfied except Assumption 13.5.1, which is also satisfied. Following Step 1 of the output feedback design algorithm, we obtain the fast observer as given by (13.5.2) with

$$L(\ell) = -\left[\, 2\ell - 2.41 \quad 1.0063\ell^2 - 2.536\ell + 5.3216 \,\right]'.$$

We note here that, with this choice of $L(\ell)$,

$$\lambda(A + L(\ell)C) = \{-\ell, -\ell\}.$$

Proceeding with Step 2 of the design, we first choose

$$\Xi = \{(\xi_1, \xi_2, \xi_3) : |\xi_1| \le 0.56, |\xi_2| \le 1.01, |\xi_3| \le 0.185\}$$

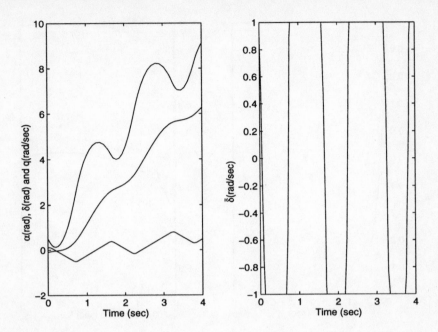

Figure 13.6.3: PLC Design $(N = 5, k = 0)$. $(\alpha(0), q(0), \delta(0), w_1(0), w_2(0)) = (-0.1, 0.5, 0.1, 0, 0.5)$ and $d = 5\sin(4t + 3)$. The left plot: states of the plant and the actuator. The right plot: the actuator rate $\dot{\delta}$.

and find,

$$\rho_0 = 9.8822, \quad \epsilon_0 = 28.5865, \quad \tilde{P}_0 = \begin{bmatrix} 2.0993 & 1.0521 & -10.7036 \\ 1.0521 & 0.7250 & -6.0107 \\ -10.7036 & -6.0107 & 82.6928 \end{bmatrix}.$$

Choosing $N = 5$ and $\Delta\rho = 0.5$, we obtain a state feedback law of the form (13.5.4) with,

$$\rho_1 = 4.9411, \quad \epsilon_1 = 15.9758, \quad \tilde{P}_1 = \begin{bmatrix} 1.6217 & 0.8010 & -7.1897 \\ 0.8010 & 0.5921 & -4.1914 \\ -7.1897 & -4.1914 & 51.6538 \end{bmatrix},$$

$$\rho_2 = 2.4705, \quad \epsilon_2 = 8.9881, \quad \tilde{P}_2 = \begin{bmatrix} 1.2857 & 0.6201 & -4.8911 \\ 0.6201 & 0.4937 & -2.9803 \\ -4.8911 & -2.9803 & 32.6997 \end{bmatrix},$$

$$\rho_3 = 1.2353, \quad \epsilon_3 = 5.0865, \quad \tilde{P}_3 = \begin{bmatrix} 1.0460 & 0.4871 & -3.3619 \\ 0.4871 & 0.4186 & -2.1552 \\ -3.3619 & -2.1552 & 20.9446 \end{bmatrix},$$

$$\rho_4 = 0.6176, \quad \epsilon_4 = 2.8930, \quad \tilde{P}_4 = \begin{bmatrix} 0.8731 & 0.3873 & -2.3302 \\ 0.3873 & 0.3596 & -1.5807 \\ -2.3302 & -1.5807 & 13.5511 \end{bmatrix},$$

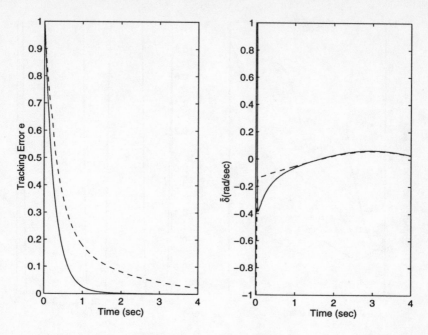

Figure 13.6.4: Transience performance: PLC/LHG ($N = 5, k = 40$, solid lines) vs. LHG ($N = 0, k = 40$, dashed lines). $(\alpha(0), q(0), \delta(0), w_1(0), w_2(0)) = (-0.1, 0.5, 0.1, 0, 0.5)$ and $d = 0$. The left plot: tracking error e. The right plot: the actuator rate $\dot{\delta}$.

$$\rho_5 = 0.3088, \quad \epsilon_5 = 1.6526, \quad \tilde{P}_5 = \begin{bmatrix} 0.7473 & 0.3113 & -1.6261 \\ 0.3113 & 0.3120 & -1.1730 \\ -1.6261 & -1.1730 & 8.8436 \end{bmatrix}.$$

Implementing the state feedback law with the fast observer state, both obtained as above, we have the output feedback law in the form of (13.5.5).

Some simulation results are shown in Fig. 13.6.5. In the simulation, $(\alpha(0), q(0), \delta(0), w_1(0), w_2(0)) = (-0.1, 0.5, 0.1, 0, 0.5)$, $d(t) = 5\sin(4t + 3)$, which are the same as in the simulations in the state feedback case, and $(\hat{x}_1(0), \hat{x}_2(0)) = (0, 0)$. With $k = 400$, $\ell = 1000$, these simulations results are almost the same as those shown in Fig. 13.6.2.

13.7. Concluding Remarks

A new design technique that combines two existing design techniques, the piecewise linear LQ control (PLC) and the low-and-high gain (LHG) design technique, both originally developed for linear systems subject to actuator magnitude saturation, has been developed for linear systems with rate-limited actuators. The new design technique yields feedback laws that cause the system output to track a desired command signal. The new design method retains the advantages of both PLC and LHG design techniques, while avoiding their disadvantages. Application of the proposed design to an F-16 fighter aircraft demonstrates the applicability, robustness and effectiveness of this novel control law.

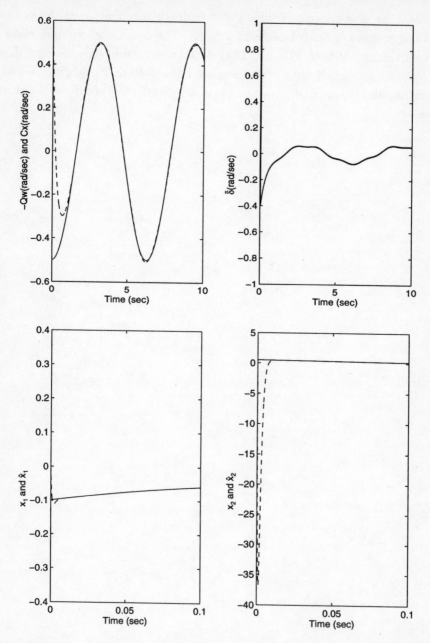

Figure 13.6.5: PLC Design ($N = 5$, $k = 400$). $(\alpha(0), q(0), \delta(0), w_1(0), w_2(0),$ $\hat{x}_1(0), \hat{x}_2(0)) = (-0.1, 0.5, 0.1, 0, 0.5, 0, 0)$ and $d = 5\sin(4t + 3)$. The upper left plot: the plant output Cx (dashed line) and the reference signal $-Qw$ (solid line). The upper right plot: the actuator rate $\dot{\delta}$. The lower plots: plant states $x(t)$ (solid line) and their estimates $\hat{x}$ (dashed line).

Appendix A

Some Technical Tools

A.1. Special Coordinate Basis

In this section, we recall the special coordinate basis (SCB) of linear systems introduced by Sannuti and Saberi [95], and Saberi and Sannuti [92]. Such a special coordinate basis has a distinct feature of explicitly displaying the finite and infinite zero structures as well as the invertibility structure of a given system. Connections between the special coordinate basis and the various invariant subspaces of geometric theory as needed for our development are also given. Although we only recall the SCB for continuous-time systems. Its discrete-time counterpart is completely analogous.

Let us consider a continuous-time linear system Σ_* characterized by the quadruple (A, B, C, D) or in the state space form,

$$\begin{cases} \dot{x} = A\,x + B\,u, \\ y = C\,x + D\,u, \end{cases} \tag{A.1.1}$$

where $x \in \mathbb{R}^n$, $u \in \mathbb{R}^m$ and $y \in \mathbb{R}^p$ are the state, the input and the output of Σ_*. It is simple to verify that there exist non-singular transformations U and V such that

$$UDV = \begin{bmatrix} I_{m_0} & 0 \\ 0 & 0 \end{bmatrix}, \tag{A.1.2}$$

where m_0 is the rank of matrix D. In fact, U can be chosen as an orthogonal matrix. Hence hereafter, without loss of generality, it is assumed that the matrix D has the form given on the right hand side of (A.1.2). One can now rewrite the system of (A.1.1) as,

$$
\begin{cases}
\dot{x} = A\,x + [B_0 \quad B_1] \begin{pmatrix} u_0 \\ u_1 \end{pmatrix}, \\[2mm]
\begin{pmatrix} y_0 \\ y_1 \end{pmatrix} = \begin{bmatrix} C_0 \\ C_1 \end{bmatrix} x + \begin{bmatrix} I_{m_0} & 0 \\ 0 & 0 \end{bmatrix} \begin{pmatrix} u_0 \\ u_1 \end{pmatrix},
\end{cases}
\tag{A.1.3}
$$

where the matrices B_0, B_1, C_0 and C_1 have appropriate dimensions. We have the following theorem.

Theorem A.1.1. *Given the linear system Σ_* of (A.1.1), there exist*

1. *Coordinate free non-negative integers n_a^-, n_a^0, n_a^+, n_b, n_c, n_d, $m_b \le p - m_0$, $m_d \le m - m_0$, r_i, $i = 1, \cdots, m_b$, and q_i, $i = 1, \cdots, m_d$, and*

2. *Non-singular state, output and input transformations Γ_s, Γ_o and Γ_i which take the given Σ_* into a special coordinate basis that displays explicitly both the finite and infinite zero structures of Σ_*.*

The special coordinate basis is described by the following set of equations:

$$
x = \Gamma_s \bar{x}, \quad y = \Gamma_o \bar{y}, \quad u = \Gamma_i \bar{u},
\tag{A.1.4}
$$

$$
\bar{x} = \begin{pmatrix} x_a \\ x_b \\ x_c \\ x_d \end{pmatrix}, \quad
x_a = \begin{pmatrix} x_a^- \\ x_a^0 \\ x_a^+ \end{pmatrix},
\tag{A.1.5}
$$

$$
x_b = \begin{pmatrix} x_{b1} \\ x_{b2} \\ \vdots \\ x_{bm_b} \end{pmatrix}, \quad
x_{b_i} = \begin{pmatrix} x_{bi1} \\ x_{bi2} \\ \vdots \\ x_{bir_i} \end{pmatrix}, \quad
x_d = \begin{pmatrix} x_1 \\ x_2 \\ \vdots \\ x_{m_d} \end{pmatrix}, \quad
x_i = \begin{pmatrix} x_{i1} \\ x_{i2} \\ \vdots \\ x_{iq_i} \end{pmatrix},
\tag{A.1.6}
$$

$$
\bar{y} = \begin{pmatrix} y_0 \\ y_d \\ y_b \end{pmatrix}, \quad
y_d = \begin{pmatrix} y_1 \\ y_2 \\ \vdots \\ y_{m_d} \end{pmatrix}, \quad
y_b = \begin{pmatrix} y_{b1} \\ y_{b2} \\ \vdots \\ y_{bm_b} \end{pmatrix},
\tag{A.1.7}
$$

$$
\bar{u} = \begin{pmatrix} u_0 \\ u_d \\ u_c \end{pmatrix}, \quad
u_d = \begin{pmatrix} u_1 \\ u_2 \\ \vdots \\ u_{m_d} \end{pmatrix},
\tag{A.1.8}
$$

and

$$
\dot{x}_a^- = A_{aa}^- x_a^- + B_{0a}^- y_0 + L_{ad}^- y_d + L_{ab}^- y_b,
\tag{A.1.9}
$$

$$\dot{x}_a^0 = A_{aa}^0 x_a^0 + B_{0a}^0 y_0 + L_{ad}^0 y_d + L_{ab}^0 y_b, \tag{A.1.10}$$

$$\dot{x}_a^+ = A_{aa}^+ x_a^+ + B_{0a}^+ y_0 + L_{ad}^+ y_d + L_{ab}^+ y_b, \tag{A.1.11}$$

for each $i = 1$ to m_b,

$$\dot{x}_{bi} = A_{r_i} x_{bi} + B_{0ib} y_0 + L_{bib} y_b + L_{bid} y_d, \tag{A.1.12}$$

$$y_{bi} = C_{r_i} x_{bi} = x_{bi1} \tag{A.1.13}$$

$$\dot{x}_c = A_{cc} x_c + B_{0c} y_0 + L_{cb} y_b + L_{cd} y_d + B_c \left[u_c + E_{ca}^- x_a^- + E_{ca}^0 x_a^0 + E_{ca}^+ x_a^+ \right], \tag{A.1.14}$$

$$y_0 = C_{0c} x_c + C_{0a}^- x_a^- + C_{0a}^0 x_a^0 + C_{0a}^+ x_a^+ + C_{0d} x_d + C_{0b} x_b + u_0, \tag{A.1.15}$$

and for each $i = 1$ to m_d,

$$\dot{x}_i = A_{q_i} x_i + B_{0id} y_0 + L_{id} y_d + B_{q_i} \left[u_i + E_{ia} x_a + E_{ib} x_b + E_{ic} x_c + \sum_{j=1}^{m_d} E_{ij} x_j \right], \tag{A.1.16}$$

$$y_i = C_{q_i} x_i, \quad y_d = C_d x_d. \tag{A.1.17}$$

Here the states x_a^-, x_a^0, x_a^+, x_b, x_c and x_d are respectively of dimensions n_a^-, n_a^0, n_a^+, n_b, n_c and $n_d = \sum_{i=1}^{m_d} q_i$, while x_i is of dimension q_i for each $i = 1, \cdots, m_d$. The control vectors u_0, u_d and u_c are respectively of dimensions m_0, m_d and $m_c = m - m_0 - m_d$, while the output vectors y_0, y_d and y_b are respectively of dimensions $p_0 = m_0$, $p_d = m_d$ and $p_b = p - p_0 - p_d$. Also for an integer $r \geq 1$,

$$A_r = \begin{bmatrix} 0 & I_{r-1} \\ 0 & 0 \end{bmatrix}, \quad B_r = \begin{bmatrix} 0 \\ 1 \end{bmatrix}, \quad C_r = [1, 0, \cdots, 0]. \tag{A.1.18}$$

Assuming that x_i, $i = 1, 2, \cdots, m_d$, are arranged such that $q_i \leq q_{i+1}$, the matrix L_{id} has the particular form

$$L_{id} = [L_{i1d} \quad L_{i2d} \quad \cdots \quad L_{ii-1d} \quad 0 \quad \cdots \quad 0]. \tag{A.1.19}$$

Also, the last row of each L_{id} is identically zero, $\lambda(A_{aa}^-) \subset \mathbb{C}^-$, $\lambda(A_{aa}^0) \subset \mathbb{C}^0$, and $\lambda(A_{aa}^+) \subset \mathbb{C}^+$. Moreover, the pair (A_{cc}, B_c) is controllable and the pair (A_{bb}, C_b) is observable. ⊺

Proof of Theorem A.1.1. See [92,95]. From the proof, we observe that when the state x_b is non-existent, $\Gamma_o = I$. Dually, when the state x_c is non-existent, $\Gamma_i = I$. The software realizations of the above decomposition in LAS and in MATLAB can be found [8] and [40], respectively. ⊠

We can rewrite the special coordinate basis of the quadruple (A, B, C, D) given by Theorem A.1.1 in a more compact form,

$$\bar{A} = \Gamma_s^{-1}(A - B_0 C_0)\Gamma_s$$

$$= \begin{bmatrix} A_{aa}^- & 0 & 0 & L_{ab}^- C_b & 0 & L_{ad}^- C_d \\ 0 & A_{aa}^0 & 0 & L_{ab}^0 C_b & 0 & L_{ad}^0 C_d \\ 0 & 0 & A_{aa}^+ & L_{ab}^+ C_b & 0 & L_{ad}^+ C_d \\ 0 & 0 & 0 & A_{bb} & 0 & L_{bd} C_d \\ B_c E_{ca}^- & B_c E_{ca}^0 & B_c E_{ca}^+ & L_{cb} C_b & A_{cc} & L_{cd} C_d \\ B_d E_{da}^- & B_d E_{da}^0 & B_d E_{da}^+ & B_d E_{db} & B_d E_{dc} & A_{dd} \end{bmatrix}, \quad \text{(A.1.20)}$$

$$\bar{B} = \Gamma_s^{-1}[\, B_0 \quad B_1 \,]\Gamma_{\mathrm{I}} = \begin{bmatrix} B_{0a}^- & 0 & 0 \\ B_{0a}^0 & 0 & 0 \\ B_{0a}^+ & 0 & 0 \\ B_{0b} & 0 & 0 \\ B_{0c} & 0 & B_c \\ B_{0d} & B_d & 0 \end{bmatrix}, \quad \text{(A.1.21)}$$

$$\bar{C} = \Gamma_o^{-1}\begin{bmatrix} C_0 \\ C_1 \end{bmatrix}\Gamma_s = \begin{bmatrix} C_{0a}^- & C_{0a}^0 & C_{0a}^+ & C_{0b} & C_{0c} & C_{0d} \\ 0 & 0 & 0 & 0 & 0 & C_d \\ 0 & 0 & 0 & C_b & 0 & 0 \end{bmatrix}, \quad \text{(A.1.22)}$$

$$\bar{D} = \Gamma_o^{-1} D \Gamma_{\mathrm{I}} = \begin{bmatrix} I_{m_0} & 0 & 0 \\ 0 & 0 & 0 \\ 0 & 0 & 0 \end{bmatrix}. \quad \text{(A.1.23)}$$

In what follows, we state some important properties of the above special coordinate basis which are pertinent to our presentation of this monograph.

Property A.1.1. *The given system Σ_* is observable (detectable) if and only if the pair $(A_{\mathrm{obs}}, C_{\mathrm{obs}})$ is observable (detectable), where*

$$A_{\mathrm{obs}} = \begin{bmatrix} A_{aa} & 0 \\ B_c E_{ca} & A_{cc} \end{bmatrix}, \quad C_{\mathrm{obs}} = \begin{bmatrix} C_{0a} & C_{0c} \\ E_{da} & E_{dc} \end{bmatrix}, \quad \text{(A.1.24)}$$

and where

$$A_{aa} = \begin{bmatrix} A_{aa}^- & 0 & 0 \\ 0 & A_{aa}^0 & 0 \\ 0 & 0 & A_{aa}^+ \end{bmatrix}, \quad \text{(A.1.25)}$$

$$C_{0a} = [\, C_{0a}^- \quad C_{0a}^0 \quad C_{0a}^+ \,], \ E_{da} = [\, E_{da}^- \quad E_{da}^0 \quad E_{da}^+ \,], \ E_{ca} = [\, E_{ca}^- \quad E_{ca}^0 \quad E_{ca}^+ \,].$$
$$\text{(A.1.26)}$$

Also, define

$$B_{0a} = \begin{bmatrix} B_{0a}^- \\ B_{0a}^0 \\ B_{0a}^+ \end{bmatrix}, \quad L_{ab} = \begin{bmatrix} L_{ab}^- \\ L_{ab}^0 \\ L_{ab}^+ \end{bmatrix}, \quad L_{ad} = \begin{bmatrix} L_{ad}^- \\ L_{ad}^0 \\ L_{ad}^+ \end{bmatrix}, \qquad (A.1.27)$$

and

$$A_{\text{con}} = \begin{bmatrix} A_{aa} & L_{ab}C_b \\ 0 & A_{bb} \end{bmatrix}, \quad B_{\text{con}} = \begin{bmatrix} B_{0a} & L_{ad} \\ B_{0b} & L_{bd} \end{bmatrix}. \qquad (A.1.28)$$

Similarly, Σ_* is controllable (stabilizable) if and only if the pair $(A_{\text{con}}, B_{\text{con}})$ is controllable (stabilizable). Ⓟ

Property A.1.2. *Invariant zeros of Σ_* are the eigenvalues of A_{aa}, which are the unions of the eigenvalues of A_{aa}^-, A_{aa}^0 and A_{aa}^+.* Ⓟ

Property A.1.3. *Σ_* has $m_0 = \text{rank}\,(D)$ infinite zeros of order 0. The infinite zero structure (of order greater than 0) of Σ_* is given by $S_\infty^\star(\Sigma_*) = \{q_1, q_2, \cdots, q_{m_d}\}$, i.e., each q_i corresponds to an infinite zero of Σ_* of order q_i.*

Moreover, the set of integers $\{q_1, q_2, \cdots, q_{m_d}\}$ is the list $\mathcal{I}_4$ of Morse structural invariants [79]. Similarly, the set of integers $\{r_1, r_2, \cdots, r_{m_b}\}$ is the list $\mathcal{I}_2$ of the Morse structural invariants. Ⓟ

Property A.1.4. *The given system Σ_* is right invertible if and only if x_b (and hence y_b) are non-existent, left invertible if and only if x_c (and hence u_c) are non-existent, and invertible if and only if both x_b and x_c are non-existent. Moreover, Σ_* is degenerate if and only if it is neither left nor right invertible, i.e., both x_b and x_c are present.* Ⓟ

By now it is clear that the special coordinate basis decomposes the state-space into several distinct parts. In fact, the state-space $\mathcal{X}$ is decomposed as

$$\mathcal{X} = \mathcal{X}_a^- \oplus \mathcal{X}_a^0 \oplus \mathcal{X}_a^+ \oplus \mathcal{X}_b \oplus \mathcal{X}_c \oplus \mathcal{X}_d. \qquad (A.1.29)$$

The following property shows interconnections between the special coordinate basis and various invariant geometric subspaces as defined in Definition 11.2.1.

Property A.1.5. *Various components of the state vector of the special coordinate basis have the following geometrical interpretations:*

1. *$\mathcal{X}_a^+ \oplus \mathcal{X}_c \oplus \mathcal{X}_d$ spans $\mathcal{S}^-(\Sigma_*)$;*

2. *$\mathcal{X}_a^- \oplus \mathcal{X}_a^0 \oplus \mathcal{X}_c \oplus \mathcal{X}_d$ spans $\mathcal{S}^+(\Sigma_*)$;*

3. *$\mathcal{X}_c \oplus \mathcal{X}_d$ spans $\mathcal{S}^*(\Sigma_*)$;*

4. $\mathcal{X}_a^- \oplus \mathcal{X}_a^0 \oplus \mathcal{X}_c$ spans $\mathcal{V}^-(\Sigma_*)$;

5. $\mathcal{X}_a^+ \oplus \mathcal{X}_c$ spans $\mathcal{V}^+(\Sigma_*)$;

6. $\mathcal{X}_a^- \oplus \mathcal{X}_a^0 \oplus \mathcal{X}_a^+ \oplus \mathcal{X}_c$ spans $\mathcal{V}^*(\Sigma_*)$.   P

The $\mathcal{S}_\lambda(\Sigma_*)$ and $\mathcal{V}_\lambda(\Sigma_*)$, as defined by (11.3.2), can also be easily obtained using the special coordinate basis. The following property was proven in [10].

Property A.1.6.

$$
\mathcal{S}_\lambda(\Sigma_*) = \mathrm{Im} \left\{ \Gamma_s \begin{bmatrix} \lambda I - A_{aa} & 0 & 0 & 0 \\ 0 & Y_{b\lambda} & 0 & 0 \\ 0 & 0 & I_{n_c} & 0 \\ 0 & 0 & 0 & I_{n_d} \end{bmatrix} \right\}, \tag{A.1.30}
$$

where

$$
\mathrm{Im}\,\{Y_{b\lambda}\} = \mathrm{Ker}\,\left[C_b (A_{bb} + L_b C_b - \lambda I)^{-1} \right], \tag{A.1.31}
$$

and where L_b is any appropriate matrix subject to the constraint that matrix $A_{bb} + L_b C_b$ has no eigenvalues at λ. We note that such an L_b always exists as (A_{bb}, C_b) is complete observable.

$$
\mathcal{V}_\lambda(\Sigma_*) = \mathrm{Im} \left\{ \Gamma_s \begin{bmatrix} X_{a\lambda} & 0 \\ 0 & 0 \\ 0 & X_{c\lambda} \\ 0 & 0 \end{bmatrix} \right\}, \tag{A.1.32}
$$

where $X_{a\lambda}$ is a matrix whose columns form a basis for the subspace,

$$
\left\{ \zeta_a \in \mathbb{C}^{n_a} \;\middle|\; (\lambda I - A_{aa})\zeta_a = 0 \right\}, \tag{A.1.33}
$$

and

$$
X_{c\lambda} = \left(A_{cc} + B_c F_c - \lambda I \right)^{-1} B_c, \tag{A.1.34}
$$

with F_c being any appropriately dimensional matrix subject to the constraint that $A_{cc} + B_c F_c$ has no eigenvalues at λ. Again, we note that the existence of such an F_c is guaranteed by the controllability of (A_{cc}, B_c).   P

A.2. A High Gain Observer Lemma

The following lemma is adapted from [115]. It is a critical tool in establishing the output feedback results of Chapters 4 and 9.

Lemma A.2.1. *Consider the nonlinear system*

$$\dot{z} = f(z, e, t), \; z \in \mathbb{R}^n, \tag{A.2.1}$$

$$\dot{e} = \ell Ae + g(z, e, t), \; e \in \mathbb{R}^m, \tag{A.2.2}$$

where $\ell > 0$ and A is an asymptotically stable matrix. Assume that for the system

$$\dot{z} = f(z, 0, t),$$

there exists a neighborhood $\mathcal{W}_1$ of the origin in $\mathbb{R}^n$ and a C^1 function V_1 : $\mathcal{W}_1 \to \mathbb{R}^+$ which is positive definite on $\mathcal{W}_1 \setminus \{0\}$ and proper on $\mathcal{W}_1$ and satisfies

$$\frac{\partial V_1}{\partial z} f(z, 0, t) \leq -\psi_1(z),$$

where $\psi_1(z)$ is continuous on $\mathcal{W}_1$ and positive definite on $\{z : \nu_1 < V_1(z) \leq c_1 + 1\}$ for some nonnegative real number $\nu_1 < 1$ and some real number $c_1 \geq 1$. Also assume that there exist positive real numbers α and β and a bounded function γ with $\gamma(0) = 0$ satisfying

$$\left. \begin{array}{l} |f(z, e, t) - f(z, 0, t)| \leq \gamma(|e|) \\ |g(z, e, t)| \leq \alpha|e| + \beta \end{array} \right\}, \; \forall (z, e, t) \in L_{V_1}(c_1 + 1) \times \mathbb{R}^m \times \mathbb{R}^+. \tag{A.2.3}$$

Let c_2 be a class $\mathcal{K}_\infty$ function satisfying

$$\lim_{\ell \to \infty} \frac{\ell}{c_2^4(\ell)} = \infty. \tag{A.2.4}$$

Let P solve the Lyapunov equation $A'P + PA = -I$. Define the function

$$V(z, e) = c_1 \frac{V_1(z)}{c_1 + 1 - V_1(z)} + c_2(\ell) \frac{\ln(1 + e'Pe)}{c_2(\ell) + 1 - \ln(1 + e'Pe)} \tag{A.2.5}$$

and the set

$$\mathcal{W} = \{z : V_1(z) < c_1 + 1\} \times \{e : \ln(1 + e'Pe) < c_2(\ell) + 1\}.$$

Then, for $\ell > 0$, $V : \mathcal{W} \to \mathbb{R}^+$ is positive definite on $\mathcal{W} \setminus \{0\}$ and proper on $\mathcal{W}$. Furthermore, for any $\nu_2 \in (0, 1)$, there exists an $\ell^*(\nu_2) > 0$ such that, for all $\ell \in [\ell^*(\nu_2), \infty)$, the derivative of V along the trajectories of (A.2.1)-(A.2.2) satisfies

$$\dot{V} \leq -\psi(z, e),$$

where $\psi(z, e)$ is positive definite on $\{(z, e) : \nu_1 + \nu_2 \leq V(z, e) \leq c_1^2 + c_2^2(\ell) + 1\}$.

☐

Notes and References

This monograph takes a unified approach to present some low gain design techniques and their applications in solving various control problems. Some of the results presented are already published, some are drastically simplified or generalized version of the published results and others are unpublished. The following notes and references give a brief description of the source of the material presented in the monograph.

Chapter 1. There is a vast literature on high gain feedback. The few references cited here are a small sample of it. This sample of references serve as a little background for introducing the recent literature on low gain feedback. The cited references on low gain feedback are related to the material to be described in the monograph.

Chapter 2. For continuous-time systems, the eigenstructure assignment based low gain design algorithm is new. Its single input case is the same as the one given in [54], where the multiple input case is somewhat cumbersome and could result in numerically ill conditioned feedback gains. For a system with n states and m inputs, the algorithm of [54] would yield a feedback gain that is polynomial matrix in the low gain parameter ε of order as high as $r_1(r_2+1)(r_3+1)\cdots(r_m+1)$, where $r_1+r_2+\cdots+r_m=n$. The new algorithm presented here however leads to polynomials of order r_i's. For $n=6$ and $m=2$, it could be ε^{12} versus ε^3. Lemma 2.2.1 is taken from [42]. Lemmas 2.2.2-2.2.5 are adapted from [49]. The ARE based low gain feedback design algorithm was originally proposed in [70]. Lemma 2.2.6 is generalized from [45,70]. For discrete-time systems, the eigenstructure assignment based low gain design algorithm is new. Its single input case is the same as the one given in [58], where multiple input case is somewhat cumbersome. Lemma 2.3.1 is taken from [42]. Lemmas 2.3.2-2.3.5 are new and have not appeared anywhere. The ARE based low gain feedback design algorithm was originally proposed in [67]. Lemma 2.3.6 is adapted and generalized from [48].

Chapter 3. For continuous-time systems, semi-global stabilizability by linear feedback of linear systems subject to actuator saturation was first established in [54] using a somewhat cumbersome eigenstructure assignment based low gain feedback design technique. The new proof given here is based on a simplified eigenstructure assignment procedure as presented in Chapter 2. This simplified eigenstructure procedure enables the elegant Lyapunov proof of the results. The alternative proof of the result by using H_2-ARE and H_∞-ARE based low gain feedback laws were later given in [70] and [113], respectively. The proof given here is based on the H_2-ARE based low gain feedback design of Chapter 2. For discrete-time systems, semi-global stabilizability by linear feedback was first established in [58] using an eigenstructure assignment based low gain feedback design technique, similar to that of [54]. The new proof given here is based on the simplified eigenstructure assignment procedure as presented in Chapter 2. This simplified eigenstructure procedure enables the elegant Lyapunov proof of the results. Alternative proof of the result by using the ARE based low gain feedback design technique, as given in Section 3.4.3, is adapted from [67].

Chapter 4. The low-and-high gain design was initiated in [60], where an eigenstructure assignment based design was developed for the class of linear systems that contain only a chain of integrators. The development of low-and-high gain design was completed in [89] for ARE based design. The extension of [89] to allow dead zone nonlinearities in the actuators was made in [43]. Section 4.4 is based on the results of [89] and [43]. The first complete, though complicated, eigenstructure assignment based low-and-high gain design was given in [61]. The eigenstructure assignment based low-and-high gain design as presented in Section 4.5 is new. The low gain based variable structure control design method as presented in Section 4.6 is based on [43].

Chapter 5. The continuous-time results are based on [70] and some unpublished notes. The discrete-time results are taken from [72]. Efforts have been made to simplify and unify the proofs by using Lyapunov analysis.

Chapter 6. The material of this chapter is taken from [68] and [69]. The semi-global finite gain L_p-stabilization was solved in [68] as the problem of D-bounded input finite gain L_p-stabilization, where the magnitude of the external input is required to be uniformly bounded by a known number $D > 0$. This result was later strengthened in [69] as the solution to the so called almost D-bounded disturbance almost disturbance decoupling problem with stability.

The presentation here unified both these results by introducing the concept of semi-globalility with respect to external input. A generalization is that the external input can be bounded either in magnitude or in energy.

Chapter 7. The presentation of the Section 7.2 is based on [44], while the material of Section 7.3 is drawn from [47]. Presentation is given in a unified framework.

Chapter 8. The continuous-time results are taken from [45]. The discrete-time results are drawn from [48]. The presentation is given in a unified framework and draws on the results of Chapter 2.

Chapter 9. The results were originally given in [59]. Both the designs and the proofs presented here have been drastically simplified, thanks to the new and simplified low gain design technique of Chapter 2 and the finer low gain feedback properties established there. In the design, the powers on both the low gain and high gain parameters are drastically reduced (see notes and references on Chapter 2 for the reduction in the power on the low gain parameter, the reduction in the power on the high gain parameter is greater). This leads to much higher low gain feedback gain and much lower high gain feedback gain and thus reduces possible numerical stiffness due to extremely small and/or extremely large numbers. The new properties of the low gain feedback established in Chapter 2 also enable us to present a much simpler proof of the results.

Chapter 10. For both continuous-time and discrete-time systems, the algorithms for explicitly constructing parameterized families of feedback laws that solve the problems of perfect regulation exist in the literature [63] and [64]. The new algorithms presented here take advantage of the new simplified low gain design techniques of Chapter 2 and result in much simpler and numerically more stable families of feedback laws. For instance, for the system in Section 10.2.3, the algorithm of [63] would result in a family of feedback laws as given by $u = -(1+1/\varepsilon^{12})x_1 - (\varepsilon+1/\varepsilon^{13})x_2$, instead of the simple $u = -2x_1 - (\varepsilon+1/\varepsilon)x_2$ as obtained here. The unified Lyapunov analysis for the low gain feedback system as developed in Chapter 2 also significantly simplifies the proofs of this chapter. Following [65,66], the design procedures of this chapter can be used to construct feedback laws that solve general H_2-suboptimal control problems.

Chapter 11. The material of Sections 11.2 and 11.3 is taken respectively from [12] and [50]. The presentation is however toward the goal of demonstrating how low gain feedback design techniques of Chapter 2 play a key role in the explicit construction of families of feedback laws that solve the general H_∞-ADDPMS. As a result, many details of [12] and [50] were not included. Similarly, Section 11.4 is aimed to provide an insight into how low gain feedback design techniques can be utilized in the solution of nonlinear H_∞-ADDPS. Further generalization of the results of this section can be found in [49].

Chapter 12. The material is taken from [62]. The design presented here includes two tuning parameters, ε and μ. A design that uses only a single parameter was carried out in [41]. The motivation for introducing a new design parameter μ is to achieve infinite gain margin of the resulting feedback laws.

Chapter 13. The presentation is based on [52] and [53].

Bibliography

[1] B.R. Barmish, M. Corless and G. Leitmann, "A new class of stabilizing controllers for uncertain dynamical systems," *SIAM Journal on Control and Optimization*, Vol. 21, pp. 255-264, 1983.

[2] J.M. Berg, K.D. Hammett, C.A. Schwartz, and S.S. Banda, "An analysis of the destabilizing effect of daisy chained rate-limited actuators," *IEEE Transactions on Control Systems Technology*, Vol. 4, pp. 171-176, 1996.

[3] D.S. Bernstein and A.N. Michel, "A chronological bibliography on saturating actuators," *International Journal of Robust and Nonlinear Control*, Vol. 5, pp. 375-380, 1995.

[4] C.I. Byrnes and A. Isidori, "A frequency domain philosophy for nonlinear systems with applications to stabilization and adaptive control," *Proceedings of the 23rd IEEE Conference on Decision and Control*, pp. 1569-1573, 1984.

[5] C.I. Byrnes and A. Isidori, "Local stabilization of minimum-phase nonlinear systems," *Systems & Control Letters*, Vol. 10, pp. 9-17, 1988

[6] C.I. Byrnes and A. Isidori, "Asymptotic stabilization of minimum phase nonlinear systems," *IEEE Transactions on Automatic Control*, Vol. 36, pp. 1122-1137, 1991.

[7] V.-S. Chellaboina and W.M. Haddad, "Fixed-order dynamic compensation for linear systems with actuator amplitude and rate saturation constraints, preprint, 1997.

[8] B. M. Chen, *Software Manual for the Special Coordinate Basis of Multivariable Linear Systems*, Washington State University Technical Report Number: ECE 0094, Pullman, Washington, 1988.

[9] B. M. Chen, *Linear Systems and Control Toolbox*, Department of Electrical Engineering, National University of Singapore, 1997.

[10] B.M. Chen, H_∞ *Control and Its Applications*, Lecture Notes in Control and Information Sciences, Springer-Verlag, to appear.

[11] B. M. Chen, J. He and Y. Chen, "Explicit solvability conditions for general discrete-time H_∞ almost disturbance decoupling problem with internal stability," *International Journal of Systems Science*, to appear.

[12] B. M. Chen, Z. Lin and C. C. Hang, "Design for general H_∞ almost disturbance decoupling problem with measurement feedback and internal stability – an eigenstructure assignment approach," *International Journal of Control*, to appear.

[13] W.P. Dayawansa, C.F. Martin and G. Knowles, "A global nonlinear tracking problem," *Proceedings of the 29th IEEE Conference on Decision and Control*, pp. 1268-1271, 1990.

[14] M.A. Dornheim, "Report pinpoints factors leading to YF-22 crash," *Aviation Week Space Technology*, pp. 53-54, Nov. 9, 1992.

[15] F. Esfandiari and H.K. Khalil, "Output feedback stabilization of fully linearizable systems," *International Journal of Control*, Vol. 56, pp. 1007-1037, 1992

[16] B.A. Francis, "The linear multivariable regulator problem", *SIAM Journal on Control and Optimization*, Vol. 15, pp. 486-505, 1977.

[17] B.A. Francis, "The optimal linear-quadratic time-invariant regulator with cheap control," *IEEE Transactions on Automatic Control*, Vol. 24, pp. 616-621, 1979.

[18] A.T. Fuller, "In the large stability of relay and saturated control systems with linear controllers," *International Journal of Control*, Vol. 10, pp. 457-480, 1969.

[19] V.M. Gavrilyako, V.I. Korobov and G.M. Skylar, "Designing a bounded control of dynamic systems in entire space with the aid of a controllability function," *Automat. Remote Control*, Vol. 11, pp. 1484-1490, 1986.

[20] P.-O. Gutman and P. Hagander, "A new design of constrained controllers for linear systems," *IEEE Transactions on Automatic Control*, Vol. 30, pp. 22-33, 1985.

[21] M. Hautus, "Linear matrix equations with applications to the regulator problem," in *Outils and Modèles Mathématique pour l'Automatique*, I.D. Landou, Ed. Paris: C.N.R.S., pp. 399-412, 1983.

[22] I.W. Horowitz, *Synthesis of Feedback Systems*, Academic Press, New York, 1963.

[23] Y.S. Hung and A.G.J. MacFarlane, "On the relationships between the unbounded asymptotic behaviour of multivariable root-loci, impulse response and infinite zeros," *International Journal of Control*, Vol. 34, pp. 31-69, 1981.

[24] A. Isidori, *Nonlinear Control Systems*, 3rd edition, Springer, Berlin, 1995.

[25] A. Isidori, "A note on almost disturbance decoupling for nonlinear minimum phase systems," *Systems & Control Letters*, Vol. 27, pp. 191-194, 1996.

[26] A. Isidori, "Global almost disturbance decoupling with stability for non minimum-phase single-input single-output nonlinear systems," Vol. 28, pp. 115-122, 1996.

[27] T. Kailath, *Linear Systems*, Prentice-Hall, Englewood Cliffs, NJ, 1980.

[28] P. Kapasouris and M. Athans, "Control systems with rate and magnitude saturation for neutrally stable open loop systems," *Proceedings of the 29th IEEE Conference on Decision and Control*, pp. 3404-3409, 1990.

[29] H.K. Khalil, *Nonlinear Systems*, 2nd Edition, Prentice Hall, Upper Saddle River, 1996.

[30] H.K. Khalil and F. Esfandiari, "Semi-global stabilization of nonlinear system using output feedback," *Proceedings of the 31st IEEE Conference on Decision and Control*, pp. 3423-3428, 1992.

[31] H.K. Khalil and A. Saberi, "Adaptive stabilization of a class of nonlinear systems using high-gain feedback," *IEEE Transactions on Automatic Control*, Vol. 32, pp. 1031-1035, 1987.

[32] H. Kimura, "A new approach to the perfect regulation and the bounded peaking in linear multivariable control systems," *IEEE Transactions on Automatic Control*, Vol. 26, pp. 253-270, 1981.

[33] P.V. Kokotovic and H.J. Sussmann, "A positive real condition for global stabilization of nonlinear systems," *Systems & Control Letters*, Vol. 12, pp. 125-134, 1989.

[34] B. Kouvaritakis and J.M. Edmund, "A multivariable root-loci: a unified approach to finite and infinite zeros," *International Journal of Control*, Vol. 29, pp. 393-428, 1979.

[35] B. Kouvaritakis and U. Shaked, "Asymptotic behaviour of root-locus of linear multivariable systems," *International Journal of Control*, Vol. 23, pp. 297-340, 1976.

[36] J. Kurzweil, "On the inversion of Lyapunov's second theorem on stability of motion," *Am. Math. Soc. Transl.*, Ser. 2, 24:19-77, 1956.

[37] H. Kwakernaak and R. Sivan, *Linear Optimal Control Systems*, John Wiley & Sons, Inc., New York, 1972.

[38] G. Leitmann, "Guaranteed asymptotic stability for some linear systems with bounded uncertainties," *Journal of Dynamic Systems, Measurement and Control Systems*, Vol. 101, pp. 212-216, 1979.

[39] J.M. Lenorovitz, "Gripen control problems resolved through in-flight, ground simulations," *Aviation Week Space Technology*, pp. 74-75, June 18, 1990.

[40] Z. Lin, *The Implementation of Special Coordinate Basis for Linear Multivariable Systems in Matlab*, Washington State University Technical Report Number ECE0100, Pullman, Washington, 1989.

[41] Z. Lin, "State feedback design for the inverted pendulum system – a linear high-and-low-gain approach," personal notes.

[42] Z. Lin, *Global and Semi-global Control Problems for Linear Systems Subject to Input Saturation and Minimum-Phase Input-Output Linearizable Systems*, Ph.D. Dissertation, Washington State University, 1994.

[43] Z. Lin, "Robust semi-global stabilization of linear systems with imperfect actuators," *Systems & Control Letters*, Vol. 29, pp. 215-221, 1997.

[44] Z. Lin, "H_∞-almost disturbance decoupling with internal stability for linear systems subject to input saturation," *IEEE Transactions on Automatic Control*, Vol. 42, pp.992-995, 1997.

[45] Z. Lin, "Semi-global stabilization of linear systems with position and rate limited actuators," *Systems & Control Letters*, Vol. 30, pp. 1-11, 1997.

[46] Z. Lin, "Almost disturbance decoupling with global asymptotic stability for nonlinear systems with disturbance affected unstable zero dynamics," *Systems & Control Letters*, Vol. 33, pp. 163-169, 1998.

[47] Z. Lin, "Global control of linear systems with saturating actuators," *Automatica*, to appear.

[48] Z. Lin, "Semi-global stabilization of discrete-time linear systems with position and rate limited actuators," *Systems & Control Letters*, to appear.

[49] Z. Lin, X. Bao and B. M. Chen, "Further results on almost disturbance decoupling with global asymptotic stability for nonlinear systems," *Proceedings of the 36th IEEE Conference on Decision and Control*, pp. 2847-2852, 1997

[50] Z. Lin and B.M. Chen, "Solutions to general H_∞ almost disturbance decoupling problem with measurement feedback and internal stability for discrete-time systems," in preparation.

[51] Z. Lin, R. Mantri and A. Saberi, "Semi-global output regulation for linear systems subject to input saturation – a low-and-high gain design," *Control – Theory and Advanced Technology*, Vol. 10, pp. 2209-2231, 1995.

[52] Z. Lin, M. Pachter, S. Banda and Y. Shamash, "Stabilizing feedback design for linear systems with rate limited actuators," *Control of Uncertain Systems with Bounded Inputs*, eds. S. Tarbouriech and Germain Garcia, Lecture Notes in Control and Information Sciences, Springer-Verlag, Vol. 227, pp. 173-186, 1997.

[53] Z. Lin, M. Pachter, S. Banda, and Y. Shamash, "Feedback design for robust tracking of linear with rate limited actuators," in preparation.

[54] Z. Lin and A. Saberi, "Semi-global exponential stabilization of linear systems subject to 'input saturation' via linear feedbacks," *Systems & Control Letters*, Vol. 21, pp. 225-239, 1993.

[55] Z. Lin and A. Saberi, "Semi-global stabilization of partially linear composite systems via feedback of the state of the linear part," *Systems & Control Letters*, Vol. 20, pp. 199-207, 1993.

[56] Z. Lin and A. Saberi, "Semi-global stabilization of minimum phase nonlinear systems in special normal form via linear high-and-low-gain state feedback," *International Journal of Robust and Nonlinear Control*, Vol. 4, pp. 353-362, 1994.

[57] Z. Lin and A. Saberi, "Semi-global stabilization of partially linear composite systems via linear dynamic state feedback," *Control - Theory and Advanced Technology*, Vol. 10, pp. 447-463, 1994.

[58] Z. Lin and A. Saberi, "Semi-global exponential stabilization of linear discrete-time systems subject to 'input saturation' via linear feedbacks," *Systems & Control Letters*, Vol. 24, pp. 125-132, 1995.

[59] Z. Lin and A. Saberi, "Robust semi-global stabilization of minimum-phase input-output linearizable systems via partial state and output feedback," *IEEE Transactions on Automatic Control*, Vol. 40, pp. 1029-1041, 1995.

[60] Z. Lin and A. Saberi, "A semi-global low-and-high gain design technique for linear systems with input saturation - stabilization and disturbance rejection," *International Journal of Robust and Nonlinear Control*, Vol. 5, pp. 381-398, 1995.

[61] Z. Lin and A. Saberi, "Low-and-high gain design technique for linear systems subject to input saturation - a direct method," *International Journal of Robust and Nonlinear Control*, Vol. 7, pp. 1071-1101, 1997.

[62] Z. Lin, A. Saberi, M. Gutmann and Y. Shamash, "Linear controller for an inverted pendulum having restricted travel - A high-and-low gain approach," *Automatica*, Vol. 32, pp. 933-937, 1996.

[63] Z. Lin, A. Saberi, P. Sannuti and Y. Shamash, "Perfect regulation of linear multivariable systems - a low-and-high gain design," *Proceedings the Workshop on Advances on Control and Its Applications, Lecture Notes in Control and Information Sciences* Vol. 208, pp. 172-193, editors: H. Khalil, J. Chow and P. Ioannou, 1996.

[64] Z. Lin, A. Saberi, P. Sannuti and Y. Shamash, "Perfect regulation for linear discrete-time systems - A low-gain based design approach," *Automatica*, Vol. 32, pp. 1085-1091, 1996.

[65] Z. Lin, A. Saberi, P. Sannuti, Y. Shamash, "A direct method of constructing H_2 suboptimal controllers – continuous-time systems," *Journal of Optimization Theory and Applications*, to appear.

[66] Z. Lin, A. Saberi, P. Sannuti, Y. Shamash, "A direct method of constructing H_2 suboptimal controllers – discrete-time systems," *Journal of Optimization Theory and Applications*, to appear.

[67] Z. Lin, A. Saberi and A. Stoorvogel, "Semi-global stabilization of linear discrete-time systems subject to input saturation via linear feedback – an ARE-based approach," *IEEE Transactions on Automatic Control*, Vol. 41, pp. 1203-1207, 1996.

[68] Z. Lin, A. Saberi and A.R. Teel, "Simultaneous L_p-stabilization and internal stabilization of linear systems subject to input saturation – state feedback case," *Systems & Control Letters*, Vol. 25, pp. 219-226, 1995.

[69] Z. Lin, A. Saberi and A.R. Teel, "Almost disturbance decoupling with internal stability for linear systems subject to input saturation – state feedback case," *Automatica*, Vol. 32, pp. 619-624, 1996.

[70] Z. Lin, A. Stoorvogel and A. Saberi, "Output regulation for linear systems subject to input saturation," *Automatica*, Vol. 32, pp. 29-47, 1996.

[71] W. Liu, Y. Chitour and E. Sontag, "On finite gain stabilizability of linear systems subject to input saturation," *SIAM Journal on Control and Optimization*, Vol. 34, pp. 1190-1219, 1996.

[72] R. Mantri, A. Saberi, Z. Lin and A. Stoorvogel, "Output regulation for linear discrete-time systems subject to input saturation," *International Journal of Robust and Nonlinear Control*, Vol. 7, No. 11, pp. 1003-1022, 1997.

[73] R. Marino, "High-gain feedback in nonlinear control systems," *International Journal of Control*, Vol. 42, pp. 1369-1385, 1985.

[74] R. Marino, W. Respondek, and A.J. van der Shaft, "Almost disturbance decoupling for single-input single-output nonlinear systems," *IEEE Transactions on Automatic Control*, Vol. 34, pp. 1013-1017, 1989.

[75] R. Marino, W. Respondek, A.J. van der Shaft, and P. Tomei, "Nonlinear H_∞ almost disturbance decoupling," *Systems & Control Letters*, Vol. 23, pp. 159-168, 1994.

[76] F. Mazenc, L. Praly and W.P. Dayawansa, "Examples and counterexamples on global stabilization by output feedback," *Systems & Control Letters*, Vol. 23, pp. 119-125, 1994.

[77] A. Megretski, "Output feedback stabilization with saturated control: making the input-output map L_2-bounded," preprint. See also "L_2 BIBO output feedback stabilization with saturated control," *Proceedings of the 13th Triennial World Congress of IFAC*, Vol. D, pp. 435-440, 1996.

[78] P. Miotto, J.M. Shewchun, E. Feron, and J.D. Paduano, "High performance bounded control synthesis with application to the F18 HARV. *Proceedings of the AIAA Guidance, Navigation and Control*, AIAA Paper 96-3697, 1996.

[79] A.S. Morse, "Structural invariants of linear systems," *SIAM Journal on Control*, Vol. 11, pp. 446-465, 1973.

[80] T. Nguyen and F. Jabbari, "Output feedback controllers for disturbance attenuation with bounded control," *Proceedings of the 36th IEEE Conference on Decision and Control*, pp. 177-182, 1997.

[81] D.H. Owens, "On structural invariants and the root-loci of linear multivariable systems," *International Journal of Control*, Vol. 28, pp. 328-337, 1978.

[82] H.K. Ozcetin, A. Saberi, and P. Sannuti, "Design for H_∞ almost disturbance decoupling problem with internal stability via state or measurement feedback – singular perturbation approach," *International Journal of Control*, Vol. 55, pp. 901-944, 1992.

[83] H.K. Ozcetin, A. Saberi and Y. Shamash, "H_∞-almost disturbance decoupling for non-strictly proper systems – a singular perturbation approach," *Control – Theory and Advanced Technology*, Vol. 9, pp. 203-245, 1993.

[84] M. Pachter, P.R. Chandler and M. Mears, "Reconfigurable tracking control with saturation," *AIAA Journal of Guidance, Control and Dynamics*, Vol. 18, pp. 1016-1022, 1995.

[85] A. Saberi, "A decentralization of large-scale systems: a new canonical form for linear multivariable systems," *IEEE Transactions on Automatic Control*, Vol. 30, No. 11, 1985

[86] A. Saberi, B.M. Chen and P. Sannuti, "Theory of LTR for non-minimum phase systems, Recoverable Target loops, Recovery in a subspace – Part II: Design, " *International Journal of Control*, Vol. 53, pp. 1117-1160, 1991.

[87] A. Saberi, P.V. Kokotovic and H.J. Sussmann, "Global stabilization of partially linear composite systems," *SIAM Journal on Control and Optimization*, Vol. 28, pp. 1491-1503, 1990.

[88] A. Saberi and Z. Lin, "Adaptive high-gain stabilization of 'minimum-phase' nonlinear systems," *Control – Theory and Advanced Technology*, Vol. 6, pp. 595-607, 1990.

[89] A. Saberi, Z. Lin and A.R. Teel, "Control of linear systems with saturating actuators," *IEEE Transactions on Automatic Control*, Vol. 41, pp. 368-378, 1996.

[90] A. Saberi and P. Sannuti, "Global stabilization with almost disturbance decoupling of a class of uncertain non-linear systems," *International Journal Control*, Vol. 47, pp. 717-727, 1988.

[91] A. Saberi and P. Sannuti, "Time-scale structure assignment in linear multivariable systems using high gain feedback," *International Journal of Control*, Vol. 49, pp. 2191-2213, 1989.

[92] A. Saberi and P. Sannuti, "Squaring down of non-strictly proper systems," *International Journal of Control*, Vol. 51, pp. 621-629, 1990.

[93] A. Saberi and P. Sannuti, "Observer design for loop transfer recovery and for uncertain dynamical systems," *IEEE Transactions on Automatic Control*, Vol. 35, pp. 878-897, 1990.

[94] A. Saberi, P. Sannuti and B.M. Chen, H_2 *Optimal Control*, Prentice Hall International, London, 1995.

[95] P. Sannuti and A. Saberi, "A special coordinate basis of multivariable linear systems–Finite and infinite zero structure, squaring down and decoupling," *International Journal of Control*, Vol. 45, pp. 1655-1704, 1987.

[96] P. Sannuti and H. Wason, "A singular perturbation canonical form of invertible systems: determination of multivariable root-loci, *International Journal of Control*, Vol. 37, pp. 1259-1296, 1983.

[97] C. Scherer, "H_∞-optimization without assumptions on finite or infinite zeros," *SIAM Journal on Control and Optimization*, Vol. 30, No. 1, pp. 143-166, 1992.

[98] W.E. Schmitendorf and B.R. Barmish, "Null controllability of linear systems with constrained controls," *SIAM Journal Control and Optimization*, Vol. 18, pp. 327-345, 1980.

[99] U. Shaked, "Design techniques for high-feedback gain stability," *International Journal of Control*, Vol. 24, pp. 137-144, 1976.

[100] J.M. Shewchun and E. Feron, "High Performance Bounded Control," *Proceedings of the 1997 American Control Conference*, pp. 2350-2354, 1997.

[101] C.A. Shifrin, "Gripen likely to fly again soon," *Aviation Week Space Technology*, pp. 72, Aug. 23, 1993.

[102] E.D. Sontag, "An algebraic approach to bounded controllability of linear systems," *International Journal of Control*, Vol. 39, pp. 181-188, 1984.

[103] E.D. Sontag and H.J. Sussmann, "Nonlinear output feedback design for linear systems with saturating controls," *Proceedings of the 29th IEEE Conference on Decision and Control*, pp. 3414-3416, 1990.

[104] R. Suarez, J. Alvarez-Ramirez and J. Solis-Daun, "Linear systems with bounded inputs: global stabilization with eigenvalue placement," *International Journal of Robust and Nonlinear Control*, to appear.

[105] R. Suarez, J. Alvarez-Ramirez, M. Sznaier, C. Ibarra-Valdez, "L_2-disturbance attenuation for linear systems with bounded controls: an ARE-Based Approach," *Control of Uncertain Systems with Bounded Inputs*, eds. S. Tarbouriech and Germain Garcia, Lecture Notes in Control and Information Sciences, Springer-Verlag, Vol. 227, pp. 25-38, 1997.

[106] H.J. Sussmann, "Limitations on the stabilizability of globally minimum phase systems," *IEEE Transactions on Automatic Control*, Vol. 35, pp. 117-119, 1990.

[107] H.J. Sussmann and P.V. Kokotovic, "The peaking phenomenon and the global stabilization of nonlinear systems'" *IEEE Transactions on Automatic Control*, Vol. 36, pp. 424-440, 1991.

[108] H.J. Sussmann, E.D. Sontag and Y. Yang, "A general result on the stabilization of linear systems using bounded controls", *IEEE Transactions on Automatic Control*, Vol. 39, pp. 2411-2425, 1994.

[109] H.J. Sussmann and Y. Yang, "On the stabilizability of multiple integrators by means of bounded feedback controls," *Proceedings of the 30th IEEE Conference on Decision and Control*, pp. 70-72, 1991.

[110] A. R. Teel, "Global stabilization and restricted tracking for multiple integrators with bounded controls," *Systems & Control Letters*, Vol. 18, pp. 165-171, 1992.

[111] A.R. Teel, *Feedback Stabilization: Nonlinear Solutions to Inherently Nonlinear Problems*, Ph.D dissertation, University of California, Berkeley, 1992.

[112] A.R. Teel, "Linear systems with input nonlinearities: global stabilization by scheduling a family of H_∞-type controllers," *International Journal of Robust and Nonlinear Control*, Vol. 5, pp. 399-441, 1995.

[113] A.R. Teel, "Semi-global stabilization of linear controllable systems with input nonlinearities," *IEEE Transactions on Automatic Control*, Vol. 40, pp. 96-100, 1995.

[114] A.R. Teel and L. Praly, "Global stabilizability and observability imply semi-global stabilizability by output feedback," *Systems and Control Letters*, Vol. 22, pp. 313-325, 1994.

[115] A.R. Teel, L. Praly, "Tools for semiglobal stabilization by partial state and output feedback", *SIAM Journal on Control and Optimization*, Vol. 33, pp. 1443-1488, 1995.

[116] H.L. Trentelman, "Families of linear-quadratic problems: continuity properties", *IEEE Transactions on Automatic Control*, Vol. 32, pp. 323-329, 1987.

[117] H. Trentelmen and J.C. Willems, "Guaranteed roll-off in a class of high gain feedback design problems", *Systems & Control Letters*, Vol. 3, pp. 361-369, 1981.

[118] F. Tyan and D.S. Bernstein, "Dynamic output feedback compensation for systems with independent amplitude and rate saturation," *International Journal of Control*, Vol. 67, pp. 89-116, 1997.

[119] M. Vidyasagar, *Nonlinear Systems Analysis*, 2nd edition, Englewood Cliffs, NJ: Princeton-Hall, 1993.

[120] W. Walter, *Differential and Integral Inequalities*, Berlin, Springer-Verlag, 1970.

[121] Q.F. Wei, Dayawansa, W.P. and W.S. Levine, Nonlinear controller for an inverted pendulum having restricted travel, *Automatica*, Vol. 31, pp. 841-850, 1995.

[122] S. Weiland, and J.C. Willems, "Almost disturbance decoupling," *IEEE Transactions on Automatic Control*, Vol. 34, pp. 277-286, 1989.

[123] J.C. Willems, "Least squares stationary optimal control and the algebraic Riccati equation", *IEEE Transactions on Automatic Control*, Vol. 16, pp. 621-634, 1971.

[124] J.C. Willems, "Almost invariant subspaces: an approach to high gain feedback design. Part I: Almost controlled invariant subspaces," *IEEE Transactions on Automatic Control*, Vol. 26, pp. 235-252, 1981.

[125] J.C. Willems, "Almost invariant subspaces: an approach to high gain feedback design. Part II: Almost conditionally invariant subspaces," *IEEE Transactions on Automatic Control*, Vol. 27, pp. 1071-1085, 1982.

[126] W. M. Wonham, *Linear Multivariable Control: A Geometric Approach*, Springer-Verlag, New York, 1979.

[127] G.F. Wredenhagen and P.R. Belanger, "Piecewise-linear LQ control for systems with input constraints," *Automatica*, Vol. 30, pp. 403-416, 1994.

[128] Y. Yang, *Global Stabilization of Linear Systems with Bounded Feedback*, Ph.D dissertation, New Brunswick Rutgers, the State University of New Jersey, 1993.

Index

Lecture Notes in Control and Information Sciences

Edited by M. Thoma

1993–1998 Published Titles:

Vol. 186: Sreenath, N.
Systems Representation of Global Climate
Change Models. Foundation for a Systems
Science Approach.
288 pp. 1993 [3-540-19824-5]

Vol. 187: Morecki, A.; Bianchi, G.;
Jaworeck, K. (Eds)
RoManSy 9: Proceedings of the Ninth
CISM-IFToMM Symposium on Theory and
Practice of Robots and Manipulators.
476 pp. 1993 [3-540-19834-2]

Vol. 188: Naidu, D. Subbaram
Aeroassisted Orbital Transfer: Guidance
and Control Strategies
192 pp. 1993 [3-540-19819-9]

Vol. 189: Ilchmann, A.
Non-Identifier-Based High-Gain Adaptive
Control
220 pp. 1993 [3-540-19845-8]

Vol. 190: Chatila, R.; Hirzinger, G. (Eds)
Experimental Robotics II: The 2nd
International Symposium, Toulouse,
France, June 25-27 1991
580 pp. 1993 [3-540-19851-2]

Vol. 191: Blondel, V.
Simultaneous Stabilization of Linear
Systems
212 pp. 1993 [3-540-19862-8]

Vol. 192: Smith, R.S.; Dahleh, M. (Eds)
The Modeling of Uncertainty in Control
Systems
412 pp. 1993 [3-540-19870-9]

Vol. 193: Zinober, A.S.I. (Ed.)
Variable Structure and Lyapunov Control
428 pp. 1993 [3-540-19869-5]

Vol. 194: Cao, Xi-Ren
Realization Probabilities: The Dynamics of
Queuing Systems
336 pp. 1993 [3-540-19872-5]

Vol. 195: Liu, D.; Michel, A.N.
Dynamical Systems with Saturation
Nonlinearities: Analysis and Design
212 pp. 1994 [3-540-19888-1]

Vol. 196: Battilotti, S.
Noninteracting Control with Stability for
Nonlinear Systems
196 pp. 1994 [3-540-19891-1]

Vol. 197: Henry, J.; Yvon, J.P. (Eds)
System Modelling and Optimization
975 pp approx. 1994 [3-540-19893-8]

Vol. 198: Winter, H.; Nüßer, H.-G. (Eds)
Advanced Technologies for Air Traffic Flow
Management
225 pp approx. 1994 [3-540-19895-4]

Vol. 199: Cohen, G.; Quadrat, J.-P. (Eds)
11th International Conference on
Analysis and Optimization of Systems –
Discrete Event Systems: Sophia-Antipolis,
June 15–16–17, 1994
648 pp. 1994 [3-540-19896-2]

Vol. 200: Yoshikawa, T.; Miyazaki, F. (Eds)
Experimental Robotics III: The 3rd
International Symposium, Kyoto, Japan,
October 28-30, 1993
624 pp. 1994 [3-540-19905-5]

Vol. 201: Kogan, J.
Robust Stability and Convexity
192 pp. 1994 [3-540-19919-5]

Vol. 202: Francis, B.A.; Tannenbaum, A.R.
(Eds)
Feedback Control, Nonlinear Systems,
and Complexity
288 pp. 1995 [3-540-19943-8]

Vol. 203: Popkov, Y.S.
Macrosystems Theory and its Applications:
Equilibrium Models
344 pp. 1995 [3-540-19955-1]

Vol. 204: Takahashi, S.; Takahara, Y.
Logical Approach to Systems Theory
192 pp. 1995 [3-540-19956-X]

Vol. 205: Kotta, U.
Inversion Method in the Discrete-time
Nonlinear Control Systems Synthesis
Problems
168 pp. 1995 [3-540-19966-7]

Vol. 206: Aganovic, Z.; Gajic, Z.
Linear Optimal Control of Bilinear Systems
with Applications to Singular Perturbations
and Weak Coupling
133 pp. 1995 [3-540-19976-4]

Vol. 207: Gabasov, R.; Kirillova, F.M.;
Prischepova, S.V.
Optimal Feedback Control
224 pp. 1995 [3-540-19991-8]

Vol. 208: Khalil, H.K.; Chow, J.H.;
Ioannou, P.A. (Eds)
Proceedings of Workshop on Advances
inControl and its Applications
300 pp. 1995 [3-540-19993-4]

Vol. 209: Foias, C.; Özbay, H.;
Tannenbaum, A.
Robust Control of Infinite Dimensional
Systems: Frequency Domain Methods
230 pp. 1995 [3-540-19994-2]

Vol. 210: De Wilde, P.
Neural Network Models: An Analysis
164 pp. 1996 [3-540-19995-0]

Vol. 211: Gawronski, W.
Balanced Control of Flexible Structures
280 pp. 1996 [3-540-76017-2]

Vol. 212: Sanchez, A.
Formal Specification and Synthesis of
Procedural Controllers for Process Systems
248 pp. 1996 [3-540-76021-0]

Vol. 213: Patra, A.; Rao, G.P.
General Hybrid Orthogonal Functions and
their Applications in Systems and Control
144 pp. 1996 [3-540-76039-3]

Vol. 214: Yin, G.; Zhang, Q. (Eds)
Recent Advances in Control and Optimization
of Manufacturing Systems
240 pp. 1996 [3-540-76055-5]

Vol. 215: Bonivento, C.; Marro, G.;
Zanasi, R. (Eds)
Colloquium on Automatic Control
240 pp. 1996 [3-540-76060-1]

Vol. 216: Kulhavý, R.
Recursive Nonlinear Estimation: A Geometric
Approach
244 pp. 1996 [3-540-76063-6]

Vol. 217: Garofalo, F.; Glielmo, L. (Eds)
Robust Control via Variable Structure and
Lyapunov Techniques
336 pp. 1996 [3-540-76067-9]

Vol. 218: van der Schaft, A.
L_2 Gain and Passivity Techniques in Nonlinear
Control
176 pp. 1996 [3-540-76074-1]

Vol. 219: Berger, M.-O.; Deriche, R.;
Herlin, I.; Jaffré, J.; Morel, J.-M. (Eds)
ICAOS '96: 12th International Conference on
Analysis and Optimization of Systems -
Images, Wavelets and PDEs:
Paris, June 26-28 1996
378 pp. 1996 [3-540-76076-8]

Vol. 220: Brogliato, B.
Nonsmooth Impact Mechanics: Models,
Dynamics and Control
420 pp. 1996 [3-540-76079-2]

Vol. 221: Kelkar, A.; Joshi, S.
Control of Nonlinear Multibody Flexible Space
Structures
160 pp. 1996 [3-540-76093-8]

Vol. 222: Morse, A.S.
Control Using Logic-Based Switching
288 pp. 1997 [3-540-76097-0]

Vol. 223: Khatib, O.; Salisbury, J.K.
Experimental Robotics IV: The 4th International
Symposium, Stanford, California,
June 30 - July 2, 1995
596 pp. 1997 [3-540-76133-0]

Vol. 224: Magni, J.-F.; Bennani, S.;
Terlouw, J. (Eds)
Robust Flight Control: A Design Challenge
664 pp. 1997 [3-540-76151-9]

Vol. 225: Poznyak, A.S.; Najim, K.
Learning Automata and Stochastic Optimization
219 pp. 1997 [3-540-76154-3]

Vol. 226: Cooperman, G.; Michler, G.;
Vinck, H. (Eds)
Workshop on High Performance Computing
and Gigabit Local Area Networks
248 pp. 1997 [3-540-76169-1]

Vol. 227: Tarbouriech, S.; Garcia, G. (Eds)
Control of Uncertain Systems with Bounded
Inputs
203 pp. 1997 [3-540-76183-7]

Vol. 228: Dugard, L.; Verriest, E.I. (Eds)
Stability and Control of Time-delay Systems
344 pp. 1998 [3-540-76193-4]

Vol. 229: Laumond, J.-P. (Ed.)
Robot Motion Planning and Control
360 pp. 1998 [3-540-76219-1]

Vol. 230: Siciliano, B.; Valavanis, K.P. (Eds)
Control Problems in Robotics and Automation
328 pp. 1998 [3-540-76220-5]

Vol. 231: Emel'yanov, S.V.; Burovoi, I.A.;
Levada, F.Yu.
Control of Indefinite Nonlinear Dynamic
Systems
196 pp. 1998 [3-540-76245-0]

Vol. 232: Casals, A.; de Almeida, A.T. (Eds)
Experimental Robotics V: The Fifth
International Symposium Barcelona, Catalonia,
June 15-18, 1997
190 pp. 1998 [3-540-76218-3]

Vol. 233: Chiacchio, P.; Chiaverini, S. (Eds)
Complex Robotic Systems
189 pp. 1998 [3-540-76265-5]

Vol. 234: Arena, P.; Fortuna, L. ; Muscato, G. ;
Xibilia, M.G.
Neural Networks in Multidimensional Domains:
Fundamentals and New Trends in Modelling
and Control
179 pp. 1998 [1-85233-006-6]

Vol. 235: Chen, B.M
H∞ Control and Its Applications
361 pp. 1998 [1-85233-026-0]

Vol. 236: De Almeida, A.T.; Khatib, O. (Eds)
Autonomous Robotic Systems
283 pp. 1998 [1-85233-036-8]

Vol. 237: Kreigman, D.J.; Hagar, G.D. ;
Morse, A.S.
The Confluence of Vision and Control
304 pp. 1998 [1-85233-025-2]

Vol. 238: Elia, N. ; Dahleh, M.A.
Computational Methods for Controller Design
176 pp. 1998 [1-85233-075-9]

Vol. 239: Wang, Q.G. ; Lee, T.H. ; Tan, K.K.
Finite Spectrum Assignment for Time-Delay
Systems
136 pp. 1998 [1-85233-065-1]